W0245863

BIOMONITORING OF TRACE AQUATIC CONTAMINANTS

BIOMONITORING OF TRACE AQUATIC CONTAMINANTS

DAVID J.H. PHILLIPS

Acer Environmental, Howard Court, Manor Park,
Daresbury, Cheshire WA7 1SJ, UK

and

PHILIP S. RAINBOW

School of Biological Sciences, Queen Mary and Westfield College,
Mile End Road, London E1 4NS, UK

ELSEVIER APPLIED SCIENCE
LONDON and NEW YORK

ELSEVIER SCIENCE PUBLISHERS LTD
Crown House, Linton Road, Barking, Essex, England IG11 8JU

WITH 31 TABLES AND 101 ILLUSTRATIONS

SOFTCOVER REPRINT OF THE HARDCOVER 1ST EDITION 1993

British Library Cataloguing-in-Publication Data

Phillips, David J. H.
Biomonitoring of Trace Aquatic
Contaminants
I. Title II. Rainbow, P. S.
628.1

ISBN 978-94-010-9131-2 ISBN 978-94-010-9129-9 (eBook)
DOI 10.1007/978-94-010-9129-9

Library of Congress Cataloging-in-Publication Data

Biomonitoring of trace aquatic contaminants / David J.H. Phillips,
Philip S. Rainbow.
p. cm.
Includes bibliographical references and index.
ISBN 978-94-010-9131-2
1. Water quality bioassay. 2. Indicators (Biology) 3. Biological monitoring.
4. Water—Analysis. I. Rainbow, P. S. II. Title.
QH90.57.B5P46 1992
628.1'61—dc20 92-18802
CIP

With love and gratitude to Lee and Mary

Preface

Twenty years ago, researchers wishing to identify contaminated areas in aquatic environments generally took water samples, and analysed them badly (as we have since discovered) for a few "pollutants" which were of topical note at the time (and which could be quantified by the methods then available).

Today, the use of aquatic organisms as biomonitors in preference to water analysis has become commonplace, and many national and international programmes exist around the world involving such studies. We believe that this trend will continue, and have complete faith in the methodology (when it is employed correctly). We hope that the following text assists in some part in attaining this goal, such that the quality of our most basic global resource – water – is adequately protected in the future.

DAVE PHILLIPS, PHIL RAINBOW
England, March 1992

Acknowledgements

Our thanks for contributions to this book are due to several individuals and groups, for varying reasons. Firstly, a co-authored book is always a triumph, and we trust that the following text is an acceptable compromise of the views of two individual authors, on a complex and developing topic.

Secondly, many of the ideas herein have crystallised over the last two decades as the field has grown, and we are individually and collectively grateful to a number of researchers for their insight and assistance. Amongst these are Mike Depledge, Geoff Moore, Des Connell, Mike Martin, Bruce Richardson, Steve White, Dayanthi Nugegoda, Victor Wong, Sian Pullen, Laurie Chan and Jason Weeks. Several of these are ex-students of P.S.R., and are already contributing markedly to the development of the field as a whole through the open literature. Others are friends and colleagues, and we trust that our efforts ring an occasional chord from long nights in distant lands.

In terms of pragmatics, the following text, figures and tables have been produced with the assistance of several individuals. Talent and time were contributed to several of the figures by Joy McDermott, Lee Phillips, Cheryl Evans, and Tony Miskiewicz and his staff. Judith Hutchinson was instrumental in producing a decipherable text from P.S.R.'s handwritten scrawl, providing an eye for detail and a calm environment whenever these were lacking.

Contents

Chapter 1

Definitions and Scope

A. BIOMONITORING

It is an unfortunate fact that many of the terms employed in studies of aquatic and terrestrial pollution are broad in nature, and have been defined inadequately and used improperly or ambiguously in the literature over some decades. As a result, there is a need to review the terms employed in the title of the present work, to define its contents and scope.

Martin and Coughtrey (1982) have provided a particularly useful discussion of the terms "biological indicators" and "biological monitors". They proposed that the two terms are in fact distinct, although many authors have employed them synonymously. *Biological indicators* (also termed 'bio-indicators') are considered to be organisms which, by their own presence or absence, indicate the existence or abundance of a particular critical factor. Thus, all organisms exhibit a defined tolerance to an environmental stimulus (whether the latter is natural or anthropogenic in nature), and can exist in particular locations only within their zone of tolerance (Fig. 1). Within this zone of tolerance, enhanced exposure to contaminants or to natural stressors (e.g. increasing or decreasing salinities or temperatures) may be met through compensation mechanisms, although signs of toxicity are likely to occur as the upper limit of the zone

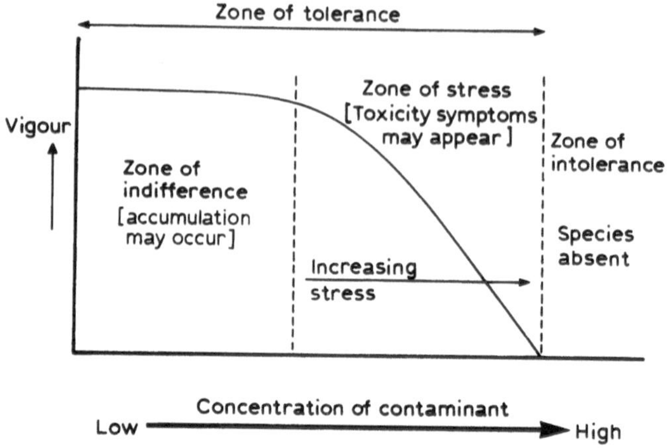

Fig. 1. Response of an organism in terms of tolerance and "vigour" to increased concentrations of an environmental contaminant. After Martin and Coughtrey (1982).

of tolerance is approached. Through its presence or absence in a particular environment, a biological indicator acts as a signal of the existence of a stimulus at or above a given threshold or critical level.

This concept is well developed in several types of environmental applications. For example, geologists and others study the distribution of plant species to indicate the likely presence or absence of critical concentrations of minerals or trace metals in soils. This may be relevant to mineral exploitation, or to agricultural activities. In aquatic environments, the degree of pollution in particular locations is commonly quantified through the analysis of benthic (or sometimes, pelagic) community structures. Alterations in both total biomass and the structure of communities are relevant, as noted by Hellawell (1986; see Fig. 2). Certain aquatic species have been widely documented as particularly useful pollution indicators; for example, the abundance of the polychaete *Capitella capitata* is often considered a reliable indicator of organic enrichment (or occasionally, of other types of impacts) in the sediments of temperate marine ecosystems. Such investigations of community structure are simply sophistications of studies of the presence and absence of single species, and both employ essentially the same indicator concept.

Environmental monitoring was considered by Chapman *et al.* (1987) to consist of repetitive data collection for the purpose of determining trends in environmental parameters. They argued that assessments which are

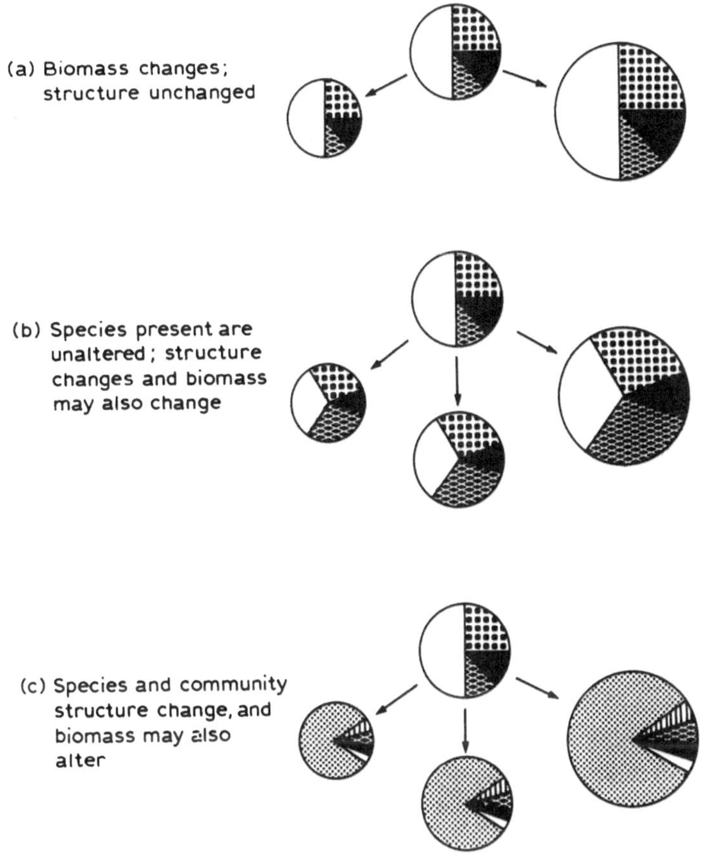
(a) Biomass changes; structure unchanged
(b) Species present are unaltered; structure changes and biomass may also change
(c) Species and community structure change, and biomass may also alter

Fig. 2. Possible alterations in the total biomass and community structure of populations due to an environmental stimulus. Biomass is represented by the size of the circles; community structure by the various types of shading. After Hellawell (1986).

possible through monitoring include both *a priori* indications of problems developing in a resource before such problems become critical, and *a posteriori* evaluations of temporal changes in particular parameters of interest. Other authors, however, have recognised more numerous reasons for conducting environmental monitoring (e.g. see Preston, 1975; Holdgate, 1979; Reay, 1979; Martin and Coughtrey, 1982). These may be paraphrased as follows:

- To screen effluents, receiving waters or biota for potentially harmful toxicants, including in certain cases the assessment of requirements for emission controls.
- To investigate the effects of environmental quality on human health or other parameters, in an attempt to elucidate cause-and-effect relationships.
- To study the sources, transport pathways and sinks for contaminants in the environment.
- To provide a historical record of emissions or of environmental quality, in order to ascertain compliance rates with relevant standards or legislation.
- To investigate the specific environmental impacts of individual or multiple developments.

Certain of these applications involve the use of organisms as types of biological monitors (or biomonitors) of the impacts of toxicants, through toxicity testing of single species, or of communities. Such studies may be undertaken in product evaluations, or in attempts to minimise the detrimental impacts of emissions on the ecosystem, or in investigations such as hazard assessments (see Cairns and van der Schalie, 1980; Cairns and Pratt, 1987). The present text does not cover any of these applications of the use of biota for toxicity testing, hazard assessment, or ecotoxicological studies.

Martin and Coughtrey (1982) distinguished *biological monitors* from biological indicators by two criteria. They proposed that biological monitors provide the means for regular surveillance, and may also be used to quantify the amount of a pollutant present in a particular environment. A flow diagram used to describe the responses of biota to contamination by trace metals was provided (Fig. 3), and this assists in defining the distinct types of biomonitors used in aquatic studies, and their inter-relationships. As shown in Fig. 3, contaminants such as metals may exert several impacts upon aquatic species or communities. These range from effects on species composition or distribution (involving the study of biological indicators; see above), through the accumulation of contaminants in the tissues of a biological monitor, to the creation of toxic symptoms in particular sensitive species or their target organs.

The present text does not seek to discuss biological indicators, i.e. those species and populations which are employed by virtue of their presence, absence or change in numbers (with consequent effects on community structure) to investigate aspects of environmental quality. Rather, it concentrates primarily on the use of organisms to measure contaminant

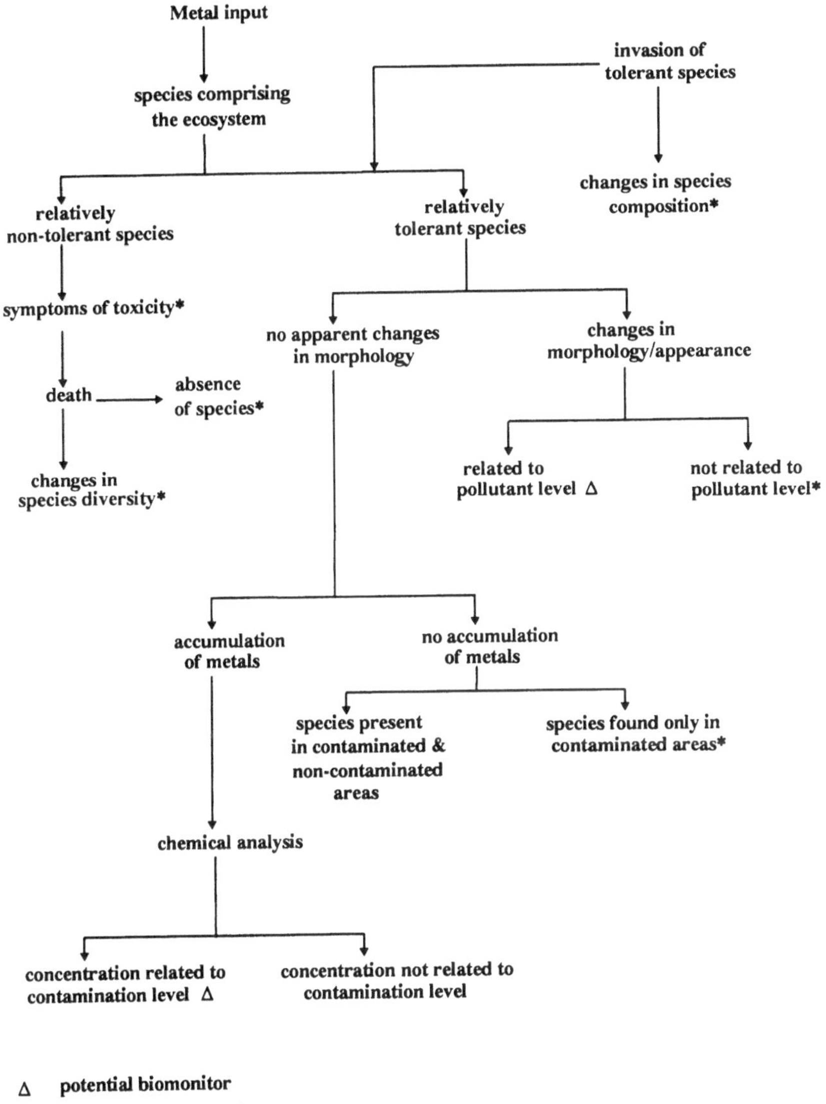

Fig. 3. Relationships between the responses of organisms and metal contamination, indicating the potential for use of organisms as biological indicators or biomonitors. After Martin and Coughtrey (1982).

abundance in aquatic ecosystems through the accumulation of these contaminants in their tissues. Goldberg *et al.* (1978) proposed the use of the term "sentinel organism" for such species, as an alternative to either biological indicator or biological monitor, but this has not enjoyed widespread usage. Hellawell (1986) employed the term "bioaccumulative indicator"; while more accurate in some senses, this term is extremely clumsy in use. The term "biomonitor" is employed in the present text as the most accurate to describe organisms which accumulate contaminants in their tissues, and may therefore be analysed to identify the abundances and bioavailabilities of such contaminants in aquatic environments.

Moreover, the authors do not seek to exhaustively cover the impacts of contaminants on aquatic organisms. However, the methodology employed in such "effects-related monitoring" is reviewed briefly, providing the reader with an introduction to this interesting and rapidly developing field. Additional information on this precise application for biomonitoring can be found in other sources (e.g. Bayne *et al.*, 1985).

B. TRACE CONTAMINANTS

Further confusion exists with respect to the terminology employed for classes of contaminants in aquatic (and terrestrial) ecosystems. Hopkin (1989) discussed the distinction between *trace metals* and *heavy metals*, and considered that neither term was adequately defined. Nieboer and Richardson (1980) proposed a chemical classification system employing Lewis acid properties for metal ions, separating these into classes A, B, or a Borderline category (Fig. 4) according to their "hardness" or "softness" as acids and bases. Class A metal ions are essentially oxygen-seeking, while those of Class B tend to bind to ligands including nitrogen or sulphur atoms; the Borderline metal ions exhibit intermediate behaviour. This classification has relevance also to whether metals are of an essential or non-essential nature, and to their toxicity and potential for aquatic pollution (see Hellawell, 1988, and Table 1). The metals which are dealt with in the present text are those with ions listed as Borderline or Class B metals in Fig. 4, and the terms trace metals or trace elements will be employed herein as a generic descriptor of these important environmental contaminants.

Both stable isotopes and their radioactive counterparts are taken up significantly by aquatic biota, and many laboratory-based studies of metal kinetics in organisms have employed radionuclides rather than stable

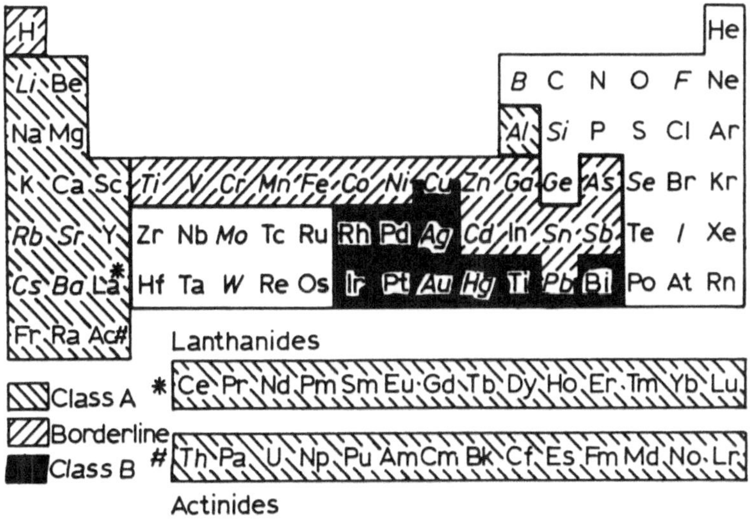

Fig. 4. The classification of metal ions into Class A, Borderline, and Class B categories based upon Lewis acid properties. Elements which are generally present only in trace quantities but are thought to be essential are shown in italics. After Nieboer and Richardson (1980).

isotopes. Such investigations have provided useful insights into the behaviour of trace metals in aquatic organisms, and are therefore included in the present text.

A further class of contaminants which exhibits significant potential for bioaccumulation in aquatic biota comprises the chlorinated lipophilic organic compounds, i.e. the organochlorines (or chlorinated hydrocarbons). These are of much greater concern in aquatic ecosystems than are

Table 1 The approximate order of toxicity of trace metals (after Hellawell, 1988)

Highly toxic →											Less toxic
Hg	Cu	Cd	Au?	Ag?	Pt?	Ba	Mn	Li	Ca	Sr	Na
		Zn	Sn	Al	Fe^{3+}		Co	K		Mg	
				Ni	Fe^{2+}						
Highly toxic →											Less toxic

Table 2 Comparisons of the environmental characteristics of synthetic insecticides (after Hellawell, 1988)

Characteristic	*Organochlorine*	*Organophosphate*	*Carbamate*
Potential for entry into freshwater	Strong	Strong	Moderate
Water solubility	Very low	Low	Low
Aquatic toxicity	High	Moderate	Moderate
Aquatic persistence	Prolonged	Short	Short
Bioaccumulation	Strong	Weak	Weak

organophosphates or carbamates, as they have a considerably enhanced potential for bioaccumulation and are of significantly greater toxicity compared to compounds of the other two classes (Table 2). Most are employed as insecticides or pesticides, but a few are employed in industrial applications only (polychlorinated biphenyls or PCBs), or are by-products of incineration or other chemical processes (e.g. the dioxins and dibenzofurans). Biomonitors have long been employed to study the distributions of these contaminants in freshwater and marine environments, and their use for this purpose is reviewed herein.

Certain types of biomonitors have also enjoyed significant use in the measurement of petroleum-derived hydrocarbon concentrations in aquatic ecosystems (Samiullah, 1990). While higher aquatic species such as fish or marine mammals may metabolise such compounds at significant rates (e.g. see Connell and Miller, 1984; Walker, 1990), invertebrates of freshwater and marine ecosystems possess considerably lower capacities for their degradation. The latter types of organisms may therefore be particularly useful in monitoring the amounts of petroleum-derived hydrocarbons in aquatic environments, and this application is also reviewed here.

C. CONTENTS AND STRUCTURE

The current interest in the contamination of aquatic environments is due in no small amount to the emergence of pollution as an item of public (and hence, political) debate. This has eventuated largely through the focusing of scientific and media attention on pollution incidents worldwide. The importance of such incidents as a driving force for the increased monitoring of contaminants in aquatic ecosystems cannot be doubted. To place contaminant monitoring in context, therefore, Chapter 2 presents a brief history of the growth of aquatic pollution and its emergence as a political issue.

The selection of the contaminants to be monitored in any particular programme is obviously of fundamental importance. Chapter 3 reviews contaminants of concern in aquatic environments, and compares the legislation relevant to this topic in Europe and the USA. An introduction to the biomonitoring concept is presented in Chapter 4, and the early use of biomonitors is reviewed therein.

The two subsequent chapters deal with the use of biomonitors to measure the abundances and distributions of particular classes of contaminants: trace metals and radionuclides (Chapter 5); and organochlorines and hydrocarbons (Chapter 6). Chapter 7 deals specifically with biomonitoring techniques for freshwater environments, which differ somewhat from those employed in estuarine or marine situations.

Chapters 8 and 9 are concerned with the development of biomonitoring programmes in particular situations. Chapter 8 addresses the concepts employed when new programmes of this nature are to be developed. Published data on monitoring the effects of contaminants in aquatic ecosystems (which is a completely distinct field from the use of biomonitors to measure contaminant abundances and distributions) are reviewed in Chapter 9.

The final chapter presents a brief overview of biomonitoring successes and failures to date, and considers the future development of the field as a whole.

Chapter 2

The Green Revolution

A. INTRODUCTION: THE GROWTH OF AQUATIC POLLUTION

(i) The Early Years

It is probable that the contamination of aquatic environments has been of at least some concern to Man throughout his existence. Early nomadic groups of *Homo sapiens* may have polluted their drinking water supplies with human wastes, and perhaps altered the local ecology of streams and estuaries. However, both their nomadic nature and the fact that they lived in groups of relatively small numbers suggest that any impacts which they may have exerted on local aquatic environments are likely to have been minor.

Inland and coastal waters came under greater threat when populations grew and Man began to live in communities of significant size. The effect of this was to both amplify and concentrate the previously widely-dispersed pollution sources, and organic wastes from sewage began to significantly challenge the assimilative capacities of the receiving waters into which they were discharged. Rivers and semi-enclosed coastal embayments would have been at greatest risk, and most of the densely-populated areas of human habitation grew up (and remain) in such environments.

The shift from a nomadic existence to one based around discrete fixed settlements of greater human densities was accompanied by the emergence of an enhanced reliance upon agriculture. This too must have created adverse impacts on certain local waters, soil erosion giving rise to increased turbidity, and animal wastes adding to the organic enrichment of waters of poor receiving capacity. Notwithstanding such trends, however, human populations in most areas remained relatively small, and any impacts on local aquatic environments are unlikely to have been significant on anything other than a highly localised scale.

(ii) The Development of the Western Nations

In Europe, this situation was radically altered by the expansion of overall populations and city dwelling habits created by the industrial revolution, which commenced in about 1760. The growth of industry not only acted as a focus for the creation of larger cities, but also had fundamental impacts on the types of aquatic contaminants finding their way into inland watercourses and to the sea. Organic wastes from sewage and agriculture were joined by other anthropogenic contaminants, enriched or created through industrial processes.

Fundamental changes occurred outside Europe at this time also, industrialisation and associated events again being critical in defining both population growth and impacts on aquatic resources. The initial settlers of the New World encountered environments which were effectively pristine, pre-existing populations being small, often semi-nomadic, and exhibiting both a respect for and understanding of wildlife protection. New settlements were established in areas offering a supply of fresh water, but in most cases population growth was slow until the mid-1800s. In San Francisco Bay, for example, the Spanish noted on their arrival in 1769 the presence of a population of some 20,000 aboriginals who were largely dependent on the (then abundant) aquatic resources of the Bay for their food (Nichols *et al.*, 1986). Populations rose little until the discovery of gold in the Sierra Nevada foothills in 1848, which gave rise to massive immigration to the area. Thereafter, pollution loads to the local streams and to the Bay itself increased dramatically, both human wastes and the hydraulic mining of gold (using large quantities of mercury) creating significant impacts on aquatic resources (Nichols *et al.*, 1986). Far-reaching changes of this nature have occurred in most of the larger estuaries of the western nations over the last century, and few can be considered to have retained many traces of their original character.

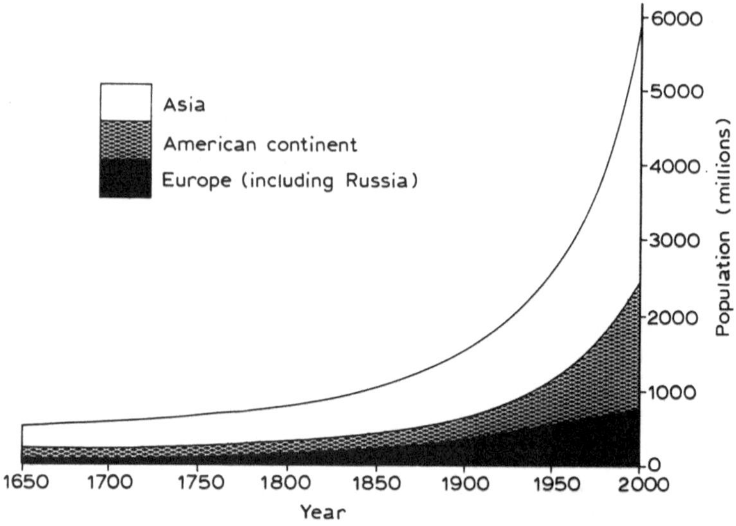

Fig. 5. Global population growth by major area between 1650 and the present, with estimates to the year 2000. Abstracted from various United Nations sources.

(iii) The Recent Emergence of the Developing Nations

This trend towards increasing populations and their reliance upon a greater degree of industrialisation has strengthened markedly over the last century, and its geographical emphasis has shifted. The global population reached one billion (10^9) in about 1840, and surpassed two billion in 1930. Over the intervening 60 years since 1930, this rate of growth has accelerated markedly, and the present human population of the world is estimated at about 5.1 billion (Fig. 5). This population increase has recently been fuelled largely by the developing nations of Latin America, Africa and the Far East; in most of the western nations, birth control has created conditions close to zero population growth over the last decade or more. This distinction between the western and the developing nations is portrayed graphically by their respective age-related population structures (Fig. 6). City-dwelling populations also reflect this trend, the conurbations of western nations growing little at present, while massive expansion continues in many cities in the developing nations (Fig. 6).

As industrial development is one of the major sources of contaminants to inland and coastal waters and also provides the basis of most trade, it is possible to draw broad conclusions on the incidence of pollution

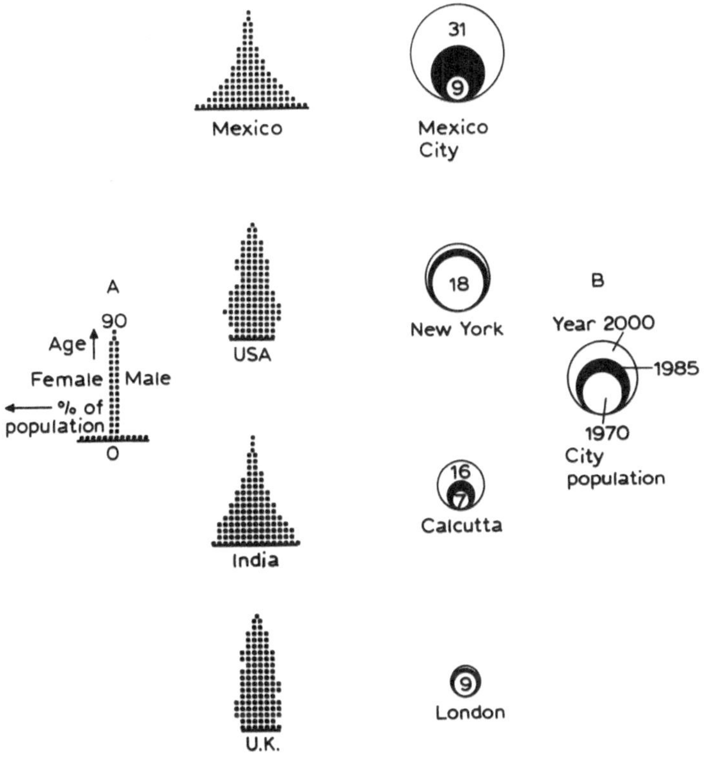

Fig. 6. Population data for selected countries and cities of the western and the developing nations. (A) Age-related population structures of countries as a whole. (B) Population change in selected cities. All population data in millions. After the *Collins Longman Atlas* (1989).

problems from economic statistics (Fig. 7; Table 3). While the western nations dominated world trade until perhaps 30 years ago, the present picture documents the emergence of the developing countries as industrial and economic forces. The massive development of an industrial trading base in Japan over the last three decades is the best example, but the gross national product of many other subtropical and tropical nations is also increasing rapidly at present.

(iv) The Current Status

Industrial development is inevitably accompanied by problems with the contamination of aquatic environments. Over the last three decades, most

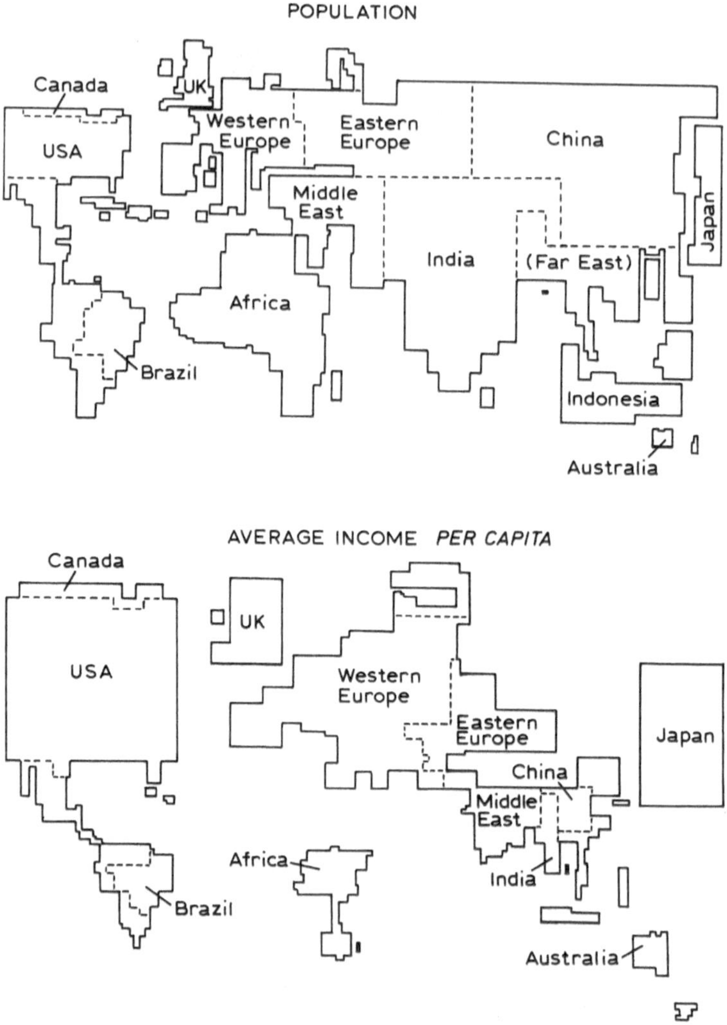

Fig. 7. Graphical protrayal of the population distribution of the world and the average annual income *per capita*. After the *Readers Digest Atlas of the World* (1987) and the *Collins Longman Atlas* (1989).

western nations have introduced controls on sewage and industrial discharges to aquatic environments. There can be no doubt that this has helped to protect their local waters; however, problems remain in many areas, these often representing a legacy of previous neglect. By contrast,

Table 3 Data for the gross national product (US$ millions) and gross national product *per capita* of population (US$) for selected western and developing nations (after the *Guinness World Data Book*, 1991)

Nation	*GNP*	*GNP per capita*
Argentina	83 040	2 640
Australia	204 446	12 390
Bangladesh	18 310	170
Botswana	1 191	1 050
Brazil	328 860	2 280
Brunei	3 317	14 120
Canada	437 471	16 760
PR China	356 490	330
Denmark	94 792	18 470
Egypt	33 250	650
Ethiopia	5 760	120
Finland	92 015	18 610
France	898 671	16 080
Germany	1 252 205	25 710
Haiti	2 240	360
Hungary	26 030	2 460
Iceland	5 019	20 160
India	271 440	330
Indonesia	76 060	580
Italy	765 282	13 320
Japan	2 576 541	21 040
Kenya	8 310	360
Madagascar	2 080	180
Malaysia	31 620	1 870
Mexico	151 870	1 820
Mozambique	1 550	100
New Zealand	32 109	9 620
Nigeria	31 770	290
Norway	84 165	20 020
Pakistan	37 153	350
Peru	29 185	1 440
Philippines	37 710	630
Poland	69 970	1 850
Senegal	4 520	630
Singapore	24 010	9 100
Somalia	970	170
Spain	301 829	7 740
Sweden	160 029	19 150
Tanzania	3 780	160
Thailand	54 550	1 000
USSR	2 310 000	8 160
United Kingdom	730 038	12 800
USA	4 863 674	19 780
Zaire	5 740	170

the developing nations possess far fewer controls of this sort in most instances (e.g. see Phillips and Tanabe, 1989), and those which do exist are often ignored (Beanlands and Si, 1988).

As a result, the rates and patterns of use of many chemicals which are of concern as aquatic contaminants (see Chapter 3) have altered in recent decades, utilisation in the developing nations in the tropics and subtropics gradually taking over from that in the westernised countries of the higher latitudes (Phillips, 1991). Perhaps the best example of this "southward tilt" concerns the use of DDT and other organochlorine pesticides, now heavily dominated by the tropical and subtropical nations of the world (Goldberg, 1975). Global utilisation rates for these compounds have not fallen subsequent to the imposition of controls on their use in western nations; demand from the developing world has increased to more than account for the reduced utilisation elsewhere.

Such patterns of increasing contamination of subtropical and tropical waters are repeated for many pollutants, the rapidly expanding populations and industrialisation of the developing nations being the principal causes (Phillips, 1991). Cross-frontier pollution problems have thus emerged as a major concern, especially for contaminants such as the organochlorines, which are sufficiently volatile to be subject to aerial transport both regionally and globally (e.g. Tatton and Ruzicka, 1967; Tanabe and Tatsukawa, 1980; Tanabe *et al.*, 1982a; Rapaport *et al.*, 1985; Ramesh *et al.*, 1989). Interestingly, while most of the DDT in Japan is now demonstrably from external sources such as India, chlordane remains in frequent use in Japan and has attained some significance recently as an aquatic pollutant (Loganathan *et al.*, 1989; Tanabe *et al.*, 1989a).

Thus, global pollution loads have both inexorably increased and shifted somewhat in geographical emphasis over the last century, being driven by a combination of increasing population and the intensification of industry. Estimates of the rates of mobilisation of metals (Table 4) serve to emphasise the scale of these anthropogenic changes. Many trace elements are now being mined at a pace which exceeds their natural mobilisation rates by greater than an order of magnitude, with clear implications for the pollution of aquatic resources, both on local and global scales. The contributions of anthropogenically-derived metals to the total metals present have been quantified for certain local environments, and are seen to be very considerable for many of the more toxic elements (Fig. 8; Förstner, 1980). In addition, rivers and coastal waters are presently exposed not only to increasing quantities of natural materials such as metals and nutrients, but also to cocktails of industrially-derived contami-

Table 4 The effects of Man on mobilisation rates of trace metals. Two estimates of the man-induced mobilisation rates are given (after Bryan, 1976)

Element	*Geological rate (10^3 t/year)*	*Man-induced rate (mining) (10^3 t/year)*		*Total in oceans (10^6 t)*
Iron	25000	319000	395000	4110
Manganese	440	1600	8150	2740
Copper	375	4460	6000	4110
Zinc	370	3930	5320	6850
Nickel	300	358	481	2740
Lead	180	2330	3200	41
Molybdenum	13	57	74	13700
Silver	5	7	9	137
Mercury	3	7	10.5	68
Tin	1.5	166	227	14
Antimony	1.3	40	65	274
Cadmium	No data	No data	17	68

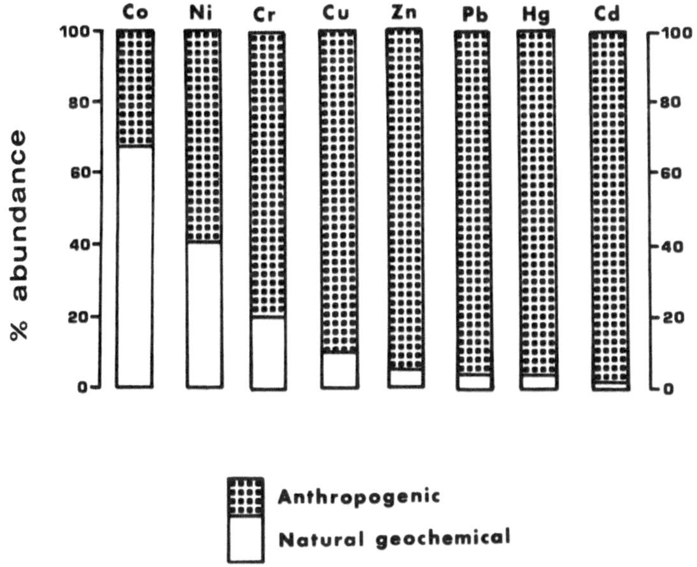

Fig. 8. Comparison of the magnitudes of natural geochemical and anthropogenic sources of trace metals in recent sediments of the lower Rhine River, Germany. After Förstner (1980).

nants, many of which exhibit significant persistence and capacities for bioaccumulation.

This increased pollution loading has been accompanied by massive physical alterations to the natural catchments of many aquatic environments, and the effects of these changes are often most pronounced in the downstream estuaries. For example, the catchment of San Francisco Bay has been radically altered through extensive urbanisation, the introduction of massive water management schemes, the agricultural development of the naturally arid Central Valley of California, and the loss of some 96% of the historical area of wetlands (Nichols *et al.*, 1986; Wright and Phillips, 1988). Each of these changes exerts significant impacts on both the hydrodynamics of the estuary and its ability to assimilate contaminants, thus providing the potential for synergistic effects between the increases in pollutant loading and other changes to the estuary.

This pattern is repeated in many parts of the world, and the changes occurring are not restricted to individual catchments, but are also evident on much greater scales. Concern over the pollution of the Mediterranean Sea gave rise to the birth of the United Nations Regional Seas Programme and the Mediterranean Action Plan (Griffiths, 1977; Waddington, 1981), and such regional programmes have since been extended to many other areas. Despite such initiatives, however, many aquatic environments continue to exhibit evidence of severe degradation, with semi-enclosed waters being at particular risk.

The northern Adriatic Sea provides an excellent example; here, nutrient loads cause extensive algal blooms and accompanying mucilage formation from diatoms in summer (Cognetti, 1989). A consideration of long-term trends in water quality suggests that these problems derive principally from agricultural development, industrialisation and urbanisation of the Po River catchment since about 1950. The impacts of organic enrichment on algal bloom incidence and on dissolved oxygen levels offshore are both significant and increasing (Justic, 1987; Degobbis, 1989; see Fig. 9). Where studies have been undertaken in the developing nations, problems similar to these are noted, both inland waters and coastal environments being threatened by the discharge of sewage and industrial wastes (e.g. see Phillips and Tanabe, 1989).

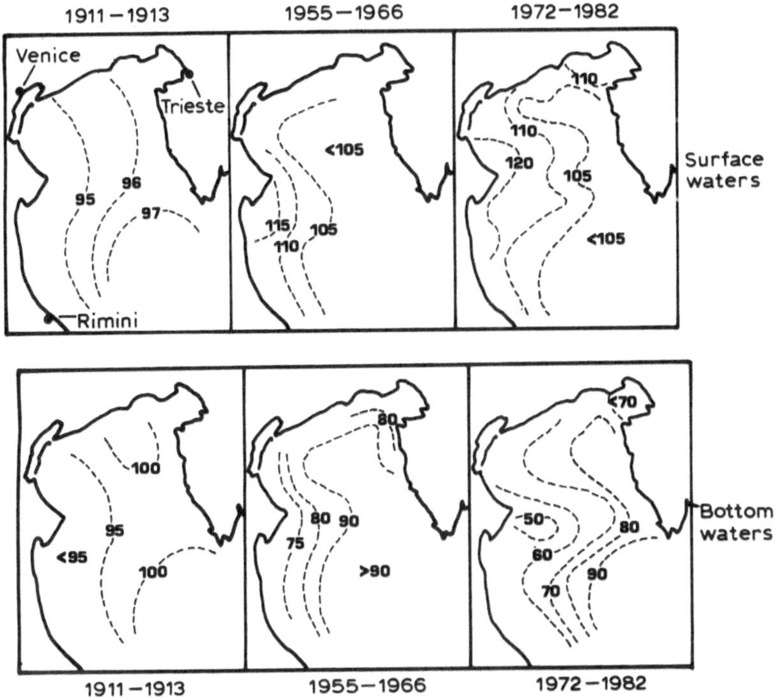

Fig. 9. Distribution of means for oxygen saturation (%) in the Northern Adriatic Sea during August and September, over the periods 1911–1913, 1955–66, and 1972–1982. Upper horizontal row relates to surface waters; lower horizontal row relates to waters 2 m above the bottom. After Justic (1987).

B. AQUATIC POLLUTION AS A POLITICAL ISSUE

The emergence of aquatic pollution as an issue of major social importance was driven initially by discrete events in which human health was of principal concern, environmental protection *per se* being seen as of lesser relevance in most instances. However, in more recent years, the protection of aquatic environments has been accorded rather higher priority than previously.

(i) Nutrients and Public Health

While the discharges of nutrients to aquatic environments and their resulting impacts are not the principal topic of the present text, concerns generated by these have contributed measurably to the emergence of aquatic pollution as a political issue. These problems are thus addressed briefly here.

The discharge of large quantities of nutrients into coastal waters from domestic sewage, industrial effluents, and agricultural sources is of increasing concern in many parts of the world. The eutrophication of inland and coastal waters which results from such discharges of nutrients may be accompanied by several problems. In certain areas, large mats of macroalgae may develop, choking rivers and estuaries. The Peel-Harvey Estuary in Western Australia provides an excellent example; here, the problem is believed to be due principally to the use of superphosphate fertilisers on farmland in the catchment (Birch *et al.*, 1986). Apart from the physical blockage of the estuary to vessel traffic, oxygen sags are experienced at night and fisheries productivity suffers as a result.

A similar nuisance arises in the northern Adriatic Sea from the production of mucilage by diatoms which bloom in summer, especially in conditions of high freshwater inflows or hot, calm weather. In both 1988 and 1989, this problem was sufficiently severe to significantly affect tourism in north-eastern Italy, as well as reducing fishery success. It appears that nutrient discharges from the Po River Estuary are principally to blame, and water quality in the area as a whole has been deteriorating over several decades (Justic, 1987; Degobbis, 1989; Fig. 9).

While these algal blooms may create nuisance and give rise to fish kills through the deoxygenation of local waters, they do not constitute a public health threat *per se*. By contrast, the proliferation of certain species of dinoflagellates may give rise to very significant impacts on public health. Dinoflagellate species of several genera produce toxins under particular conditions, and these toxins are passed through the food chain to affect higher consumers, including humans. The resulting poisoning is classified in several different fashions, for example as paralytic shellfish poisoning (PSP), diarrhetic shellfish poisoning (DSP), venerupin poisoning, or ciguatera. Each type of poisoning tends to be associated with specific species of dinoflagellates and with particular routes of exposure to humans.

The incidence of such problems varies considerably between different parts of the world. In general, the subtropical and tropical zones suffer the greatest incidence of PSP and ciguatera. This is not to say, however, that PSP is unknown in temperate locations. Ayres (1975) produced an interesting review of PSP in Britain from 1814 to 1968, including a discussion of the poisoning episode affecting 78 persons in May 1968, which was triggered by the consumption of mussels (*Mytilus edulis*). PSP toxins were again found to be present in the north-east of England in the summer of

1990, possibly correlating to a particular combination of high nutrient inflows from rivers and unusually warm and sunny weather, which may have promoted the growth of particular species of toxic dinoflagellates.

In Australia, the incidence of PSP is quite low. Hallegraeff and Sumner (1986) noted the existence of blooms of the toxic dinoflagellate *Gymnodinium catenatum* in Tasmania in early 1986, and local oyster and mussel farms were closed temporarily as a result. Ciguatera poisoning, due to the transfer of toxins from the dinoflagellates *Gambierdiscus toxicus* and *Ostreopsis siamensis* to fish, is more common in the Great Barrier Reef region than elsewhere in Australia, as might be anticipated from its global incidence, which is heavily skewed to tropical areas (Gillespie, 1984).

In Hong Kong, algal blooms were relatively rare prior to the late 1970s (Morton, 1985, 1989; Phillips, 1985a). However, the development at that time of two new town areas in the north-east of the Territory gave rise to massive increases in nutrient loading to Tolo Harbour, a semi-enclosed embayment. A gradual increase through the 1980s in algal blooms in Tolo Harbour was recorded (Phillips, 1985a; Morton, 1989; see Fig. 10), and in the mid-1980s, the incidence of PSP also began to rise. It has been theorised that the altered nitrogen:phosphorus ratios in discharges to the estuary were a contributory factor to the shift in species dominance from diatoms to the potentially toxic dinoflagellates (Holmes and Lam, 1985).

Elsewhere in the Indo-Pacific, PSP has been recognised as a considerable public health problem. Reviews of the incidence of PSP in this area have been published by Maclean and co-workers (Maclean, 1984, 1989; Maclean and Temprosa, 1984; Maclean and White, 1985). They have concluded that, while *Noctiluca scintillans* and other non-toxic species may attain bloom proportions and create anoxia and fish kills in the region, the principal toxic dinoflagellate species of importance is *Pyrodinium bahamense* var. *compressa*. There is evidence that this dinoflagellate species is spreading geographically through the Indo-Pacific, and it is known to have been the cause of several significant PSP incidents, involving both illness and death on numerous occasions (Fig. 11). In many of these instances, the consumption of shellfish (sometimes of the cultured variety) was the primary cause of the poisoning outbreak. Given the acknowledged link between the nutrient enrichment of coastal waters and the incidence of algal blooms, this is of no little concern.

(ii) Trace Metals and Public Health

The first widely-documented instance of a public health impact resulting from aquatic contamination by a trace metal occurred at Minimata Bay

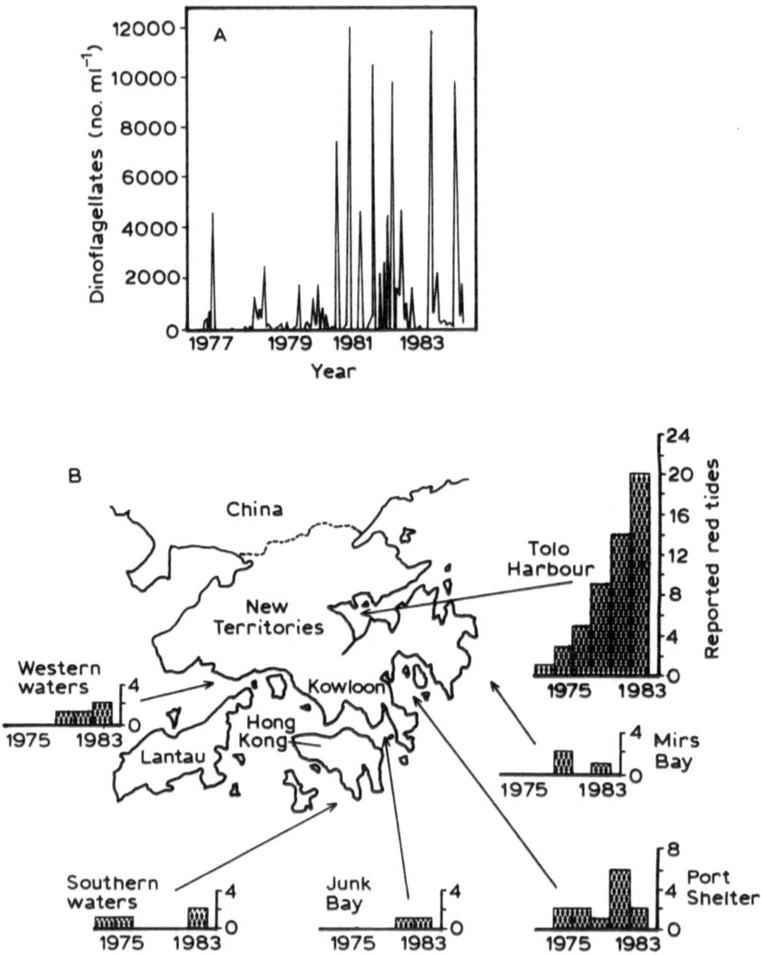

Fig. 10. Eutrophication in Hong Kong coastal waters. (A) Numbers of dinoflagellates recorded between 1977 and 1984 in Tolo Harbour, to the north-east of the Territory. (B) Records of red tides in the Territory as a whole between 1977 and 1984. After Holmes and Lam (1985) and Morton (1989).

in Japan, commencing in 1953. Minimata Bay lies on the eastern aspect of the Shiranui Sea, off south-western Kyushu (Fig. 12). These waters are partially landlocked, and their assimilative capacity for pollutants is therefore poor. Fish kills and bird mortalities in the area were noted in the early 1950s, and were followed in 1953 by the outbreak of a "dancing

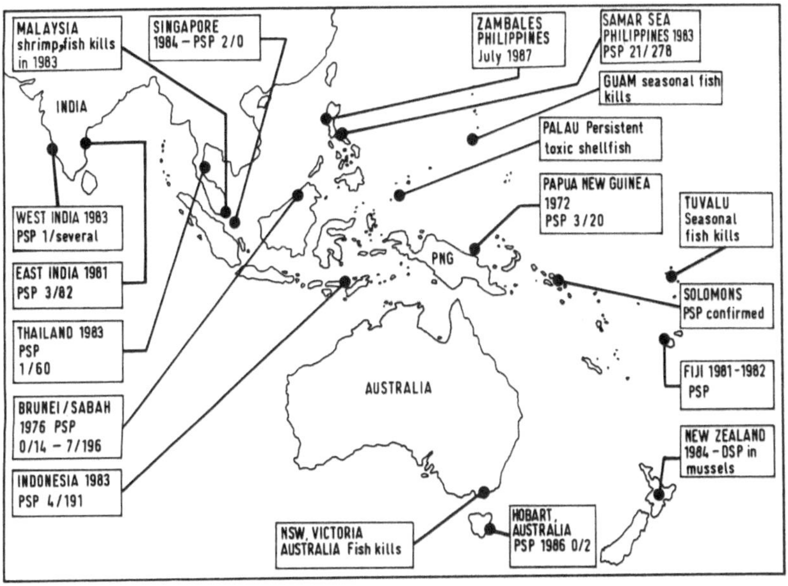

Fig. 11. Sites where red tides, fish kills and Paralytic Shellfish Poisoning have been recorded in the Indo-Pacific area. Numbers and dates in the boxes refer to the incidence of deaths/illnesses and the timing of the reported incidents. Harmless red tides have been omitted. After Maclean (1989).

disease" in the local cat population (Harada and Smith, 1975; Takizawa, 1979). This affliction involved coordination problems and convulsions in the affected cats, death frequently resulting. Dogs and pigs were also affected, but in smaller numbers.

In April 1956, a 5-year-old girl admitted to the hospital of the Chisso Chemical Corporation in Minimata was diagnosed to have brain damage. The symptoms displayed by the girl included disturbances of gait and speech, and delirium. Her younger sister and four members of a neighbouring family were soon found to be suffering from similar symptoms, and in early May 1956, Dr Hosokawa of the hospital reported to the Minimata Public Health Department that an outbreak of an "unclarified disease of the central nervous system" had occurred in the Minimata area.

The disease was rapidly found to be non-infectious, and a poisoning origin was suspected, rather than any microbiological cause. A research group set up at the Kumamoto University Medical School began to investigate the Chisso Chemical Corporation factory in Minimata as a

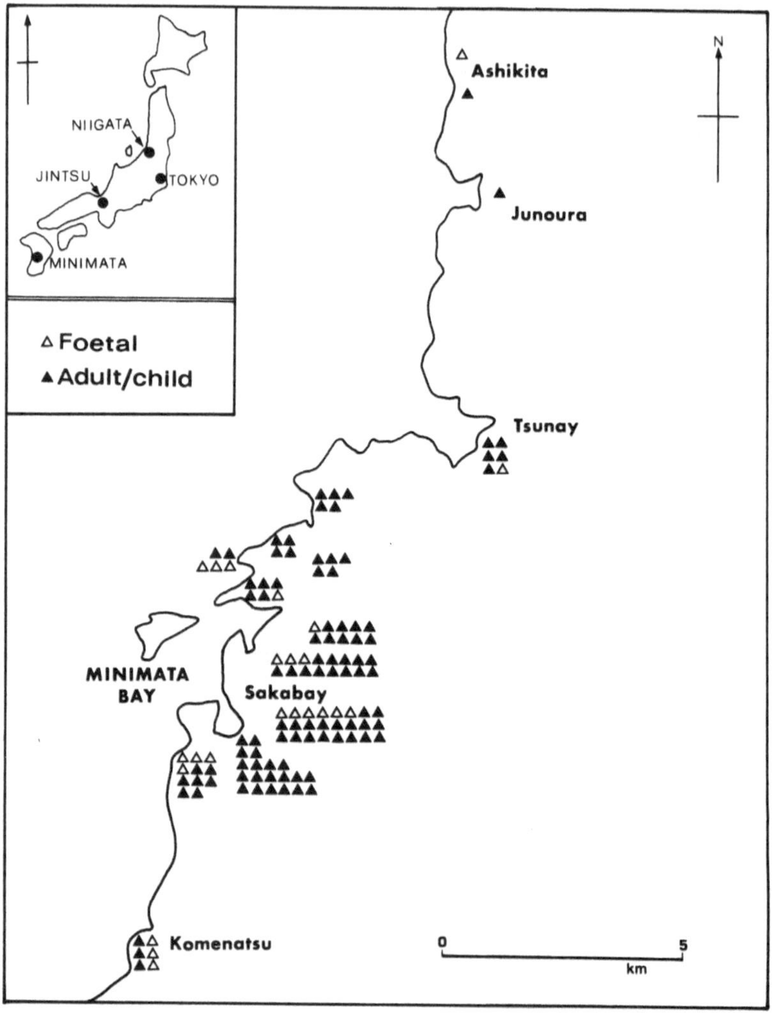

Fig. 12. Instances of foetal poisoning cases and those affecting children or adults in the Minimata area, Japan. The geographical distribution of the 121 cases officially recognised by the end of 1962 is shown. After Takeuchi (1972).

possible source of the toxins. The Chisso factory had been constructed in 1907, and had originally manufactured fertilisers and carbide products, expanding later into petrochemicals and plastics. The production of acetaldehyde had commenced in 1932, and ceased in 1968; vinyl chloride was

Table 5 Clinical symptoms displayed in 34 cases of acute and sub-acute Minimata disease studied in the 1950s (after Harada and Smith, 1975)

Symptoms	*Percentage*
Disturbance of sensation:	
superficial	100
deep	100
Constriction of the visual field	100
Disturbances of coordination:	
adiadochokinesia	93.5
ataxic gait	82.4
dysmetria	80.6
Dysarthria	88.2
Impairment of hearing	85.3
Tremor	75.8
Mental disturbances	70.6
Romberg's sign	42.9
Tendon reflex:	
exaggerated	38.2
weak	8.8
Salivation	23.5
Pathological reflexes	11.8

produced between 1941 and 1971. Both of these processes employed mercury salts as catalysts, although in very different amounts. Recent estimates suggest that about 80 tonnes of mercury were lost to the environment from the process manufacturing acetaldehyde, whereas only 0.2 tonnes were derived from the vinyl chloride production process (Taylor, 1982).

Effluents from the factory contained significant quantities of several trace metals, including arsenic, copper, lead, manganese, selenium, and thallium. All of these were investigated in the mid-1950s as possible causative agents of the Minimata symptoms, but were eventually discounted. Meanwhile, the incidence of the disease was increasing steadily, and congenital cases began to be reported, the latter involving children with severe brain damage born to apparently healthy mothers. Clinical symptoms of the disease were fully identified (Table 5), and appeared to be almost completely neurologically-related.

In late 1958, Professor T. Takeuchi of the Kumamoto University team noted that the clinical and pathological findings from the Minimata victims were virtually identical to those reported by Hunter and Russell (1954) for occupational poisoning of workers in a factory in England

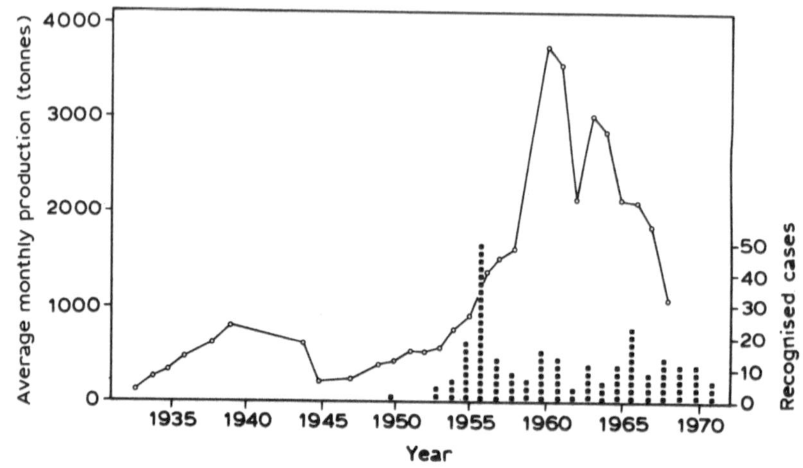

Fig. 13. The time courses of acetaldehyde production (left axis) at the Chisso Chemical Corporation factory at Minimata, Japan, and of the onset of Minimata disease (histograms; right axis). After Harada and Smith (1975).

producing methylmercury. Studies in 1959 confirmed that the disease in cats could be reproduced by feeding the animals with methylmercury, and the same contaminant was found to be present at high concentrations in the fauna and sediments of Minimata Bay. Later investigations established that both methylmercury and inorganic mercury were present in the effluents from the Chisso Chemical Corporation facility, and that methylation of inorganic mercury could occur in sediments in the environment. The principal route of mercury exposure of both the cats and humans affected in Minimata was considered to be the consumption of contaminated fish and shellfish from Minimata Bay (Jernelöv, 1969; Fujiki, 1972; Takeuchi, 1972; Taylor, 1982).

By the end of 1962, 121 cases of Minimata disease had been officially recognised (Fig. 12), and 46 of these patients had died. Astonishingly, the Chisso factory continued to discharge effluents contaminated by mercury until 1968, and the number of cases continued to increase. By late 1974, 107 deaths had occurred amongst some 798 officially verified patients, and 2,800 additional inhabitants of the area had applied for official verification. The time courses of acetaldehyde production by the factory and of the reported cases of Minimata disease are shown in Fig. 13, and the temporal correlation between these is clear. It might also be noted here that a further 59 cases of Minimata disease were reported by the

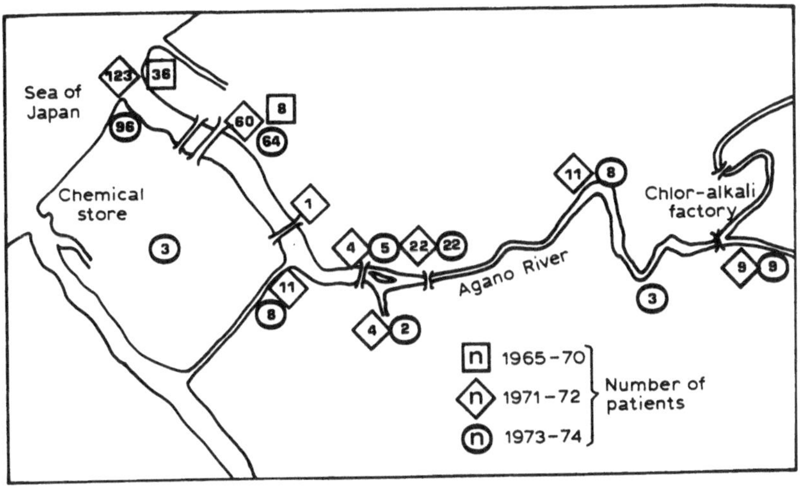

Fig. 14. The geographical distribution of mercury poisoning in Niigata, Japan. See the insert to Fig. 12 for the location of Niigata in Japan. After Takizawa (1979).

Kumamoto University research team in 1973, these being inhabitants of the city of Goshonoura, which lies opposite Minimata on the north-western side of the Shiranui Sea (see Förstner and Wittmann, 1983).

During the course of the protracted Minimata incident, mercury contamination was becoming recognised elsewhere as a public health threat. In 1964–65, some 47 cases of Minimata disease (with 6 deaths) were reported from Niigata, to the north-west of the island of Honshu in Japan (Fig. 14). These were caused by the discharge of mercury in the effluents of the Showa Electric industrial plant, and the resulting contamination of fish in the Agano River, which flows through Niigata. The ingestion of contaminated fish was once again the principal cause of the poisoning.

In the 1960s, problems had also been noted in Sweden, where the use of organomercurial dressings as fungicide agents for grain and other seeds was widespread. Bird populations were severely affected, and mercury accumulation by fish led to the instigation of bans in 1967 on fishing in many inland lakes and rivers (D'Itri, 1972). Similar concerns arose in Canada and the United States in the late 1960s and early 1970s, although in these instances the principal sources of the mercury were chlor-alkali plants and other industrial concerns, rather than the use of organomercurial seed dressings (Fig. 15). Legislation was introduced on the

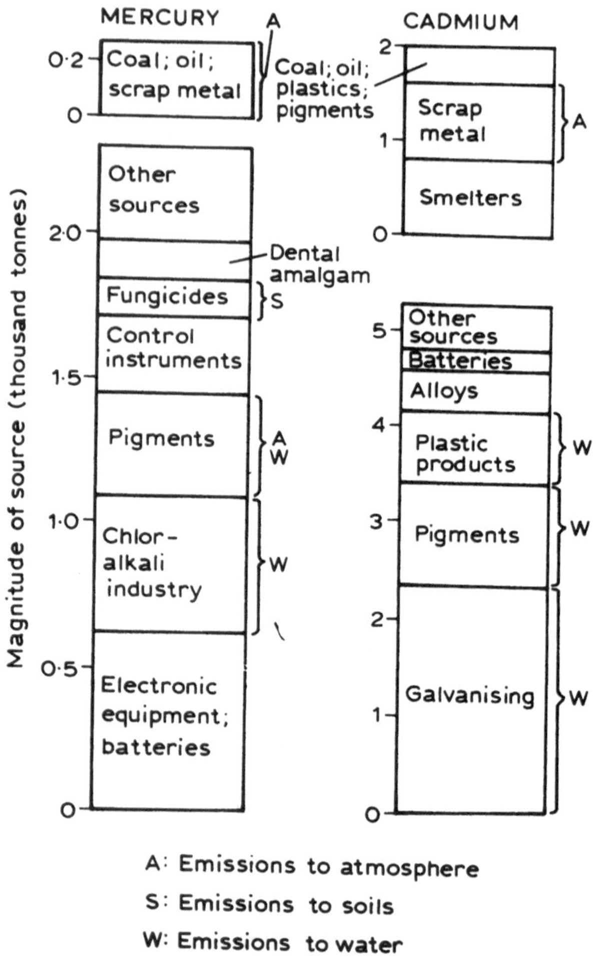

Fig. 15. Sources of mercury and cadmium emissions in the United States. Where emissions affect principally the atmosphere, soils, or water, these tendencies are shown. After Förstner and Wittmann (1983).

maximum permissible concentrations of mercury in fish products, and the swordfish and tuna fisheries in the USA were severely affected. An American female of age 44 was found to have contracted mercury poisoning through eating excessive amounts of fresh swordfish as a method of dieting (D'Itri, 1972).

More recently, Hansen (1990) concluded that mercury ingested principally in whale meat may pose a significant threat to the Eskimo populations of Greenland, at least in terms of foetal exposure. An isolated report of Minimata disease in the village of Muara Angke in Indonesia also exists (Webb, 1983), but does not appear to have been followed up to date.

Further poisoning episodes involving mercury were reported in the 1970s, although not all of these were strictly environmentally-related. Bakir *et al.* (1973) reported that the population in northern Iraq had suffered a massive poisoning episode due to the ingestion of wheat seed dressed with methylmercury. The wheat seed had been imported from Mexico for planting, but some had been used to make bread for human consumption. In an attempt to protect the population, the Iraqi authorities stated that possession of the treated seed would be considered an offence liable to prosecution involving the death sentence. This caused panic, and the treated seed was dumped in local rivers and lakes, thereby severely contaminating large areas. Estimates of fatalities due to the combined effects of these events ranged from 5,000 to 50,000; inhabitants suffering permanent disabilities were thought to number between 100,000 and 500,000 (Förstner and Wittmann, 1983). Similar episodes involving the poisoning of local populations through the ingestion of mercury-treated grain products had occurred in Iraq previously, as well as in Guatemala in the mid-1960s, in Pakistan in 1969 (Bakir *et al.*, 1973), and in Yalovi village in the Volta region of Ghana in the early 1970s, where 144 persons were poisoned by eating mercury-treated maize (Derban, 1974).

Other trace metals have also been the cause of environmental poisoning outbreaks. In 1947, an unusual disease of an overtly "rheumatic nature" broke out in the Jintsu River basin in Toyama Prefecture in Japan (Yamagata and Shigematsu, 1970; Kobayashi, 1971; Friberg *et al.*, 1974). Some 44 cases of this affliction were recognised initially, and it became known as "itai-itai" disease (translated literally as "ouch-ouch" disease, from the cries of pain made by those suffering from the affliction). An accurate estimate of the numbers affected by the disease has not been produced, but it is considered that about 100 deaths occurred due to itai-itai disease among a total of about 200 patients prior to 1966 (Förstner and Wittmann, 1983; Mance, 1987; see Fig. 16). The symptoms of the disease included a yellow discolouration of the teeth, loss of the sense of smell, a reduction in red blood cell numbers, lumbar and leg pains, softening of the skeletal bones, pseudo-fractures of these bones, and (in severe cases) skeletal deformation. In addition, albumin and various other

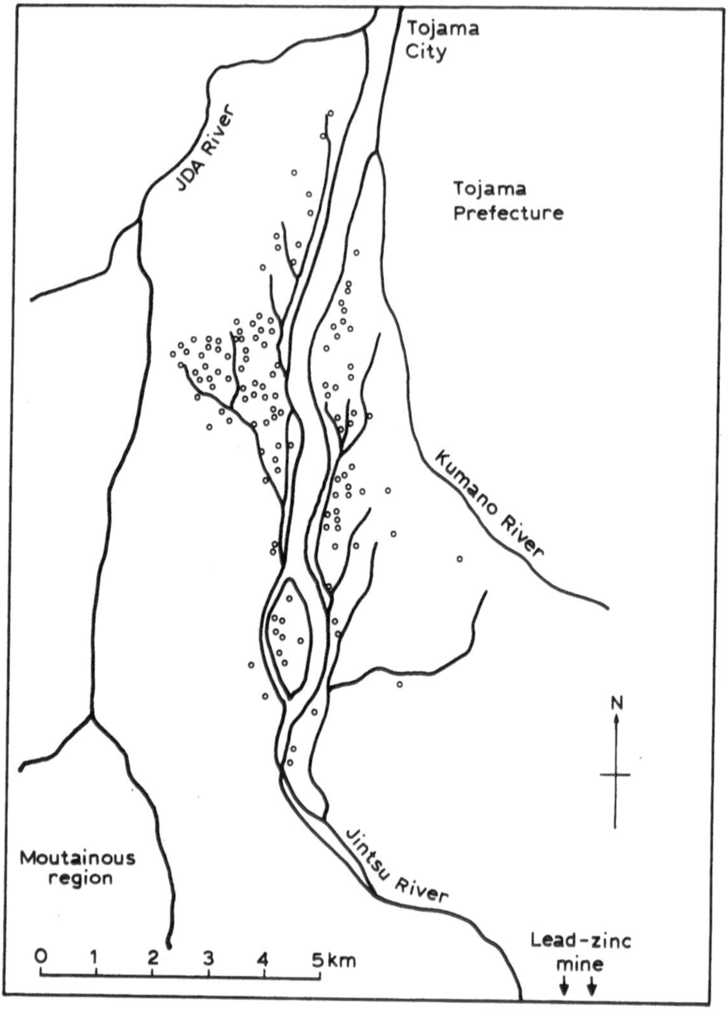

Fig. 16. The distribution of cases of itai-itai in the lower Jintsu River basin, Japan. See the insert to Fig. 12 for the location of the Jintsu basin in Japan. After Kobayashi (1971).

proteins were found in the urine of affected patients, due to kidney damage.

The cause of itai-itai disease was uncertain until the early 1960s, when Hagino and Yoshioka (1961) postulated that it was due to chronic cadmium poisoning. This was later officially confirmed by the Japanese Ministry of Health and Welfare, but controversy has continued to surround the matter, particularly as various factors such as sex, pregnancy incidence, ageing, and calcium deficiency appear to have contributed to the formation of the symptoms.

It is now believed that the primary cause of the itai-itai syndrome was the discharge of cadmium-rich effluents from a zinc mine operated by the Makioki Company, situated adjacent to the Jintsu River about 50km upstream from the affected area. Production at the mine increased in the 1940s, but the effluents and flotation sludge produced were discharged untreated to the Jintsu River. In the downstream river basin, the waters of the river were employed for the irrigation of rice fields, and the rice crop produced was severely contaminated. Concentrations of cadmium in rice from the affected area approximated 0.7 $\mu g\ g^{-1}$ wet weight, which is an order of magnitude greater than levels recorded for rice elsewhere in Japan (Friberg *et al.*, 1974). Cadmium uptake from ingested rice was thought to be the principal cause of the disease in those affected. In 1955, a retaining dam was constructed at the site of the mine, and the discharge of contaminated sludge was reduced. Thereafter, the incidence of itai-itai disease diminished rapidly in the downstream area.

Global concern over the potential for adverse public health impacts from cadmium has continued, however. The provisional tolerable weekly intake for cadmium in humans is 400–500 μg (FAO/WHO, 1972), and it is known that populations in many areas of the world approach or exceed this amount (Friberg *et al.*, 1974; Nriagu, 1988). In certain locations, cadmium contamination of soils occurs due to previous industrial activities or because of the presence of shales and other geological strata which are naturally cadmium-rich (e.g. at Shipham in south-west England; see Sherlock, 1983). High-risk groups ingesting unusually large amounts of the element in particular types of seafoods or offal undoubtedly exist in some populations, and subclinical effects (mostly involving the liver and kidney cortex) may occur (CEC, 1978; Nriagu, 1988). As a result of such concerns, cadmium has been included with mercury on the so-called "Black List" of contaminants for controls in Europe (see Chapter 3). It might also be noted here that cadmium came under some suspicion as the possible cause of vomiting after the consumption of Pacific oysters (*Crassostrea gigas*) in Tasmania, Australia in the early 1970s (Thrower and Eustace, 1973). It is now accepted, however, that the high concentrations of zinc in these oysters were the primary cause of the emetic action observed.

Lead is also of great concern as an environmental contaminant, particularly with respect to possible impacts on children. While controversy continues (at least partly due to analytical difficulties; see Patterson and Settle, 1976a,b), the evidence for a general global increase in lead concentrations over the last two centuries now appears convincing (e.g. see Grandjean, 1981; Wolff, 1990). The principal sources of lead uptake in humans are pulmonary in most instances, but significant accumulation of the metal also occurs in occupational exposure; from the ingestion of pica by children; from glazes on crockery; and from the dissolution of the element in drinking water distribution systems. Cases of lead poisoning from each of these causes have been reported (e.g. see Förstner and Wittmann, 1983; Nriagu, 1988). The most commonly observed symptoms include nephritis (often after a long incubation period) and neurological disorders; however, Nriagu (1988) lists large numbers of subsidiary effects which may be anticipated to occur in either acute or chronic exposures to the element.

Both arsenic and chromium have also caused significant public health impacts. In its inorganic form, arsenic is highly toxic to biota, and its toxicity to humans is evidenced by its historical popularity as a deliberately employed poison. In 1955, a mass outbreak of arsenic poisoning occurred in Japan, caused by the ingestion by children of milk powder containing arsenic-contaminated sodium phosphate, which had been added to the powder as a stabiliser. Some 12,000 cases were reported, 130 of these being fatal. A study undertaken 14 years later reported the existence of persistent impacts on the central nervous system. Tsuchiya (1977) has provided a useful review of this incident and other arsenic poisoning episodes in Japan, the latter mostly involving occupational exposure to the element.

In Czechoslovakia, arsenic emissions from a coal-fired power plant are considered to have given rise to respiratory problems and hearing loss in local children (WHO, 1981). The contamination of drinking water by arsenic has been of concern in Taiwan, Argentina and Chile (Tseng, 1977; Förstner and Wittmann, 1983; Nriagu, 1988). In Taiwan, an increased incidence of skin cancer has resulted in the local population, and isolated cases of Blackfoot disease (peripheral gangrene caused by vascular problems) have also occurred due to this contamination (Tseng, 1977; Calabrese, 1983). Both arsenic and lead have been associated with poisoning of humans in the USA through the consumption of "moonshine" (illicit whisky); the principal impacts are neurological and kidney-related (Gerhardt *et al.*, 1980; Harrington, 1980).

Arsenic is present in seafoods in appreciable concentrations, and this has generated considerable public health-related concern over the last two decades in particular. Phillips (1990a) has reviewed the present situation, noting in particular the dependence of the toxicological impacts of arsenic in seafoods on the chemical speciation of the element in marine biota. While total arsenic levels are commonly found to be high in marine organisms (and these concentrations would certainly be of great concern if the element were present in the inorganic form), the principal forms of arsenic found in marine organisms are organic in nature. The present evidence suggests that these organic forms of arsenic (which include arsenobetaine and arseno-sugars, in addition to several other chemical species) are of relatively low toxicological impact to humans (Phillips, 1990a).

By contrast, chromium is of considerable toxicity to humans, especially in its hexavalent form. The Kiryama factory of the Nippon-Denko Concern in Hokkaido, Japan, has been established to be the principal source of airborne chromium in that area, and a high local incidence of perforation of the nasal septum and of lung cancer has resulted, with many mortalities. In addition, Förstner and Wittmann (1983) noted that severely contaminated wastes from the Nippon Chemical Industrial Company have apparently been employed in construction materials in Tokyo and Chiba Prefecture, and the local groundwater is known to be severely contaminated by chromium. They considered this situation to "...appear to herald the beginning of an unparalleled, scandalous disaster..." (Förstner and Wittmann, 1983).

Finally here, it should be noted that the acute poisoning episodes discussed above may be accompanied by a much greater incidence of sublethal impacts due to the chronic exposure of populations to rather lower concentrations of metals. Nriagu (1988), in a thought-provoking review of such possible effects, has proposed lead, cadmium, mercury and arsenic as the four elements of greatest probable impact on human health. However, he notes that several other metals (such as aluminium and thallium) may also be implicated in significant impacts, and that many trace elements interact with other factors and hence may contribute to the incidence of disease. There can be no doubt that further epidemiological evidence is the most urgent requirement for resolving such concerns.

(iii) Radionuclides and Public Health

The contamination of aquatic and terrestrial environments by radioactive elements engenders enormous political concern. While it may be argued that relatively few incidents involving radionuclides have occurred which have directly affected public health, the spectre of long-term radioactive contamination is perhaps the most active and passionate area of political debate among environmental concerns.

Goldberg (1976) cites three groups of radionuclides which may be introduced in significant quantities to aquatic environments. These are nuclides which are present in nuclear fuels (e.g. ^{235}U and ^{238}Pu); fission products from such fuels (e.g. ^{137}Cs); and activation products arising from the irradiation of metals in nuclear reactors or nuclear weapons (e.g. ^{65}Zn and ^{55}Fe). Radium (^{226}Ra) is of importance in mining and milling operations, and is unusually toxic and hence of importance as an environmental contaminant. Isotopes of uranium are critical to fuel fabrication and enrichment processes, whereas plutonium is the most important nuclide in the recycling of nuclear fuels. Isotopes emitted from reactors include (in order of their relative importance) ^{137}Cs, ^{134}Cs, ^{58}Co, ^{60}Co, ^{54}Mn, and ^{3}H; the last of these is emitted in significant quantity from reprocessing plants (Förstner, 1980).

In general, it is believed that radioactive isotopes of trace elements exhibit similar environmental cycling to their corresponding stable isotopes. However, their potential toxicological impacts on organisms are enhanced compared to stable isotopes by their emission of radioactivity, as alpha, beta or gamma radiation. Subsequent to their accumulation by biota, alpha emitters (e.g. ^{238}U; ^{239}Pu) are considered to have much greater radiation-related impacts than either beta or gamma emitters (e.g. ^{58}Co; ^{137}Cs), due to their higher specific ionisation (Förstner, 1980).

Low-level emissions of radioactivity occur from nuclear reactors as a matter of course. Thus, for example, the Hanford reactors in the Columbia River in Washington State, USA, discharged significant quantities of several radionuclides between 1940 and 1971. The principal isotopes present in these effluents were ^{51}Cr, ^{65}Zn and ^{60}Co. The Columbia River plume was significantly enriched by these nuclides, and their presence could be tracked for large distances seaward. Examples of the contamination produced include that in oceanic migrating hake, *Merluccius productus* (Cutshall *et al.*, 1977). The seaward transport of sediment from the estuary could also be traced, by monitoring the spatial distribution of ^{65}Zn (Förstner, 1980).

The nuclear reactors at Windscale (now renamed Sellafield) in Cumbria in the north-west of England have also constituted a highly significant source of radionuclides to the marine environment. Isotopes which are known to have been discharged in quantity from this facility include ^{95}Zr, ^{95}Nb, ^{106}Ru, ^{144}Ce, and ^{137}Cs. While the annual inventories of radionuclides

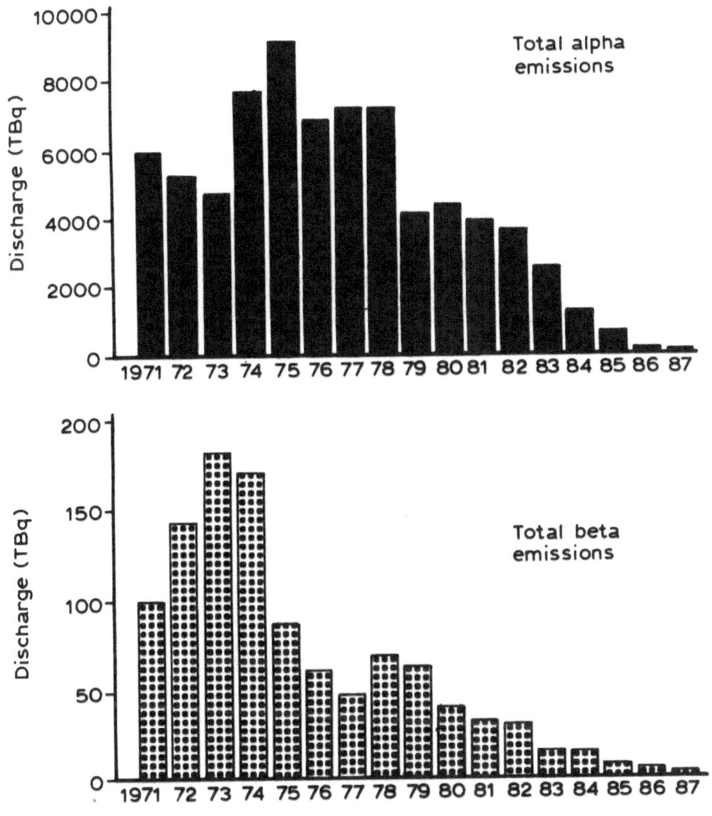

Fig. 17. Annual inventories of total alpha and total beta radioactivity (total becquerels, TBq) emitted from Sellafield (Windscale) nuclear power station between 1971 and 1987. After ISSG (1990).

released have diminished in recent decades (Fig. 17), emissions from facilities such as Sellafield may be traced for very long distances (see review by ISSG, 1990). Studies of the distribution of ^{137}Cs have shown that radionuclides derived from Sellafield are dispersed northwards along the west coast of Scotland, and then southwards through the North Sea (Fig. 18). Additional (relatively minor) discharges of ^{137}Cs are evident in Fig. 18 from the nuclear reprocessing plant at La Hague in France, which is also a documented source of ^{239}Pu and ^{240}Pu (Förstner, 1980).

In addition to the "normal" emission of radionuclides from nuclear reactors and reprocessing facilities, significant amounts of radioactive

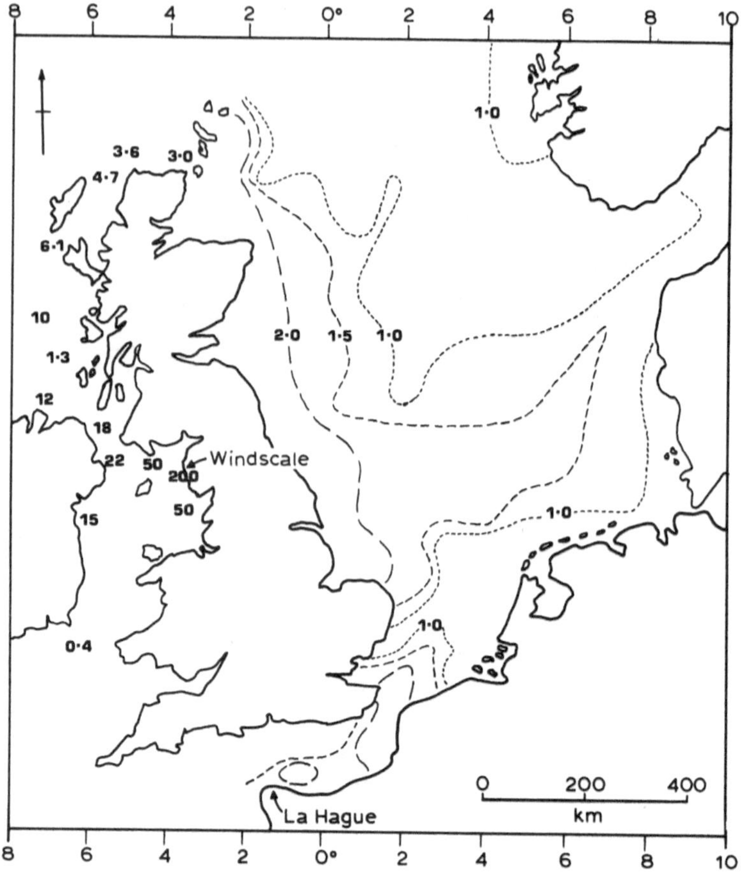

Fig. 18. Distribution of ^{137}Cs (pCi kg^{-1}) in surface waters around the United Kingdom in the 1970s. Isolines are shown only for intervals of 1.0, 1.5 and 2.0 pCi kg^{-1}, while individual values are provided to the north-west of England and Scotland. Note the predominant dispersion of nuclides from Windscale (Sellafield) in a clockwise direction, and the minor impact of the Cap La Hague installation. After Förstner (1980).

isotopes enter aquatic environments from the fallout from nuclear weapons testing and from accidents involving nuclear reactors. The latter are of particular concern, attracting huge media attention and much political debate. Eisenbud (1963) has provided an interesting review of several early case studies of nuclear testing events and accidents which gave rise to significant environmental contamination. Among these are the thermonuclear detonation of March 1954 at Bikini Atoll, which contaminated large areas of the Marshall Islands as well as affecting the crew of the Japanese fishing vessel *Fukuru Maru*, which was inadvertently not cleared from the test area.

An incident at Windscale reactor number one in October 1957 was, however, of even greater concern. This reactor was an air-cooled graphite-moderated vessel employed in the production of plutonium from uranium. An attempt to anneal the reactor graphite on 7 October 1957 led to the production of excessive heat in the reactor core, and the consequent temperature rise was not detected due to insufficient core instrumentation. A substantial fire resulted which lasted 4 days, with highly significant quantities of nuclides (especially ^{131}I, ^{137}Cs, ^{89}Sr and ^{90}Sr) being released into the atmosphere. The resulting contamination of milk by ^{131}I created considerable concern (Fig. 19), as did gamma radiation levels in the area.

Other accidents in either reactors or radiochemical facilities followed, several of these being of significant potential for public health effects. In November 1959, a chemical explosion occurred in a radiochemical processing pilot plant at Oak Ridge National Laboratory in the USA. The cause of the explosion was the presence of phenol in a decontaminating agent, which was later mixed with nitric acid. A release of ^{239}Pu resulted (with smaller amounts of ^{95}Zr and ^{95}Nb also being involved), and a lengthy and expensive decontamination and fixation process was required in the buildings of the complex and the surrounding streets (Eisenbud, 1963).

Further incidents have occurred regularly in the nuclear industry since this time, including major events at Three Mile Island in the USA and at Chernobyl in the USSR. The latter involved a super-critical event which led to an explosion of the number four reactor core at Chernobyl in April 1986, followed by the burning of the graphite moderator on its exposure to the atmosphere. The initial explosion and fire resulted in the production of a radioactive plume, which moved towards the north and north-west, contaminating areas *en route* through the USSR, the Baltic nations, and much of central and western Europe. The principal radionuclides of concern were ^{131}I, ^{134}Cs and ^{137}Cs, but their distributions subsequent to the accident were found to be highly heterogeneous (INSAG, 1986).

Attempts were made to extinguish the fire and reduce the levels of radioactive particulates released, involving the dropping of about 5,000 tonnes of clay, sand, lead and other materials onto the reactor from the air. Emissions of radioactivity were reduced by about 60% in four days, and by early May 1986, emissions had been reduced to relatively low levels (INSAG, 1986). The reactor was sealed by the construction of a sarco-

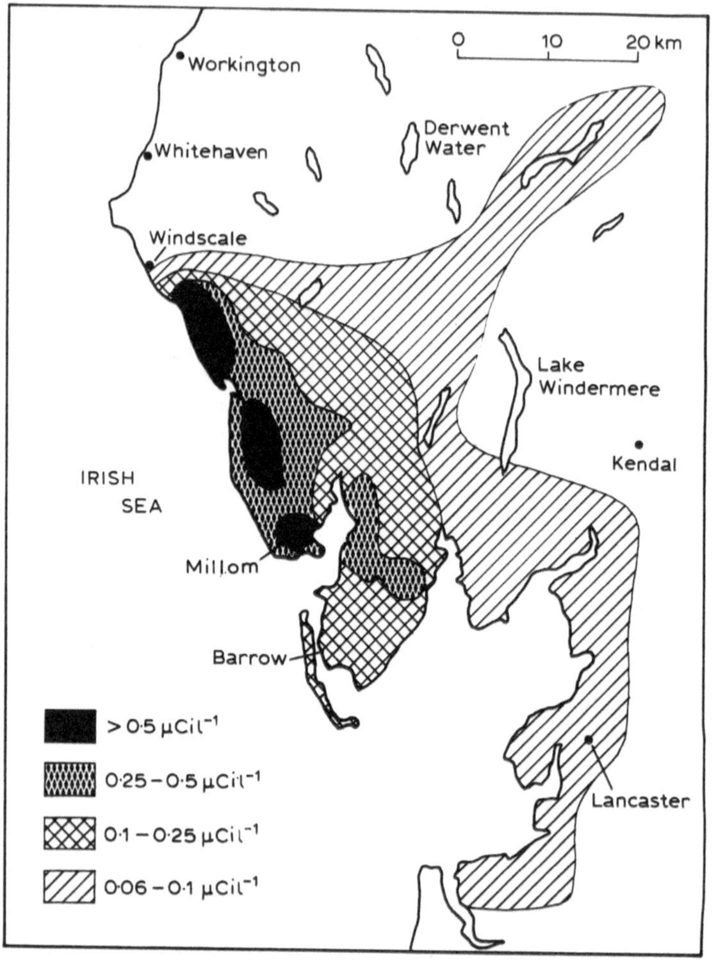

Fig. 19. The geographical distribution in concentration of ^{131}I in the milk of cattle around Windscale (Sellafield) 5 days after the commencement of the 1957 fire in reactor number one. After Eisenbud (1963).

phagus around its periphery in September of the same year, although concerns remain over possible continued releases of radioactive particulates.

Such incidents, while comparatively rare, have had a very considerable impact on the growth of the nuclear industry as a whole. Their impacts on human health remain a matter of great controversy, and there can be no doubt that concerns over the environmental effects of nuclear radiation have assisted considerably in the international development of environmental awareness.

(iv) Organochlorines and Public Health

The direct toxicity of many organochlorines to humans is low compared to their toxic impacts on other animals. Thus, instances of direct effects on public health have been rare for most organochlorines, whilst examples of their adverse impacts on aquatic resources are commonplace. The latter have occurred almost unabated from the early discovery of the effects of DDD at Clear Lake, California (see reviews by Hunt and Bischoff, 1960; Herman *et al.*, 1969), through the impacts of DDT on reproduction in birds, creating eggshell thinning (e.g. Moriarty, 1988), to the recent reports concerning the impacts of polychlorinated biphenyls (PCBs) and other contaminants on the reproduction of various fish species and of harbour seals (Reijnders, 1986; Cross and Hose, 1988; Spies and Rice, 1988; Casillas *et al.*, 1991). However, a few groups of organochlorines are of considerable significance in terms of their potential toxicity to humans. These include the dioxins (especially the most potent dioxin, which is 2,3,7,8-tetrachloro dibenzo-*p*-dioxin or TCDD), the dibenzofurans, and certain of the PCBs (Fig. 20).

In 1968, a poisoning outbreak occurred in Japan which subsequently became known as the Yusho (translated as "rice oil") affair. More than 1,500 persons were affected in total, and the symptoms displayed included chloracne, ocular and liver damage, immunological impacts, and neurological effects (Kuratsune, 1980). Investigations of the incident concluded that the cause was the contamination by a commercial PCB preparation (Kanechlor 400) of a particular brand of rice oil produced by the Kanemi Rice Oil Company. The PCBs were employed as a heat exchange fluid in the production of the rice oil, and had contaminated the product through their accidental transfer to the oil from the heat exchanger.

Originally, it was considered that the PCBs themselves were responsible for the toxic impacts observed on the victims. However, more recent research employing isomer-specific analytical techniques has demonstrated that the principal causative agent was in fact a polychlorinated dibenzofuran (2,3,4,7,8-pentachloro dibenzofuran; see Tanabe *et al.*, 1989b), which was produced through heating the Kanechlor 400 found in the affected oil. Certain coplanar PCBs also contributed to the toxic impacts of the rice oil, however, and polychlorinated quaterphenyls (which are dimers of PCBs) may also have been implicated (Mochiike *et al.*, 1986; Tanabe *et al.*, 1989b). A similar poisoning event occurred in Taichung in

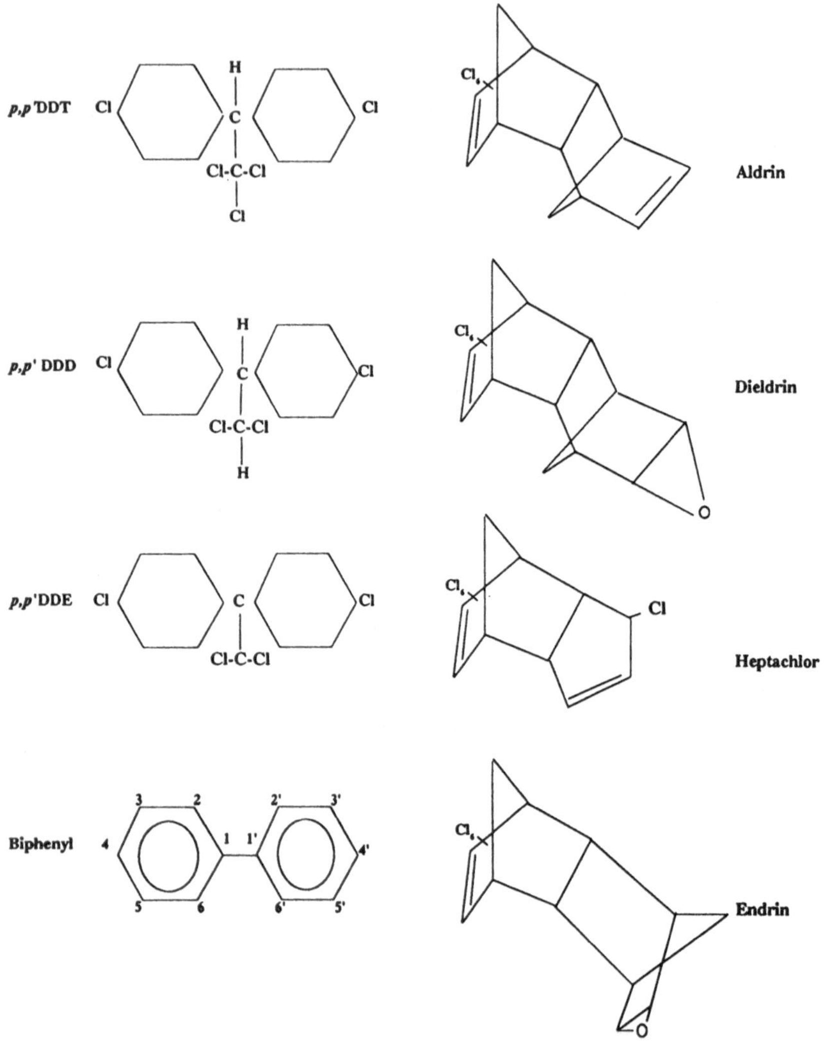

Fig. 20. Chemical structures of various organochlorine pesticides and of the biphenyl ring which forms the basis of PCBs.

Taiwan in 1979, due to the contamination of rice-bran oil by PCBs (Chen *et al.*, 1981; Chen and Hsu, 1986).

Perhaps the best-known incident of widespread environmental contamination involving dioxins occurred at Seveso in northern Italy in 1976. The ICMESA plant at that location (owned by the pharmaceutical giant Hoffmann-La Roche, through Givaudan in Switzerland) had been producing trichlorophenol subsequent to about 1970, apparently in breach of their industrial licence (Margerison *et al.*, 1980; Strigini, 1983). On 10 July 1976, an exothermic reaction occurred in a chemical reactor producing trichlorophenol. A safety valve broke under the pressure from the reaction, and a cloud of contaminated material was released into the atmosphere. It was later established that the material released contained significant quantities of dioxins, including the extremely toxic 2,3,7,8-tetrachloro-dibenzo-*p*-dioxin.

The most severely contaminated area ("Zone A") measured 108 ha and included the residences of 670 local inhabitants. The less heavily contaminated "Zone B" to the south-west of the ICMESA plant was 270 ha in area, and contained almost 5,000 inhabitants. A "Warning Zone R" was also established on the periphery of these two affected areas, this being 1,430 ha in area with a population exceeding 32,000 (Strigini, 1983). The last of these zones was initially considered to be a control area, but was later shown also to contain traces of dioxins.

Zone A was evacuated in late July 1976, and was fenced off under the surveillance of the Italian army. Pregnant women and all children were evacuated from Zone B also, and warnings against the consumption of local crops and animals were promulgated in all zones, including Zone R. Agricultural activities in the three zones were ordered to cease, and all animals were destroyed. Later attempts at decontamination included the removal and replacement of 20cm of topsoil in particularly contaminated areas; the removal and disposal of vegetation; the dismantling and removal of some buildings and their contents; the washing of other buildings and contents with solvents; and the eventual incineration of the most contaminated material in Switzerland.

Epidemiological studies on the local population were unfortunately poor, and the precise impacts of the Seveso accident are difficult to estimate (Strigini, 1983). Some 193 cases of chloracne were documented among school children, and these may be considered to be a direct result of the dioxin contamination. In addition, 600 cases of birth defects were recorded in the exposed population, the most common of which were haemangiomas (vascular malformations visible as skin spots, and usually benign in nature). The incidence of pregnancy was reduced after the dioxin exposure, and that of spontaneous abortion increased. Whether such effects are indicative of social changes in response to the contamination or of a teratogenic or other toxic effect on the population is unknown (Strigini, 1983), although there is evidence from the use of the defoliant Agent Orange in Vietnam that dioxins cause teratogenic effects in mammals, including humans.

Dioxins and dibenzofurans are also produced through the pyrolysis of PCBs, and several cases of local contamination by these compounds have been recorded, these being due to either the illegal burning of PCBs at low temperatures, or to fires involving PCB-filled transformers (e.g. USCDC, 1985; Thompson *et al.*, 1986). In such instances, decontamination is always a lengthy affair, and in some cases is not possible. Further sources of both dioxins and dibenzofurans have only very recently been documented (Anon., 1991a), and these include the production and use of chlorine gas; the chlorine bleaching of pulp; the iron and steel industry; and various photochemical reactions. There is an urgent need for further work on these exceptionally toxic compounds.

Other instances also exist of contamination by pesticides or similar types of organic contaminants, some of these being of significant public health impact. In 1973, a fire retardant composed principally of polybrominated biphenyls (PBBs) was accidentally substituted for a cattle feed supplement in Michigan, USA. PBBs rapidly appeared in both milk and meat products, and by 1978, were estimated to be present in 97% of the human population of the State (Schnare *et al.*, 1984). The precise chronic effects of PBBs on humans are largely unknown, but appear to be principally related to impacts on the liver and sensory nerves; follow-up studies are continuing in this case. The contamination of the James River and other parts of Virginia State, USA, by the pesticide Kepone was originally discovered in 1975 due to the impacts of Kepone on workers at the Life Sciences Products plant in Hopewell, Virginia, which produced the pesticide. While relatively few cases of human intoxication by Kepone were documented, the pesticide exerted significant impacts in certain rivers in the USA (Huggett and Bender, 1980). The contamination of the Love Canal area near Niagara Falls in New York State also merits mention, as the Superfund in the USA grew from this event and is still employed to fund clean-up actions in environmentally degraded areas (Dickson, 1982).

It should be noted here that the analysis of breast milk provides an excellent indication of the abundance of organochlorines in a human population (see review by Jensen, 1983). Data for organochlorines in breast milk samples from western populations generally indicate a decrease in the concentrations of DDT, hexachlorocyclohexane (HCH), PCBs and other organochlorines since the introduction of controls in western nations (mostly in the 1970s) on the use of these substances. By contrast, there is evidence from parts of the developing world for severe contamination. Thus, for example, high levels of DDT and its metabolites have been reported in breast milk samples from Guatemala, El Salvador and Iraq (Winter *et al.*, 1976; de Campos and Olszyna-Marzys, 1979; Al-Omar *et al.*, 1986). Interestingly, DDT and its metabolites and isomers of HCH are present at high concentrations in the breast milk of ethnic Chinese in Hong Kong (Ip, 1983; Ip and Phillips, 1989). It is thought that the latter pattern is due principally to the ingestion of organochlorines in seafoods in Hong Kong (Ip and Phillips, 1989). There is a need for further studies of this type in the developing nations.

(v) Oil Pollution

While oil pollution of aquatic environments generally does not affect public health directly, the scale and severity of the impacts from oil spills on aquatic resources generate very considerable political awareness.

The first large-scale oil spill from a tanker occurred on 18 March 1967. At about 09.00 hours, the *Torrey Canyon*, bound for Milford Haven from the Persian Gulf, struck the Pollard Rock of the Seven Stones, some 24 km west of Land's End, England. The tanker was carrying a total of about 115,000 tonnes of Kuwait crude oil in 18 storage tanks, and was travelling at 17 knots when she ran aground. Six of the storage tanks were immediately ruptured, and over the following several weeks (during which attempts to refloat the vessel failed, she broke her back, was bombed, and eventually sank) essentially all her oil escaped, to heavily contaminate the English and French coasts on both sides of the English Channel. The *Torrey Canyon* spill was by far the largest recorded to that time, and not only created massive international media attention, but led to significant marine research into the impacts of oil pollution (Smith, 1968). Unfortunately, with the global trend towards ever larger bulk carriers, worse was to come.

Almost exactly 11 years after the *Torrey Canyon* spill, the supertanker *Amoco Cadiz* was wrecked some 3km off the coast of Portsall in Brittany, France. The *Amoco Cadiz* was carrying a total of over 220,000 tonnes of oil (about 45% being light Arabian crude, and the remainder light Iranian crude) when she ran aground in high seas on 16 March 1978, creating the largest single spill ever recorded from a bulk carrier, and dwarfing the

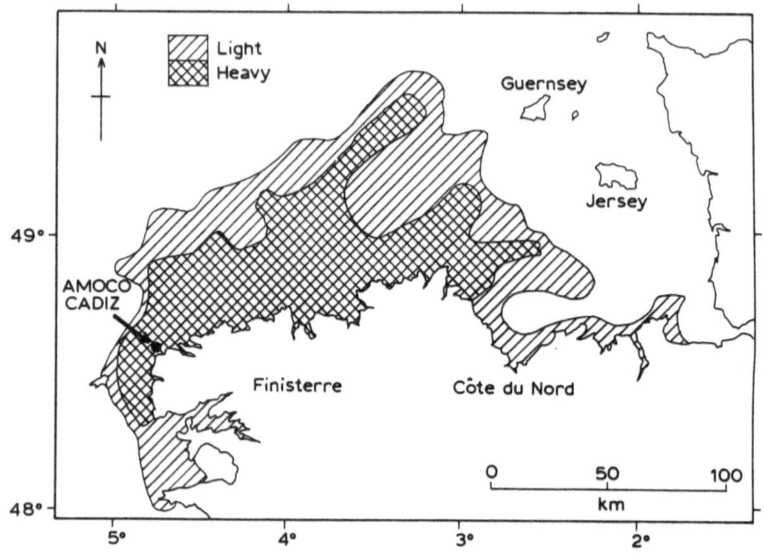

Fig. 21. The extent of the areas in which heavy and light slicks occurred off the coast of north-west France following the wreck of the *Amoco Cadiz*. After Spooner (1978).

Torrey Canyon spill in terms of volume. The complete cargo was lost over some 14 days, creating a very large area of slicks and surface sheens (Fig. 21). The rough weather, the high proportion of light aromatic fractions in the oil spilled, and the use of relatively less toxic dispersants in the *Amoco Cadiz* event rendered its impacts quite distinct from those of the *Torrey Canyon* spill (Laubier, 1978; Spooner, 1978). It is likely that about 40% of the oil spilled by the *Amoco Cadiz* was volatilised quite rapidly, to be followed by photodegradation. However, some of the light fractions were undoubtedly taken into solution, and acute toxicity was documented (e.g. Cabioch *et al.*, 1978). Importantly, this spill also afforded an excellent opportunity to study the long-term impacts of major catastrophic events of this nature, and it has become apparent that the recovery of aquatic ecosystems is slow for at least certain of the affected communities (e.g. see Conan, 1982; Dauvin and Gentil, 1990).

Highly sensitive marine environments have also suffered. On 24 March 1989, the US-flagged tanker *Exxon Valdez* ran aground on Blye Reef in Prince William Sound, Alaska. The largest spill recorded to date in American waters followed, some 268,000 barrels of oil being lost from the vessel. Oil spill contingency plans had been drafted some years before the event, but (despite the existence of unusually good weather for 3 days after the stranding of the vessel) the deployment of booms, skimmers and dispersants was extremely slow. Prince William Sound is not only an area of extreme natural beauty and high wildlife diversity, but is also heavily visited by tourists to Alaska. The precise impacts of the spill are still being debated (both in the courts and elsewhere), but its occurrence and the paucity of any rapid response should be seen as nothing short of an international scandal. Costs to date to Exxon have been estimated recently as well in excess of US$1 billion, and litigation is proceeding (Anon., 1991b).

The incidents related above may appear to be isolated events. However, this is not the case. Figure 22 and Table 6 provide a synthesis of major tanker-related spills from 1974 to early 1989. Over this period, 150 spills in excess of 10,000 barrels have been recorded. It is notable, however, that despite the increasing tonnage of oil transported in international waters with time, the improvement of operational procedures and the introduction of more stringent regulations have served to decrease the oil spilled from marine transportation, from about 2.4 million tonnes *per annum* in the early 1970s to about 1.5 million tonnes *per annum* in the early 1980s.

While oil spills of large volume from tankers frequently dominate media attention and create severe impacts in the environment local to the spill, this is by no means the major source of petroleum hydrocarbons entering the marine environment, accounting in fact for only an estimated 10% of the total. Spills of relatively small volume are extremely commonplace, as shown by Fig. 23. Oil sources of significance other than vessel-related spills include operational discharges from tankers (the cleaning of tanks and discharge of ballast water), municipal discharges and stormwater runoff, hydrocarbon inputs from the atmosphere, and natural seeps. It is accepted by the scientific community that the chronic inputs of hydrocarbons are of much greater magnitude than the acute spill-related inputs, although the latter receive much wider political and media attention (NAS, 1975; Burns and Smith, 1982; Whittle *et al.*, 1982).

The largest single oil spill recorded to date did not, however, derive from a grounded vessel, but from an armed conflict. Shortly before the liberation of Kuwait in April 1991, several major oil slicks appeared in the northern Gulf off Kuwait. It later transpired that these were derived from four main sources: damaged tankers at the loading terminals in Kuwait; discharges from the Al Ahmadi loading terminal in Kuwait and the Mina al Bakr terminal in Iraq; and damaged storage tanks in Khafji, Saudi Arabia (Pellew, 1991). Much of the discharge was of a deliberate nature,

Table 6 Compendium of major oil spills from vessels between 1974 and early 1989. All spills over 10,000 barrels are shown (after the *Times Atlas and Encyclopaedia of the Sea*, 1989)

1974

1. Metula[a]
2. Yuyo Maru No. 10[a]
3. Cherry Vinstra[b]
4. Theodoros V[b]
5. ABC 23118[c]
6. Eleftheria[c]
7. Elias[c]
8. Ercole[c]
9. Esso Garden State[c]
10. Hanetia[c]
11. John Colocotronis[c]
12. Key Trader[c]
13. Peter Maersk[c]
14. Saint Mary[c]
15. Sea Spray[c]
16. Transhuron[c]
17. Trojan[c]
18. Universe Leader[c]
19. Visahakit 1[c]

1975

20. British Ambassador[a]
21. Corinthos[a]
22. Epic Colocotronis[a]
23. Jakob Maersk[a]
24. Forthfield[b]
25. Master Stathios[b]

42. St. Peter[a]
43. Urquiola[a]
44. Argo Merchant[b]
45. LSCO Petrochem[b]
46. Nanyang[b]
47. Bohlen[c]
48. Diego Silang[c]
49. Mysella[c]
50. Sansinema II[c]
51. Sealift Pacific[c]

1977

52. Caribbean Sea[a]
53. Hawaiian Patriot[a]
54. Borag[b]
55. Irenes Challenge[b]
56. URSS 1[b]
57. Venoil[b]
58. Agip Venezia[c]
59. Al Rawdatain[b]
60. Dauntless Colocotronis[c]
61. Ethel H[c]
62. Maria B[c]
63. Oswego Glory[c]
64. Oswego Tarmac[c]
65. Venpet[c]

1978

82. Burmah Agate[a]
83. Gino[a]
84. Independenta[a]
85. Ionnis Angelicoussis[a]
86. Cys Dignity[b]
87. Gunvor Maersk[b]
88. Aegean Captain[c]
89. Antonio Gramsci[c]
90. Aviles[c]
91. Betelgeuse[c]
92. Chevron Hawaii[c]
93. Fortune[c]
94. Hitra[c]
95. Kurdistan[c]
96. Master Michael[c]
97. Messiniaki Frontis[c]
98. Sea Valiant[c]
99. Skyron II[c]
100. Vera Berlingieri[c]

1980

101. Jaun A. Lavalleja[a]
102. Georgia[c]
103. Gogo Rambler[c]
104. Irenes Serenade[a]
105. Lima[c]
106. Princess Anne Marie[c]
107. Tanio[c]

120. Honam Jade[c]
121. Monemvasia[c]
122. SF172[c]
123. Sivand[c]

1984

124. Almak[c]
125. Alvenus[c]
126. Arianna[c]
127. Glory[c]
128. Puerto Rican[c]

1985

129. Nova[a]
130. Grand Eagle[c]
131. Marina[c]
132. Napo[c]
133. Patmos[c]

1986

134. Amazon Venture[c]
135. Brotas[c]
136. Maysun[c]
137. Orleans[c]
138. Valparaiso[c]

1987

139. Cabo Pilar[c]

26. Princess Anne Marie[b]
27. Spartan Lady[b]
28. Athenian Star[c]
29. Barge 10[c]
30. Globtik Sun[c]
31. Heiwa Maru[c]
32. Olympic Alliance[c]
33. Pacific Colocotronis[c]
34. Scapbay[c]
35. Shell Barge No. 2[c]
36. Showa Maru[c]
37. Tarik Ibn Ziyad[c]

1976
38. Cretan Star[a]
39. Ellen Conway[a]
40. Grand Zenith[a]
41. Scorpio[a]

66. Amoco Cadiz[a]
67. Tadotsu[a]
68. Andros Patria[b]
69. Brazilian Marina[c]
70. Cabo Tamar[c]
71. Chrissi[c]
72. Christos Bitas[c]
73. Eleni V[c]
74. Interstate 19[c]
75. Kountouriotis[c]
76. Lima[c]
77. Ocean 250[c]
78. Peck Slip[c]
79. Sealift Mediterranean[c]
80. Takase Maru[c]

1979
81. Atlantic Empress[a]

1981
108. Cavo Cambanos[b]
109. Energy Endurance[c]
110. Globe Asimi[c]
111. Olympic Glory[c]

1982
112. Arkas[c]

1983
113. Assimi[a]
114. Castillo de Bellver[a]
115. Pericles G.C.[a]
116. Bellona[b]
117. Al Duriyah[c]
118. Da Qing 236[c]
119. Feoso Ambassador[c]

140. El Hani[c]
141. Odysseas[c]
142. Stuyvesant[c]
143. Stuyvesant[c]

1988
144. Athenian Venture[a]
145. Odyssey[a]
146. Amazzone[c]
147. Barge 283[c]
148. Esso Puerto Rico[c]
149. Nord Pacific[c]

1989
150. Exxon Valdez[a]

---

[a] Over 200 000 barrels.
[b] 100 000–200 000 barrels.
[c] 10 000–100 000 barrels.

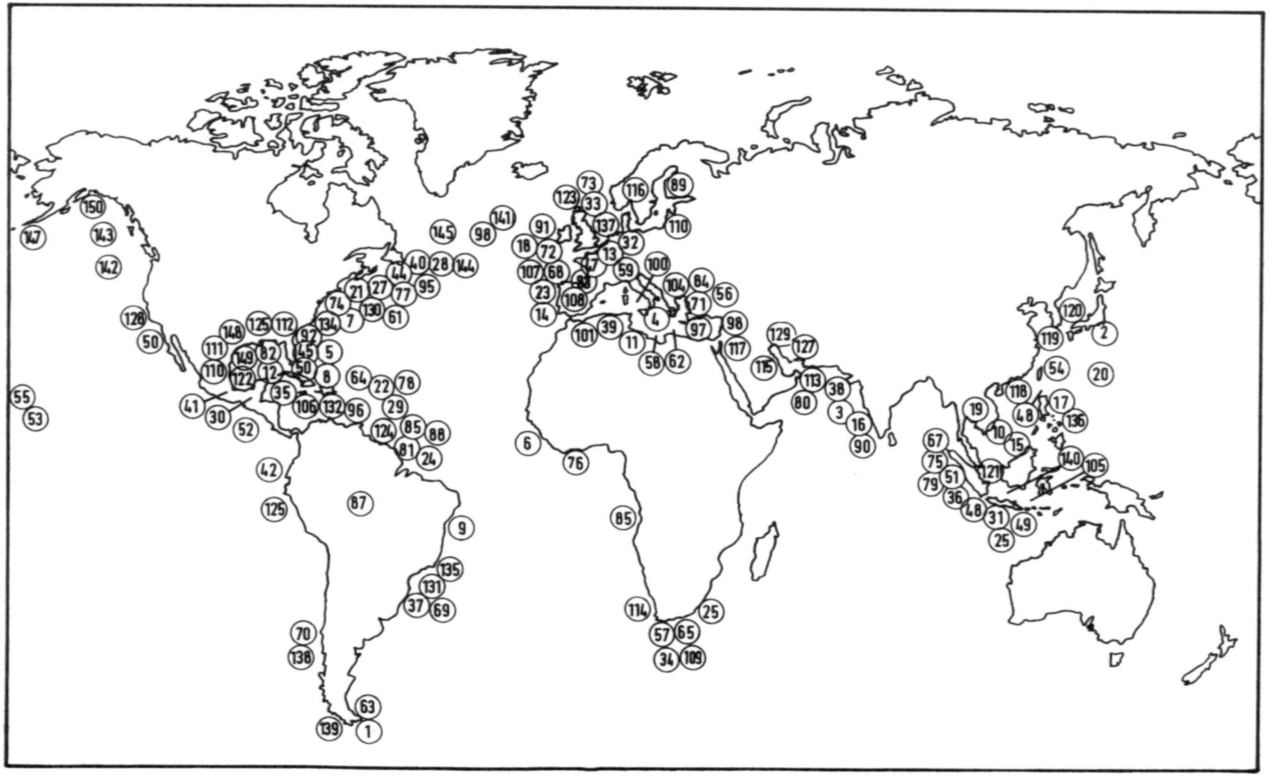

Fig. 22. The locations of tanker spillages greater than 10,000 barrels in volume, between 1974 and early 1989. See Table 7 for a key to the individual spills. After the *Times Atlas and Encyclopaedia of the Sea* (1989).

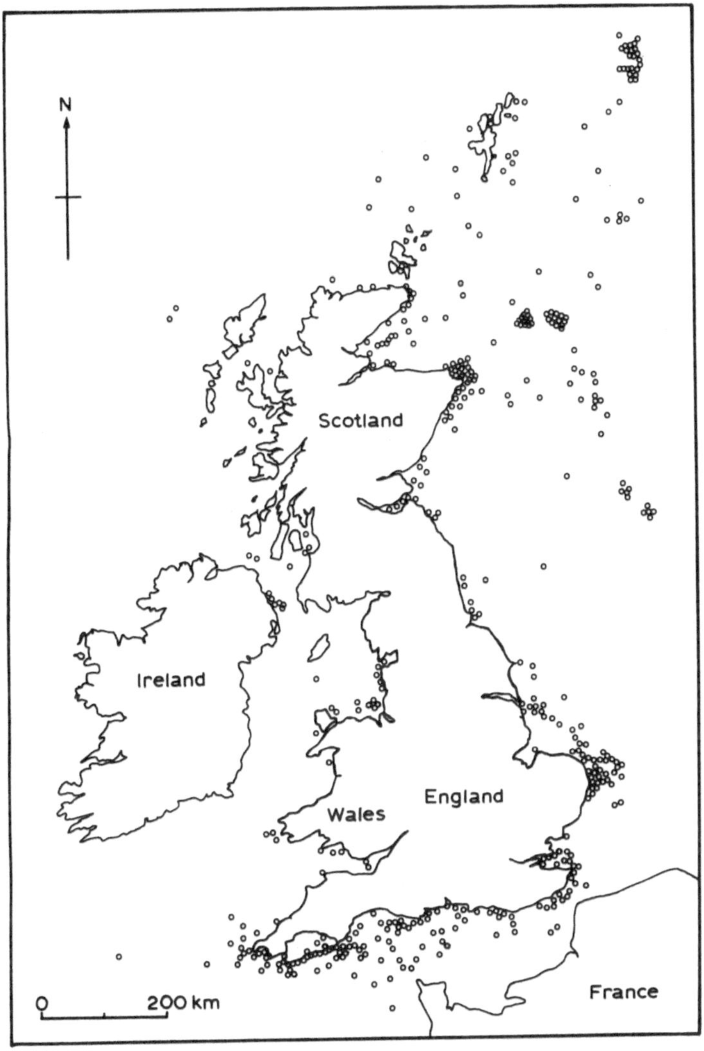

Fig. 23. The distribution of marine oil spills around Britain between 1974 and mid-1981. After Whittle *et al.* (1982).

and an estimated total of between 7 and 8 million barrels of oil was involved, the majority being derived from the tankers and loading terminal in Kuwait. This oil spill is approximately twice the magnitude of any single previous spill, and gave rise to severe oiling of at least 460km of the coastlines of southern Kuwait, Saudi Arabia, Bahrain, Qatar and the United Arab Emirates (Pellew, 1991). At the time of writing, its effects have not been fully documented.

C. THE NEED FOR MONITORING

It may be concluded from the above that the threat to inland and coastal waters of the world from contamination is at its most significant at present, and is increasing. The monitoring of aquatic contamination plays a central role in its control; unless the degree of contamination can be estimated, any attempt at regulating the impacts of pollution is doomed to failure.

However, such monitoring is a relatively new and emerging science, for several reasons. Firstly, the currently available analytical techniques are inadequate to accurately quantify many of the contaminants of potential importance in aquatic ecosystems (see chapter 3). There has been insufficient emphasis on aquatic chemistry to provide a driving force for the development and perfection of such techniques, at least until recently. Secondly, there can be little doubt that Man has for too long blithely ignored the threat posed to aquatic environments by contaminants, preferring to envisage the rivers and oceans as waste disposal facilities of limitless capacity. Finally, new techniques for the monitoring of aquatic contamination have needed to be developed, and this has taken time.

As a result of the search for and development of new monitoring techniques, the basic methods employed to monitor aquatic environments for contamination have undergone a revolution over the last two decades. Traditional methods of water sampling and their analysis for minute concentrations of contaminants have largely been replaced by a reliance on organisms to monitor pollution. The following chapters provide an overview of the initial development and present use of biomonitors for measuring the extent and degree of contamination of aquatic environments by trace contaminants.

Chapter 3

Aquatic Contaminants of Concern

A. INTRODUCTION

It should be emphasised here that any present list of contaminants thought to be of potential for exerting significant adverse impacts in aquatic environments will undoubtedly be incomplete, principally due to the shortcomings in aquatic science alluded to in the last section of the previous chapter. Thus, for example, polychlorinated biphenyls (PCBs), now thought to be of very considerable significance as a coastal and oceanic pollutant (e.g. see Waid, 1986; Tanabe *et al.*, 1987a; Tanabe, 1988) were not discovered in aquatic biota until 1966, some 37 years after their initial use by industry (Anon., 1966; Jensen, 1972). There is therefore no doubt that, as analytical techniques improve, new contaminants of concern will emerge.

Several authors have suggested novel methods of defining such contaminants, using their physical and chemical properties to provide clues to their potential for exerting impacts in aquatic environments. The use of octanol-water partition coefficients to estimate the capacity of organic contaminants for bioaccumulation has attracted a copious literature (e.g. see Ernst, 1980; Gossett *et al.*, 1983; Miller *et al.*, 1985; Connell, 1988), and other types of partition coefficients have also been considered (Chiou,

1985). There can be no doubt that these methods are valuable in certain cases; however, they are not of universal application (e.g. superlipophilic substances do not conform to the general rules; see Hawker and Connell, 1985; Connell, 1988; Phillips, in press). The more recent concept of fugacity also shows considerable promise (Mackay and Hughes, 1984; Clark *et al.*, 1988). Other authors have proposed that compounds derived from processes similar to those which create dioxins or related contaminants (i.e. products of "fire chemistry") merit further study (Tanabe *et al.*, 1989a).

The regulations surrounding the registration of new chemicals are also being strengthened over time. In both the USA and Europe, for example, new requirements are being introduced for the initial testing of chemicals prior to their use (e.g. see EEC, 1991). Such requirements include the testing of the toxicities and bioaccumulation properties of chemicals in standard test protocols relevant to both aquatic and terrestrial environments. The data produced through adherence to these legislative initiatives will undoubtedly improve our understanding of the relative toxicities of different chemical classes in aquatic ecosystems.

Contaminants which are presently considered to be of concern in aquatic environments have been identified and discussed by several authorities. This chapter provides an overview of the regulations produced in Europe and the USA to date; further detail may be found elsewhere (e.g. see Crathorne and Dobbs, 1990; Haigh, 1992).

B. EUROPEAN LEGISLATION

The initial legislative efforts to control the discharge of contaminants to aquatic environments in Europe grew out of the Paris Convention on the Prevention of Marine Pollution from Land-Based Sources. This Convention (adopted in 1974) covers the North Sea and the north-east Atlantic area, and is essentially concerned with contaminants discharged from land-based sources either to inland waters or directly to marine environments (see Bates, 1991). Contaminants were assigned to classes (which became known as "Black List" and "Grey List" substances), and the stated intention was to eliminate pollution by the former, while "strictly limiting" pollution from Grey List substances. Parallel legislation dealing with the dumping and incineration of wastes in the north-east Atlantic was introduced by the Oslo Convention, which was signed in 1972 and entered into force in 1974. Further relevant legislation also exists, such as the

London Dumping Convention of 1972, dealing with the dumping of waste at sea; and the Convention for the Protection of the Rhine, adopted in late 1976.

These early moves to limit the contamination of aquatic environments were later integrated into European legislation under the auspices of the European Economic Community (EEC). The initial regulations on these matters from the EEC, promulgated in the Council Framework Directive 76/464/EEC in 1976, contained so-called "List I" and "List II" contaminants, which were based substantially on the Paris Convention lists produced earlier (EEC, 1976). These listed contaminants also became known as the "Black List" and the "Grey List" respectively, and this more informal nomenclature is employed here in relation to the EEC Directive.

Contaminants on the EEC Black List were considered to be of significant toxicity, persistence and bioaccumulation in aquatic environments. The List as originally published included both general classes of contaminants (such as organohalogens; carcinogenic compounds) and a total of 13 specific toxicants (Table 7). The EEC sought to set limit values or emission standards (the latter set by reference to Environmental Quality Objectives, or EQOs) for Black List substances at the Community level, and to promulgate daughter Directives on individual substances.

The vexed question of whether the limit value or the EQO approach should be employed, and the political disagreement between Britain and other EEC Member States on this matter, have been reviewed recently by Haigh (1992). While this provides interesting insights into the interface between scientific realities and political life, it is considered beyond the scope of the present review.

The Framework Directive stated that toxicants included in the original Black List would only become liable to full controls as List I substances when Daughter Directives were promulgated by the EEC, being treated until that time as List II substances (Haigh, 1992). To date, limit values have been set for each of the 13 contaminants specifically included in the original Black List (see EEC, 1980, 1982a, 1983a, 1984, 1986a), and for a further four specific contaminants included in the original Black List under generic descriptors (Haigh, 1992).

The specific compounds in each of the general classes of contaminants included in the original Black List were not actually enumerated by the Commission until 1982, when a list of 129 substances which were considered to be of particular concern in aquatic environments was drawn up (EEC, 1982b; see Table 8). These chemicals were selected as potential Black List substances on the basis of their production volumes, in addition

Table 7 List I or "Black List" substances as originally published by the European Economic Commission (after EEC, 1976)

List I contains certain individual substances which belong to the following families and groups of substances, selected mainly on the basis of their toxicity, persistence and bioaccumulation, with the exception of those which are biologically harmless or which are rapidly converted into substances which are biologically harmless:

1. organohalogen compounds and substances which may form such compounds in the aquatic environment,
2. organophosphorus compounds,
3. organotin compounds,
4. substances in respect of which it has been proved that they possess carcinogenic properties in or via the aquatic environment,
5. mercury and its compounds,
6. cadmium and its compounds,
7. persistent mineral oils and hydrocarbons of petroleum origin,

and for the purposes of implementing Articles 2, 8, 9 and 14 of the Directive:

8. persistent synthetic substances which may float, remain in suspension or sink and which may interfere with any use of the waters.

Specific contaminants included in List I:

Mercury[a]	Pentachlorophenol[d]
Cadmium[b]	'Drins' (aldrin, dieldrin, endrin, and isodrin)[d]
Hexachlorocyclohexane[c]	Hexachlorobenzene[d]
Carbon tetrachloride[d]	Hexachlorobutadiene[d]
DDT[d]	Chloroform[d]

[a] See EEC (1982a) for limit values
[b] See EEC (1983a) for limit values
[c] See EEC (1984) for limit values
[d] See EEC (1986a) for limit values

to considerations of their toxicity, persistence and bioaccumulation properties in aquatic ecosystems. A further two contaminants have been added to the Black List more recently, bringing the total number of toxicants on the Black List at present to 131. Certain of these contaminants have recently been included in a list proposed for priority in the setting of further limit values (EEC, 1990).

Contaminants included in the original EEC Grey List (Table 9) were considered to exert deleterious impacts on aquatic environments, but it was thought that these could be confined to specific areas (i.e. that these contaminants were of local concern, rather than affecting ecosystems on regional or global scales). Member States are required to prepare action programmes to attempt to control the discharge of these contaminants to

Table 8 Contaminants considered as candidate chemicals for List I (after EEC, 1982b)

Chlorinated hydrocarbons	— Aldrin,[a] dieldrin[a], chlordane, chlorobenzene, dichlorobenzenes, chloronaphthalene, chloroprene, chloropropene, chlorotoluenes, chlorotoluidene, endosulfan,[a] endrin,[a] heptachlor, hexachlorobenzene,[a] hexachlorobutadiene,[a] hexachlorocyclohexane,[a], hexachloroethane, PCBs,[a] tetrachlorobenzenes, trichlorobenzenes.
Chlorophenols	— Monochlorophenols, 2,4-dichlorophenol, 2-amino-4-chlorophenol, pentachlorophenol,[a] 4-chloro-3-methylphenol, trichlorophenol.
Chloroanilines and nitrobenzenes	— Monochloroanilines, 1-chloro-2,4-dinitrobenzene, dichloroanilines, 4-chloro-2-nitrobenzene, chloronitrobenzenes, chloronitrotoluenes, dichloronitrobenzenes
Polycyclic aromatic hydrocarbons	— Anthracene, biphenyl, naphthalene, PAH.
Inorganic chemicals	— Arsenic and compounds, cadmium and compounds,[a] dibutyltin compounds, mercury and compounds,[a] tetrabutylin.
Solvents	— Benzene, carbon tetrachloride, chloroform, dichloroethane,[a] dichloroethylene, dichloromethane, dichloropropane, dichloropropanol, dichloropropene, ethylbenzene, toluene, tetrachloroethylene, trichloroethane, trichloroethylene.
Others	— Benzidine, benzyl chloride, benzylidene chloride, chloral hydrate, chloroacetic acid, chloroethanol, dibromoethane, dichlorobenzidine, dichloro diisopropylether, diethylamine, dimethylamine, epichlorohydrin, isopropylbenzene, tributyl phosphate, trichlorotrifluoroethane, vinyl chloride, xylenes.
Pesticides	— Azinphos ethyl and methyl,[a] coumaphos, cyanuric chloride, 2,4-D and derivatives, 2,4,5-T and derivatives, DDT, demeton, dichlorprop, dichlorvos,[a] dimethoate, disulfoton, fenitrothion,[a] fenthion, linuron, malathion,[a] MCPA, mecoprop, methamidiophos, mevinphos, monolinuron, omethoate, oxydemeton-methyl, parathion, phoxim, propanil, pyrazon, simazine,[a] triazophos, tributyl tin oxide,[a] trichlorofon, trifluralin,[a] triphenyltin compounds.[a]

[a] Also on the Red List of the United Kingdom.

Table 9 List II or "Grey List" substances of the European Economic Commission (after EEC, 1976)

List II contains:

- substances belonging to the families and groups of substances in List I for which the limit values referred to in Article 6 of the Directive have not been determined,
- certain individual substances and categories of substances belonging to the families and groups of substances listed below,

and which have a deleterious effect on the aquatic environment, which can, however, be confined to a given area and which depend on the characteristics and location of the water into which they are discharged.

Families and groups of substances referred to

(1) The following metalloids and metals and their compounds:

1. zinc[a]	6. selenium	11. tin	16. vanadium[a]
2. copper[a]	7. arsenic[a]	12. barium	17. cobalt
3. nickel[a]	8. antimony	13. beryllium	18. thallium
4. chromium[a]	9. molybdenum	14. boron[a]	19. tellurium
5. lead[a]	10. titanium	15. uranium	20. silver

(2) Biocides and their derivatives not appearing in List I.

(3) Substances which have a deleterious effect on the taste and/or smell of the products for human consumption derived from the aquatic environment, and compounds liable to give rise to such substances in water

(4) Toxic or persistent organic compounds of silicon and substances which may give rise to such compounds in water, excluding those which are biologically harmless or are rapidly converted in water into harmless substances.

(5) Inorganic compounds of phosphorus and elemental phosphorus.

(6) Non-persistent mineral oils and hydrocarbons of petroleum origin.

(7) Cyanides, fluorides.

(8) Substances which have an adverse effect on the oxygen balance, particularly:
ammonia, nitrates.

[a] United Kingdom national standards set for these substances.

inshore and coastal waters. The control of Grey List substances thus occurs on a national level rather than at the Community level, and operates through the setting of national emission standards designed to meet Environmental Quality Objectives, rather than through the imposition of Community-wide limit values. Several papers have been produced dealing with specific contaminants on the Grey List, such as chromium and titanium (e.g. see EEC, 1983b, 1986b).

This European legislation was intended to act as a broad framework for national controls on water pollution in each of the Member States of the

Table 10 The initial priority Red List of the United Kingdom (after DoE, 1985)

Mercury and its compounds
Cadmium and its compounds
γ-Hexachlorocyclohexane
DDT
Pentachlorophenol
Hexachlorobenzene
Hexachlorobutadiene
Aldrin
Dieldrin
Endrin
Polychlorinated biphenyls
Dichlorvos
1,2-Dichloroethane
Trichlorobenzene
Atrazine
Simazine
Tributyltin compounds
Triphenyltin compounds
Trifluralin
Fenitrothion
Azinphos-methyl
Malathion
Endosulphan

Community. Although (unusually) the EC did not lay down specific dates for the implementation of the original Framework Directive, it was suggested that this should be completed by 1986. In practice, this timetable has proven to be highly optimistic, and much remains to be done (in Britain and elsewhere) to formally introduce these controls in their entirety.

In Britain, for example, a consultation paper was produced by the Department of the Environment and the Welsh Office in 1985, setting out proposals for the strengthening of controls on the discharge of dangerous substances to aquatic environments in response to the original 1976 EEC Directive (DoE, 1985). This gave rise to the publication of the so-called Red List of contaminants in 1988, which included 23 priority contaminants, in addition to a number of further substances considered as priority candidates for eventual inclusion. The Red List has since been slightly amended, and the current version (known as the "initial priority Red List") is shown in Table 10. This is intended as a list of contaminants of specific concern to the United Kingdom, but is based upon the original Black and Grey Lists of the EEC. In addition, it may be noted that national standards for the discharge of several EEC Grey List toxicants have been set by the United Kingdom (see Table 9).

Finally here, it may be noted that the controls applied to trace contaminants in aquatic ecosystems in the United Kingdom have been strengthened recently by additional national legislation, including both the Water Act 1989 and the Environmental Protection Act 1990. The latter calls for the instigation of a system of Integrated Pollution Control (IPC), to be used to minimise all emissions of Red List (and certain other) toxicants. New Statutory Water Quality Objectives (SWQOs) are also in draft at present, and these will act as the future national framework for the protection of inland and coastal waters and groundwaters. However, it should be noted that the proposed dates for the implementation of these controls stretch to the mid-1990s and beyond.

C. LEGISLATION IN THE USA

In the USA, attempts to define contaminants of concern in aquatic environments date back to a court settlement of June 1976, involving the US Environmental Protection Agency (EPA) and a variety of "environmentally concerned" organisations. Suits were brought against the EPA by these organisations for failing to implement portions of the Federal Water Pollution Control Act, and one result of this was a requirement for the EPA to draw up a list of pollutants for which discharge limits would be required (Keith and Telliard, 1979). The criteria employed for the inclusion of pollutants were toxicity, the availability of standards for analysis, documentation of the presence of the pollutants in effluents or natural waters, and chemical production data. It is notable that these criteria differ somewhat from those of the EEC (see above), and do not include any specific cognisance of the tendency of a contaminant to bioaccumulate in aquatic environments.

A list of 65 contaminants (the "Toxic Pollutants List") was originally devised, but this was later expanded to 129 substances, principally through the citation of specific compounds which had been previously included only in generic classes of contaminants. The list of 129 contaminants of concern (Table 11) became known as the Priority Pollutants List, although it was never formally published in the US Federal Register.

In 1981, three contaminants (dichlorodifluoromethane, trichlorofluoromethane and *bis*-[chloromethyl] ether) were deleted from the Priority Pollutants List, as they were considered on further reflection by the EPA not to constitute a significant threat to aquatic environments or human health. The number of contaminants on the Priority Pollutants List has remained unchanged since that time, at 126 (Table 11), despite further court actions concerning specific inclusions (ethylbenzene, phenol, 2,4-dichlorophenol, trichlorophenol, pentachlorophenol, monochlorophenyl phenyl ether and chlorodifluoromethane; see US Federal Register, 1981).

D. THE NEED FOR IMPROVEMENTS

While the lists of contaminants produced by both the EEC and the US EPA coincidentally numbered 129 at one stage, several interesting differences are evident between them, and these merit brief discussion here.

Both lists of contaminants (and the UK-derived Red List) include a mixture of single contaminants and classes of contaminants, although the treatment of contaminant classes is different in the various cases. For example, the EEC Black List and the UK Red List both cite PCBs as a simple generic class of compounds, whereas the US EPA list includes a total of seven citations to different forms of Aroclors, which are commercially-marketed forms of PCBs, themselves constituting a mixture of different isomers and homologues. Neither of these approaches can be considered to be scientifically justifiable in the light of present knowledge. Several authors have argued that PCBs should not be treated simply as a generic class of contaminants, but that individual isomers should be quantified in environmental samples and should be considered as specific contaminants in their own rights (e.g. see Tanabe *et al.*, 1987a; Duinker *et al.*, 1988a,b; Tanabe, 1988; Zitko, 1989).

There can be no doubt of the correctness of this approach in the light of the more recent findings on the extreme toxicities of the non-*ortho* coplanar PCBs, for example (Tanabe *et al.*, 1987a). It may therefore be argued that the more toxic PCBs (such as the non-*ortho* coplanar isomers) should be specifically cited in lists of contaminants thought to be of particular concern, whereas less toxic PCB isomers should be relegated to subsidiary lists, such as the Grey List of the EEC, for example. Indeed, certain PCB isomers (such as some of the dichlorobiphenyls) probably do not merit inclusion even on the Grey List, as they are of relatively low toxicity and potential for bioaccumulation (e.g. see Phillips, 1986; Tanabe *et al.*, 1987b).

Table 11 The priority pollutants list of the US EPA

1. Acenaphthene
2. Acrolein
3. Acrylonitrile
4. Benzene
5. Benzidine
6. Carbon tetrachloride (tetrachloromethane)
7. Chlorobenzene
8. 1,2,4-Trichlorobenzene
9. Hexachlorobenzene
10. 1,2-Dichloroethane
11. 1,1,1-Trichloroethane
12. Hexachloroethane
13. 1,1,-Dichloroethane
14. 1,1,2-Trichloroethane
15. 1,1,2,2-Tetrachloroethane
16. Chloroethane
17. *Bis*(chloromethyl) ether[a]
18. *Bis*(2-chloroethyl) ether
19. 2-Chloroethyl vinyl ether (mixed)
20. 2-Chloronaphthalene
21. 2,4,6-Trichlorophenol
22. Parachlorometacresol
23. Chloroform (trichloromethane)
24. 2-Chlorophenol
25. 1,2-Dichlorobenzene
26. 1,3-Dichlorobenzene
27. 1,4-Dichlorobenzene
28. 3,3′-Dichlorobenzidine
29. 1,1-Dichloroethylene
30. 1,2-*Trans*-dichloroethylene
31. 2,4-Dichlorophenol
32. 1,2-Dichloropropane
33. 1,3-Dichloropropylene
34. 2,4-Dimethylphenol
35. 2,4-Dinitrotoluene
36. 2,6-Dinitrotoluene
37. 1,2-Diphenylhydrazine
38. Ethylbenzene
39. Fluoranthene
40. 4-Chlorophenyl phenyl ether
41. 4-Bromophenyl phenyl ether
42. Bis(2-chloroisopropyl) ether
43. Bis(2-chloroethoxy) methane
44. Methylene chloride (dichloromethane)
45. Methyl chloride (chloromethane)
46. Methyl bromide
47. Bromoform (tribromomethane)
48. Dichlorobromomethane
49. Trichlorofluoromethane[a]
50. Dichlorodifluoromethane[a]
51. Chlorodibromomethane
52. Hexachlorobutadiene
53. Hexachlorocyclopentadiene
54. Isophorone
55. Naphthalene
56. Nitrobenzene
57. 2-Nitrophenol
58. 4-Nitrophenol
59. 2,4-Dinitrophenol
60. 4,6-Dinitro-*o*-cresol
61. N-nitrosodimethylamine
62. N-nitrosodiphenylamine
63. N-nitrosodi-*n*-propylamine
64. Pentachlorophenol
65. Phenol (4APP method)
66. *Bis*(2-ethylhexyl) phthalate
67. Butyl benzyl phthalate
68. Di-*n*-butyl phthalate
69. Di-*n*-octyl phthalate
70. Diethyl phthalate
71. Dimethyl phthalate
72. Benzo(*a*)anthracene (1,2-benzanthracene)
73. Benzo(*a*)pyrene (3,4-benzopyrene)
74. 3,4-Benzofluoroanthene
75. Benzo(*k*)fluoranthane (11,12-benzofluoranthene)
76. Chrysene
77. Acenaphthylene
78. Anthracene
79. Benzo(*ghi*)perylene (1,12-benzoperylene)
80. Fluorene
81. Phenanthrene
82. Dibenzo(*a*,*h*)anthracene

Table 11—*contd.*

83. Indeno(1,2,3-*cd*)pyrene	107. PCB-1254 (Aroclor 1254)
84. Pyrene	108. PCB-1221 (Aroclor 1221)
85. Tetrachloroethylene	109. PCB-1232 (Aroclor 1232)
86. Toluene	110. PCB-1248 (Aroclor 1248)
87. Trichloroethylene	111. PCB-1260 (Aroclor 1260)
88. Vinyl chloride (chloroethylene)	112. PCB-1016 (Aroclor 1016)
89. Aldrin	113. Toxaphene
90. Dieldrin	114. Antimony (Total)
91. Chlordane (tech. mixture & metabolites)	115. Arsenic (Total)
92. 4,4′-DDT	116. Asbestos (Fibrous)
93. 4,4′-DDE (p,p′DDX)	117. Beryllium (Total)
94. 4,4′-DDD (p,p′-TDE)	118. Cadmium (Total)
95. α-endosulfan	119. Chromium (Total)
96. β-endosulfan	120. Copper (Total)
97. Endosulfan sulfate	121. Cyanide (Total)
98. Endrin	122. Lead (Total)
99. Endrin aldehyde	123. Mercury (Total)
100. Heptachlor	124. Nickel (Total)
101. Heptachlor epoxide	125. Selenium (Total)
102. α-BHC	126. Silver (Total)
103. β-BHC	127. Thallium (Total)
104. γ-BHC (Lindane)	128. Zinc (Total)
105. δ-BHC	129. 2,3,7,8-Tetrachloro-dibenzo-*p*-dioxin (TCDD)
106. PCB-1242 (Aroclor 1242)	

[a] Deleted in 1981

Similar arguments may be made in respect of other groups of contaminants. For example, the US EPA list includes specific mention of all four isomers of hexachlorocyclohexane (cited as BHC rather than HCH; there is a need for standardisation of nomenclature here, in any event), whereas the EEC Black List simply cites HCH as a chemical class, and the UK Red List includes only γ-HCH (Lindane). The various HCH isomers differ considerably in their environmental persistence and toxicity to biota, and merit individual consideration as specific contaminants in the different lists, just as noted previously for the PCBs.

This argument may also be extended to the inorganic contaminants. The US EPA Priority Pollutants List includes 13 trace elements, but does not cite any specific forms of these elements as being of particular concern. Tin is not included in this list in any form, principally because the list was developed prior to the considerable concerns over the aquatic toxicity of tributyl tin employed as an anti-fouling agent. The original EEC Black

List and its later expanded version include arsenic, cadmium and mercury and their compounds, and several forms of organotins, cited specifically (but sometimes in strange categories; see Table 8). Other trace elements are included in the EEC Grey List, although several differences exist between the elements cited here and those in the US EPA list (e.g. nine elements are found on the EEC list which are not included on the US EPA list). The UK Red List differs from either of these, including only cadmium and mercury and their compounds, together with tributyl tin and triphenyl tin compounds.

Other specific inclusions differ substantially between the various lists. For example, the Priority Pollutants List cites many specific forms of polynuclear aromatic hydrocarbons (PAHs), while only three PAHs are specifically included on the EEC Black List and no PAHs are found on the UK Red List. Perhaps most surprising of all, dioxins are mentioned only once in any of these lists, being included specifically as the most toxic isomer (2,3,7,8-tetrachlorodibenzo-*p*-dioxin) on the Priority Pollutants List, but being absent completely from the EEC and UK lists. While it might be argued that dioxins would be included under one of the generic classes of compounds in the EEC Black List, the lack of any specific citation of this most toxic of trace contaminants is an astonishing omission from both the EEC and the UK lists.

Several points may be made here. There can be no doubt that each of the lists of contaminants of concern is now outdated. Our understanding of the distributions and toxicities of trace contaminants has expanded significantly since either the US or EEC lists were produced, a decade or more ago. There is clearly an urgent need to update the lists, and their consolidation may also be considered.

This is important for several reasons. Firstly, many regulatory programmes rely upon such lists of legislative derivation to define the contaminants to be monitored in particular situations. While each of the lists is clearly incomplete in certain aspects, a case can also be made for the deletion of certain contaminants from the lists (at least in particular circumstances). Secondly, there has been no attempt to define the relevant media in which the various contaminants may be found at measurable levels. This gives rise to a situation (particularly in programmes in the USA dealing with health-related issues) in which attempts are made to analyse all the listed contaminants in each sample taken. The result of this is the unnecessary expenditure of limited research and monitoring funds on the analysis of samples for contaminants which do not accumulate therein, and hence will not be found above detection limits. There is a clear need for the rationalisation of the lists, and for the provision of additional guidance on the relevant media for analysis for each of the contaminants cited.

E. THE PRESENT REALITY

In reality, less than 100 of the many thousands of chemicals which are commonly discharged to fresh and salt waters globally are included in most regular monitoring programmes (see Table 12 for an example). Specific analyses may be undertaken on an *ad hoc* basis for perhaps a further 100 contaminants in particular environments, including many of those on the various lists discussed previously in this chapter. For most of the remaining thousands of chemicals discharged regularly to aquatic ecosystems, analytical techniques of the required sensitivity have not been developed, or (in a few cases) exist but are simply not used. It may therefore be legitimately claimed that the monitoring of aquatic contaminants undertaken at present, even in the most sophisticated of laboratories, is woefully inadequate to define the existing degree of pollution and its effects.

Jernelöv (1974) stated the following:

> ... the total number of compounds to be included in the list of marine contaminants could ... be counted in tens of thousands ... even out of those which have been looked for and shown to be present, there is only a small number of compounds of which we have more than isolated or scattered data ... Our ignorance is thus monumental and our ability to evaluate the present situation and future trend regarding total marine contamination is close to non-existent.

While this statement was made almost two decades ago, it remains largely true today.

Thus, much greater efforts are required in the development of sensitive and robust analytical methodologies and in the use of theoretical methods to assist in determining which contaminants may exert significant effects in aquatic environments, either through direct toxic impacts or because of their high bioavailabilities and/or tendency to bioaccumulate. There is also a need to reassess the lists of contaminants which are considered of importance in aquatic ecosystems in the light of recent research, and to

Table 12 Contaminants for which mussels are analysed in the California State Mussel Watch Program (after Stephenson *et al.*, 1986)

Aldrin	δ-HCH
Chlorbenside	Heptachlor
Cis-chlordane	Heptachlor epoxide
Trans-chlordane	Hexachlorobenzene
α-chlordene	Methyl parathion
γ-chlordene	PCB (Aroclor) 1248
Cis-nonachlor	PCB (Aroclor) 1254
Trans-nonachlor	PCB (Aroclor) 1260
Oxychlordane	Total PCBs
Total chlordane	Total phenol
Chlorpyrifos	Pentachlorophenol
Dacthal	Tetrachlorophenol
o,p′ DDD	Tedion
p,p′ DDD	Toxaphene
o,p′ DDE	Ronnel
p,p′ DDE	Tetradifon
p,p′ DDMS	
p,p′ DDMU	Cadmium
o,p′ DDT	Chromium
p,p′ DDT	Copper
Total DDT	Mercury
Methoxychlor	Manganese
Diazinon	Lead
Dieldrin	Zinc
Endosulfan 1	Arsenic
Endosulfan 2	Nickel
Endosulfan sulphate	Selenium
Total endosulphan	Titanium
Endrin	Barium
Ethyl parathion	Cobalt
α-HCH	Silver
β-HCH	Tributyl tin
γ-HCH	Aluminium

define the most appropriate methods of monitoring for the various types of contaminants thought to be of significance (e.g. see Chapman *et al.*, 1982). Biomonitors have a significant role to play in such studies. Biomonitors are also, however, of great use in defining the degree of contamination of aquatic environments by chemicals which are known to be of concern, and their utilisation for such a purpose is the principal topic of the remainder of the present text.

Chapter 4

The Early Use of Biomonitors

The advantages of the use of organisms over other methods for monitoring aquatic contamination have been discussed at length by several authors (Butler *et al.*, 1971; Haug *et al.*, 1974; Bryan, 1976; Phillips, 1976a, 1977, 1978, 1980, 1990b, 1991; Bryan *et al.*, 1985; Waldichuk, 1985; Phillips and Segar, 1986); their conclusions are summarised here. The three basic methods employed for monitoring aquatic ecosystems for contamination will be dealt with in turn.

A. THE ANALYSIS OF WATER

Several fundamental problems exist with the traditional methods of monitoring for contaminants by the analysis of natural waters. The concentrations of most contaminants of concern in either fresh or salt waters are very low, ranging from around the nanogram per litre (part per trillion or 10^{12}) level to about the milligram per litre (part per million or 10^{6}) level, depending on the contaminant involved. Despite this, in some cases (for example, the analysis of nutrient concentrations), robust and reliable analytical techniques have been developed, and the levels of certain of the contaminants of concern in fresh or salt waters can be accurately determined.

However, in other instances (such as in studies of radionuclides, pesticides or hydrocarbon derivatives), analysis at such low concentrations as are present in natural waters is extremely challenging and often impossible. Thus, for example, problems with the analysis of coastal waters for the radioisotopes ^{54}Mn and ^{65}Zn (and later, other radionuclides) were a driving force for the first use of biomonitors (Folsom *et al.*, 1963; Folsom and Young, 1965; Young and Folsom, 1967). Although analytical techniques to quantify the minute concentrations of contaminants such as radionuclides and organochlorines in natural waters have been developed since this time, they remain rarely used.

Most stable isotopes of trace metals lie in the intermediate portion of this spectrum of natural abundance and analytical challenge, and their concentrations may be determined with acceptable accuracy and precision through the use of lengthy preconcentration techniques and/or suitably sensitive analytical methods, undertaken in purpose-built "clean laboratory" facilities (see below). Many studies continue to be published on the analysis of trace metals in natural waters; however, the use of organisms for monitoring is becoming increasingly widespread.

The fact that most contaminants are found at very low concentrations in natural waters not only renders their analysis in such waters technically difficult, but may also introduce inadvertent errors in collection and analytical procedures. These occur due to either sample contamination or to the loss of contaminants during sample handling. The best examples of such problems derive from work on trace metals in water. Patterson and Settle (1976a,b) were the first to report the existence of very significant errors of this type, in this case with reference to the analysis of lead at trace levels. These authors reported that the actual concentrations of lead present in natural waters were orders of magnitude lower than had been previously described, approximating only 10–50 ng $litre^{-1}$ in most aquatic environments. They considered that all previous analyses of lead in such waters were grossly in error, due to inadvertent sample contamination by extraneous sources of the element.

Later work documented similar problems for the analysis of other trace elements in seawater, and it is now widely accepted that data on metals in seawater produced prior to the mid-1970s should not be relied upon (Bruland *et al.*, 1978a,b; Gordon *et al.*, 1982; Knauer *et al.*, 1982; Bruland, 1983; Martin *et al.*, 1983; Wong *et al.*, 1983; Burton and Statham, 1990). Purpose-built "clean laboratory" facilities are now thought to be mandatory for the analysis of trace metals in seawater (at least for open ocean sites), to avoid the problems caused by the extraneous contamination of samples. In addition to such problems with inadvertent contamination, it may be noted that trace metals tend to adsorb to the walls of containers used for water sampling, leading in certain cases to significant losses of the elements of interest prior to analysis.

Even if such analytical problems can be overcome (as is possible with modern techniques for at least the nutrients and certain of the trace metals), the quantification of contaminants in natural waters suffers from additional disadvantages with respect to its use as a regular monitoring tool. In many aquatic environments, the concentrations of contaminants present vary widely with time (Phillips, 1980; Frenet, 1981; Habib and Minski, 1982; Carpenter and Huggett, 1984; see Fig. 24). This is particularly the case in rivers and estuaries, where contaminant abundance may be affected by a wide variety of interacting factors (flow intensity; the intermittency of contaminating discharges; tidal and current-related effects). To overcome this temporal variation and produce reliable estimates of the mean concentrations of contaminants in such waters requires protracted study over a wide range of conditions, preferably employing time-integrating water samplers (or the compositing of grab samples). While not impossible, this demands considerable effort, and costs inevitably increase.

Perhaps the greatest disadvantage of the analysis of natural waters for contaminants, however, is the lack of any useful correlation between the concentrations of contaminants present and their biological availability. This has been discussed at some length by Phillips (1977, 1980) and Waldichuk (1985). Both authors concluded that if the impacts of contaminants on biota are the principal rationale for undertaking monitoring, the bioavailabilities of the contaminants are of fundamental interest and importance. As contaminant bioavailability cannot be inferred from the analysis of water samples, it may be concluded that organisms should be preferred for most monitoring tasks.

B. THE ANALYSIS OF SEDIMENTS

Many authors have employed sediments to monitor the contamination of aquatic environments, and there can be little doubt that this technique offers certain advantages over the analysis of natural waters (see the recent review by Luoma, 1990). Most contaminants of concern in aquatic ecosys-

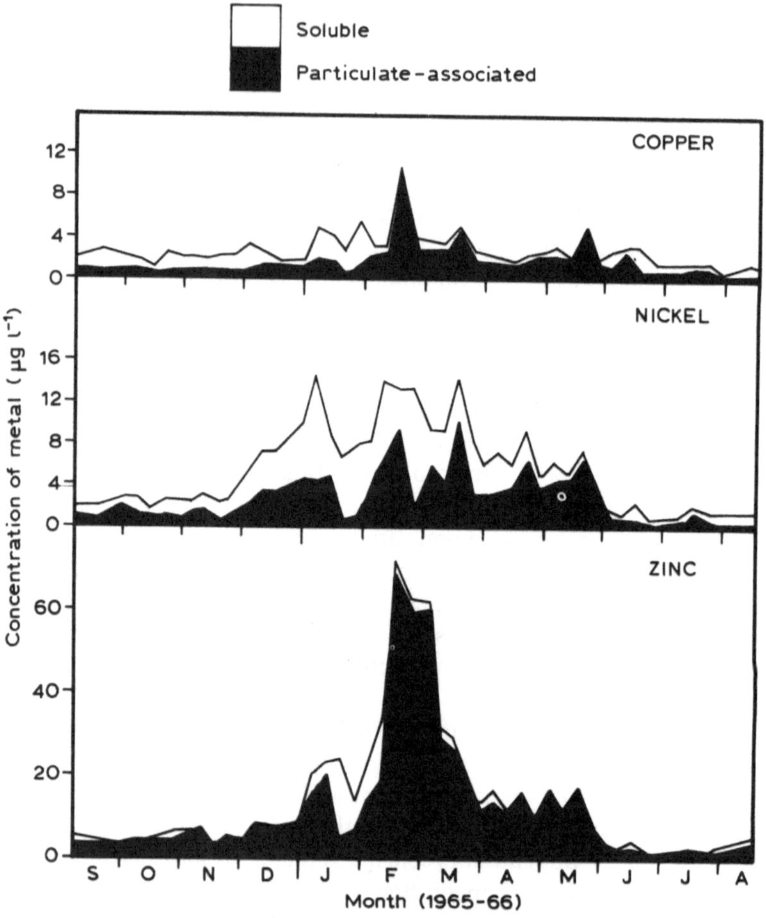

Fig. 24. Temporal variations in the concentrations of copper, nickel and zinc in surface waters of the Susquehanna River, as it enters Chesapeake Bay, USA. Data for soluble and particulate-associated metals are provided separately. After Carpenter and Huggett (1984).

tems tend to associate preferentially with suspended particulate material rather than being maintained in solution, although this behaviour varies in extent between individual contaminants.

Among the trace metals, aluminium, iron, lead and manganese are generally found to be particularly heavily associated with particulates, only very low concentrations remaining in solution under most conditions.

Table 13 Aqueous solubilities of selected organic contaminants

Compound	*Type*	*Solubility in water (μg litre^{-1})*	*Reference*
DDT	Organochlorine	1	Bowman *et al.* (1960)
DDT	Organochlorine	17	Biggar *et al.* (1966)
PCB[a]	Organochlorine	12	Courtney and Langston (1978)
PCB[b]	Organochlorine	25	Wiese and Griffin (1978)
Dieldrin	Organochlorine	90	Biggar *et al.* (1966)
Aldrin	Organochlorine	105	Biggar *et al.* (1966)
Endrin	Organochlorine	130	Biggar *et al.* (1966)
Lindane	Organochlorine	2 150	Biggar *et al.* (1966)
Parathion	Organophosphate	20 000	Martin (1963)
Malathion	Organophosphate	145 000	Martin (1963)
Dichlorvos	Organophosphate	10 000 000	Martin (1963)

[a] Refers to Aroclor 1254 and seawater; produced by air-lift pump system. Other mixtures likely to differ considerably; few precise data appear to be available.
[b] Equilibrium concentration for Aroclor 1254 in membrane-filtered, charcoal-filtered seawater at 16.5°C. Solubilities of different components shown to differ, components of lower chlorination dissolving more rapidly.

By contrast, significant proportions of the arsenic, cadmium and selenium in natural waters are generally present in solution. Other elements tend to display intermediate partitioning, and conditions such as salinity and temperature are important in determining the chemical species present and the differences in partitioning between the solution and particulate phases in distinct environments.

Pesticides and hydrocarbons also vary widely in their solubilities in aqueous media and in their partitioning between solution and suspension in natural waters. As a general rule, organochlorine pesticides and most hydrocarbons are distinctly hydrophobic in nature, and therefore tend to be found almost completely in the particulate-associated fraction. By contrast, organophosphate compounds and certain short-chain hydrocarbons (for example, the monocyclic aromatic compounds) exhibit much higher aqueous solubilities, and are maintained in solution to a greater extent. However, highly significant differences in solubilities exist within the three groups of compounds (Table 13), and the cycling of both pesticides and hydrocarbons in aquatic environments is greatly affected by these differences.

This physical behaviour leads to the preferential sequestration in sediments of many of the contaminants of concern, and the researcher may take advantage of this in the design of monitoring schemes. Because such contaminants generally accumulate in sediments over time, the problems of short-term temporal variations (associated with the analysis of water for contaminants) tend to be reduced. Thus, in most situations, contaminant levels in sediments alter little over short time periods, and reflect average conditions of contamination over a period of some weeks or months (e.g. see Boehm and Farrington, 1984; Thomson *et al.*, 1984; Luoma, 1990). Concentrations of contaminants in sediments are also several orders of magnitude greater in most cases than are those in natural waters, and analysis is therefore often simpler and less prone to the problems discussed above, caused by extraneous sources of contamination during sample handling or the preparation of samples for analysis.

As a result, there can be little doubt that the use of sediments for some types of monitoring of aquatic contamination is highly successful. Thus, sediments are frequently most useful in defining the locations of contaminant sources, gradients of increasing concentration being found towards significant sources in many cases. This method can be surprisingly sensitive under some circumstances. For example, Knauer (1976, 1977) noted increases in the concentrations of cobalt, mercury and nickel in sediments close to a smelter in Queensland, Australia, less than one month after the commencement of operation of the facility. The analysis of sediment cores can also provide "geochronological" information, on the historic levels of contaminants in particular locations (e.g. Eisenreich *et al.*, 1979; Goldberg *et al.*, 1979; Essink, 1980; Smith and Loring, 1981; Hungspreugs and Yuangthong, 1983; Baker and Harris, 1991; Barcellos *et al.*, 1991; see Fig. 25). Studies of this type rely upon the selection of areas of net sediment deposition for sampling, and the use of radioisotope dating techniques in concert with trace element analysis.

However, several problems exist with any reliance upon sediments for monitoring contaminants in aquatic ecosystems. Firstly, concentrations of contaminants in sediments do not simply reflect the absolute magnitudes of contaminant abundance at the sampling point (as assumed by many authors), but are in fact a complex function of the relative fluxes of contaminants and suspended particulates in the system (Bertine and Goldberg, 1977; Phillips, 1977). Thus, any comparison of the absolute concentrations of contaminants in the sediments of two or more sites should take account of the rates of sedimentation of particulates, if the intention is to compare the sites for their overall degree of contamination (i.e. for the magnitudes of contaminant inputs or loads). While this may be achieved in studies of historical metal loading to a system (e.g. see Smith and Loring, 1981), it is rarely attained in other investigations.

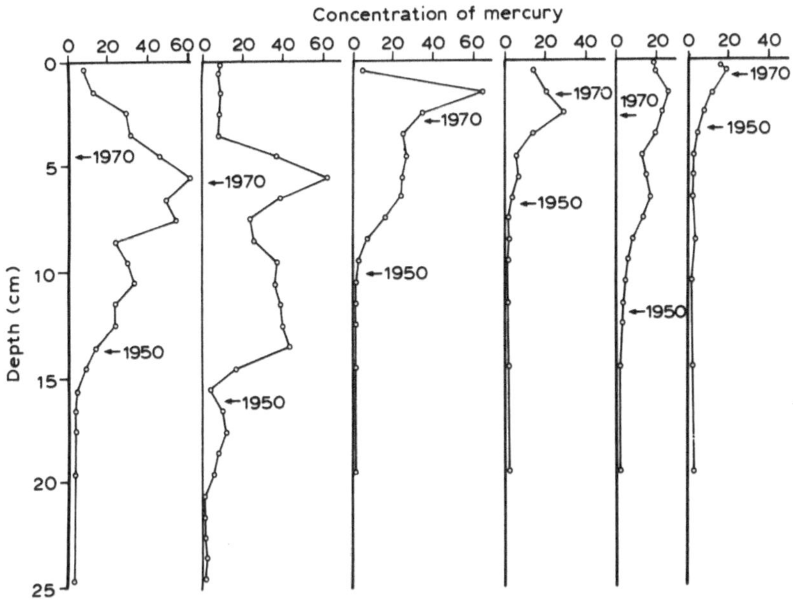

Fig. 25. The variation in mercury concentration (μg g^{-1} dry weight) with depth in sediments from the Saguenay Fjord, Quebec. Time-stratigraphic horizons for 1950 and 1970 were determined by ^{210}Pb and ^{137}Cs dating; these differ for each site due to variations in sedimentation rates with location in the fjord. After Smith and Loring (1981).

Secondly, the concentrations of contaminants in sediments are affected very significantly by two major factors characterising the sediment: grain size and organic carbon content. As a general rule, contaminant concentrations vary inversely with grain size and directly with the organic carbon content of sediments, although the precise relationships found vary between locations and with changes in ambient conditions (Luoma, 1990). The impacts of grain size are thought to be mediated by the availability of binding sites for contaminants on the surface of sediment particles, smaller particles providing a greater surface area/volume ratio, and hence tending to accumulate higher concentrations of contaminants. This tendency holds true for trace metals, organochlorines and hydrocarbons, although it is clear that the detailed chemistry of the surface binding is distinct for the different classes of contaminants.

Mayer and Fink (1980) undertook extensive studies on the impacts of grain size and other factors on chromium concentrations in sediments,

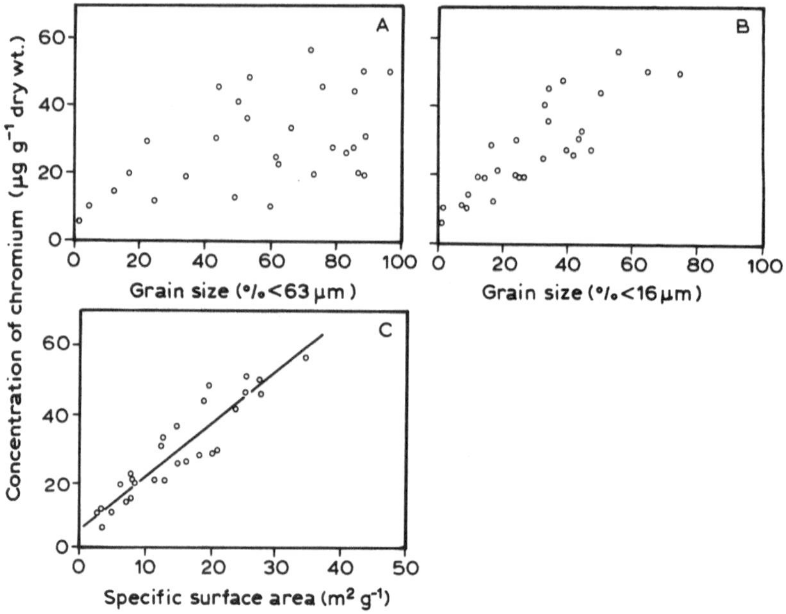

Fig. 26. Relationships between concentrations of chromium ($\mu g\,g^{-1}$ dry weight) in sediments from three estuaries in Maine, USA, and grain size or specific surface area. (A) Percentage less than 63 μm. (B) Percentage less than 16 μm. (C) Specific surface area. After Mayer and Fink (1980).

using samples from four estuaries in Maine, USA. They showed that the specific surface area of sediments is the factor which truly controls metal accumulation, rather than grain size *per se*; relationships between chromium levels and grain size or specific surface area are shown in Fig. 26. The use of specific surface area as a normalisation factor for chromium in sediments allowed them to identify areas where excess metal was present, indicating the existence of local sources of chromium. Baldi and Bargagli (1982) also employed this technique, using it in concert with chemical leaching methods (see below) to investigate the fate of mercury discharged from rivers in north-western Italy to the Tyrrhenian Sea; similar work has also been carried out on the northern Adriatic Sea (Donazzolo *et al.*, 1984). There can be no doubt that the use of specific surface areas as a normalising factor for metals in sediments is far superior to techniques involving simple sieving of samples prior to analysis. However, sieving techniques have been most frequently employed to date, providing a significant source of error in attempts to compare sediment contamination profiles from different locations.

In an effort to introduce some consistency into the interpretation of data on contaminant concentrations in sediments from distinct locations, certain authors have proposed that the concentrations found should be quoted in relation to the organic carbon content of the sediment, rather than in relation to sediment mass (see Luoma, 1990). While this has some merit, it assumes that all forms of organic carbon offer a similar binding capacity for contaminants, and the method has not as yet gained wide acceptance. There is thus no simple method available to account for any effects that differences in the organic carbon contents of sediments may have on their accumulation of contaminants.

The issue of contaminant bioavailability is once again, however, of greatest significance in defining the usefulness of sediments for regular monitoring of the contamination of aquatic environments. It is known that sediments act as both intermediate and final sinks for contaminants, and that while some remobilisation may occur (especially of organic substances; see for example Neff, 1984; Knezovich *et al.*, 1987), contaminants may also be lost to the biotic portion of the ecosystem through their irreversible binding in sediments, and eventual burial.

Unfortunately, it is not possible to accurately identify the bioavailable portion of any contaminant present in sediments using only chemical methods of analysis. Some authors have used sequential digestion or leaching techniques to attempt to define the bioavailabilities of trace metals in sediments (e.g. see Guy *et al.*, 1978; Loring, 1981; Badri and Aston, 1983; Jones, 1986). The precise techniques employed by different authors vary somewhat, although the intention is to sequentially strip metals from sediments by their treatment with chemicals of gradually increasing strength (e.g. ammonium salts, acetic acid, hydrogen peroxide, and dilute and strong nitric acid). The metals released are considered to have been bound in different forms on the sediment particles, and to be more or less "exchangeable". It is then assumed that the strength or type of binding of the metals to the sediment particles has a definite relationship to the bioavailabilities of the elements.

While such methods undoubtedly improve our understanding of the potential for the remobilisation of contaminants from sediments, they should be seen to provide only crude estimates of the bioavailabilities of contaminants, relying on many assumptions. No study has documented a reliable correlation between the true bioavailability of any contaminant and its removal from sediments through the use of leaching agents, and the accurate prediction of the bioavailability of contaminants present in sediments remains as elusive as that for contaminants in solution. Thus, while the analysis of sediments is of considerable use to identify sources of contaminants or to elucidate long-term trends in contamination at a site through the analysis of cores, the issue of contaminant bioavailability remains unsolved.

C. THE ANALYSIS OF ORGANISMS

Aquatic organisms have long been known to accumulate significant quantities of contaminants in their tissues. In the case of hydrophobic contaminants such as the organochlorine pesticides and most hydrocarbons, uptake by biota appears in almost all cases to be passive, driven by the physicochemical process of lipid-water partitioning (e.g. see Hamelink *et al.*, 1971; Pavlou and Dexter, 1979). The degree of accumulation and retention of such contaminants by either whole organisms or their individual tissues is therefore primarily dependent on the overall content and distribution of lipids in the organisms concerned (Phillips, 1978, 1980, 1986).

The accumulation of trace metals by aquatic biota is a more complex phenomenon, however. The degree to which organisms take up and retain trace metals varies markedly between phyla, and may also differ significantly between individual species within the different phyla (Eisler, 1981). Such variations are thought to reflect the evolution of distinct strategies for detoxifying trace metals (Phillips and Rainbow, 1989; Rainbow, 1990). Thus, the concentrations of certain trace metals (particularly those of an "essential" nature, i.e. those required for normal metabolic functions, such as copper and zinc) are regulated by some aquatic organisms within particular limits (see Fig. 27 for an example). By contrast, other species accumulate very large quantities of metals, sequestering these in particular tissues, often in insoluble or detoxified forms (Brown, 1982; Simkiss *et al.*, 1982; Viarengo, 1989). Between these two extremes lies a spectrum of "partial regulation", in which organisms employ several different methods to sequester metals (Phillips and Rainbow, 1989). One of the most important of these is the binding of trace elements to metallothioneins (Roesijadi, 1980/81, 1992).

This accumulation and sequestration of contaminants by organisms provides an opportunity to short-circuit the traditional methods for monitoring contamination in aquatic environments through the analysis of

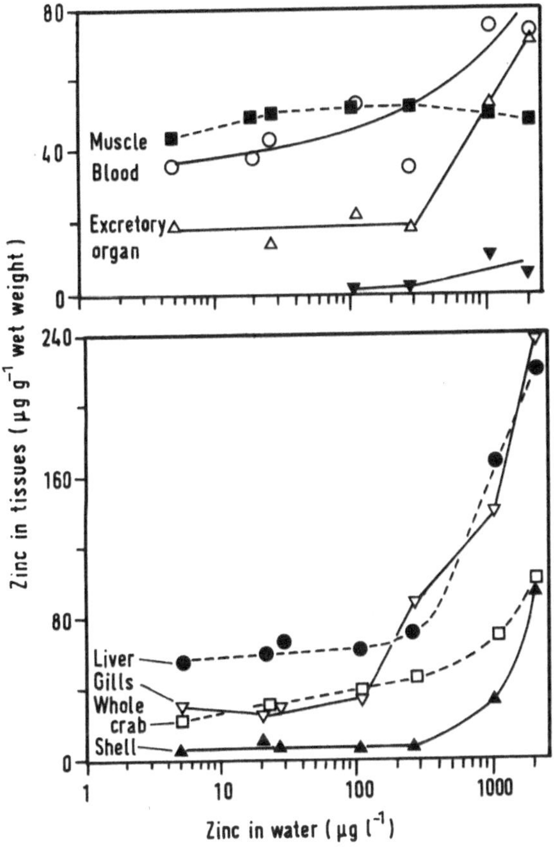

Fig. 27. Mean concentrations of zinc ($\mu g\ g^{-1}$ wet weight) in tissues of the crab *Carcinus maenas* exposed to various levels of the element in solution for 32 days at 13°C, showing the regulation of zinc accumulation. Each point represents a mean of at least two individuals. Note the difference in the vertical scales. After Bryan (1966, 1976).

water, and to employ organisms themselves for such monitoring. Such a proposition was first investigated by Folsom and co-workers, in attempts to quantify the contamination of Californian waters by radionuclides (Folsom *et al.*, 1963; Folsom and Young, 1965; Young and Folsom, 1967).

In the late 1960s and early 1970s, other authors also began to study the possibilities of using organisms to monitor conservative contaminants in aquatic ecosystems. Butler (1966, 1969a,b, 1971, 1973) headed a national programme in the USA, using bivalve molluscs to monitor pesticides in estuaries. This programme was ahead of its time, and in many ways constituted a forerunner to the later "Mussel Watch Program" in the USA, which covered radionuclides, stable isotopes of metals, and hydrocarbons in addition to pesticides (Goldberg *et al.*, 1978, 1983; Farrington *et al.*, 1982, 1983). Butler *et al.* (1971) suggested that the organisms employed in such monitoring programmes, here termed "biomonitors", should possess the following attributes:

- Contaminants should be accumulated without lethal impacts to the species employed.
- Biomonitors should be sedentary, so as to be representative of the area in which they are sampled.
- The organisms employed should be abundant throughout the study area, and sufficiently long-lived to allow the sampling of more than one year class, if desired.
- The organisms used should be easy to sample, hardy enough to survive under laboratory conditions (to permit laboratory-based studies of contaminant kinetics), and should provide sufficient tissue for contaminant analysis.

Haug *et al.* (1974) added the following requirements, based upon their studies of macroalgae as biomonitors of trace metals on Norwegian coasts:

- Biomonitors should tolerate brackish water, permitting studies in estuaries (which are often the most contaminated areas of coastal waters).
- A simple correlation should exist between the contaminant content of a biomonitor and the average contaminant concentration in its ambient environment.

Phillips (1977, 1978, 1980, 1990b) has discussed a variety of refinements to these basic criteria for an organism to act as an efficient biomonitor of conservative contaminants, but the original criteria are certainly adequate as an initial assessment of the potential of a species for monitoring.

The use of biomonitors to quantify the degree of contamination of aquatic environments has a number of theoretical and practical advantages over the analysis of either natural waters or sediments in monitoring programmes. Because of their considerable capacities for the accumulation of contaminants, most biomonitors exhibit contaminant concentrations which permit relatively simple measurement compared to the techniques required for water analysis. The problems of inadvertent contamination of samples, which bedevilled attempts at water analysis for many decades (see above), are also much reduced. Biomonitors also provide a time-integration capacity, their accumulated concentrations of contaminants reflecting an average of the short-term temporal fluctuations in contaminant abundance in the ambient waters.

The extent of this time-integration of contaminant abundance varies between species and with the particular contaminant considered. For example, the time-integration of ambient concentrations of trace metals by most biomonitors is generally longer than that for organochlorines or hydrocarbons, because of their relative kinetics of uptake and excretion (Phillips, 1980; Phillips and Segar, 1986). The accumulation of individual PCB isomers and homologues by biota provides an excellent example of such variations (see Tanabe *et al.*, 1987b). It is possible artificially to shorten or lengthen the time-integration capacity of a biomonitor in any particular study, either through the transplantation of samples or by the use of "time-bulking" techniques (Phillips and Segar, 1986).

However, the greatest advantage of the use of biomonitors is that the bioavailabilities of contaminants are measured directly, without recourse to the assumptions employed in the other methods; thus, a contaminant present in a biomonitor is by definition bioavailable (Phillips, 1980; Waldichuk, 1985). As Waldichuk (1985) stated:

> ... it is not yet possible to predict unequivocally the biological availability of metals based on physical and chemical characteristics of either the seawater or the sediments ... At the present time, there is no true substitute for chemical analysis of the tissues of exposed marine organisms for metal concentrations, if one wishes to determine the biological availability of metals at a given site.

This statement is certainly also true in terms of the monitoring of organic contaminants such as hydrocarbons and organochlorines, which are even more difficult to quantify with accuracy in natural waters than are trace metals.

This does not necessarily imply that all biomonitoring species will display equivalent patterns of relative contamination in monitoring programmes at given locations, as the bioavailability of a contaminant may vary between different types of organisms. Thus, for example, macroalgae accumulate contaminants from solution in the main, whereas bivalve molluscs take up contaminants from ingested organic and inorganic particulate material in addition to those from solution (e.g. see Bryan and

Hummerstone, 1977; Phillips, 1979a). Such differences between species do not necessarily detract from the benefits derived from biomonitoring; rather, the use of several species in any given monitoring survey can provide valuable clues as to the reasons for differences in contaminant abundance at the locations studied.

Over the last two decades, the use of biomonitoring has become standard in local, national and international programmes for monitoring aquatic contaminants, and biomonitoring techniques are now the method of choice in most situations. However, this should not be taken to imply that the technique has no disadvantages (Phillips, 1980; White, 1984). Several factors have been identified which interfere with the uptake of contaminants by biomonitors, and hence with their successful use to monitor aquatic pollution. These factors have been reviewed in detail by Phillips (1977, 1978, 1980, 1986, 1990b), and will be discussed in the following chapters as relevant.

Chapter 5

The Biomonitoring of Trace Metals and Radionuclides

A. INTRODUCTION

Natural waters contain a complex mixture of stable isotopes of trace metals and radionuclides. As noted in Chapter 2, the stable trace elements are derived from both geological weathering processes and anthropogenic sources, and in most cases, the latter predominate. The radioisotopes present in fresh and salt waters also fall into naturally-occurring and anthropogenically-derived categories (Table 14). Nuclides of anthropogenic origin in the aquatic environment are predominantly from the fallout from nuclear weapons testing (most of which ceased in 1963 following test ban treaties), nuclear fuel reprocessing plants, and the disposal of radioactive waste at sea (Woodhead, 1984; Clark, 1989).

The bioaccumulation of these contaminants depends on their biochemical properties and on the individual accumulation strategies of organisms for each element or isotope. Generally, radioactive isotopes of metals follow the metabolic pathways of their stable counterparts (where these exist), although atomic weight differences may cause slight differences in relative rates of reaction and hence in metabolic kinetics. These differences tend to be greater for the isotopes of low atomic weight, such as those of hydrogen and carbon. However, in most cases, radioisotopes will be

Table 14 Natural and fallout radionuclides in surface seawater (after Woodhead, 1984)

Radionuclide	*Half-life (years)*	*Concentration (mBq litre^{-1})*
Natural radionuclides:		
Potassium-40	1.28×10^9	11,800
Rubidium-87	4.80×10^{10}	107
Uranium-234	2.45×10^5	48
Uranium-238	4.47×10^9	44
Tritium	12.3	22–110
Carbon-14	5.73×10^3	5.9–6.7
Uranium-235	2.45×10^5	1.9
Radium-226	1.60×10^3	1.3–3.1
Lead-210	22.3	0.4–5.0
Radon-222	1.05×10^{-2}	~0.7
Polonium-210	3.79×10^{-1}	0.19–3.7
Radium-228	5.76	0.04–3.7
Thorium-228	1.91	0.007–0.12
Thorium-230	8.00×10^4	0.002–0.05
Thorium-232	1.4×10^{10}	0.004–0.03
Fallout radionuclides:[a]		
Strontium-90	28.0	0.74–19
Caesium-137	30.0	1.1–30
Tritium	12.3	1,100–2,700
Carbon-14	5.73×10^3	0.37–1.5
Plutonium-239	2.44×10^4	0.011–0.044

[a] Data for fallout radionuclide concentrations relate to the North Atlantic Ocean. Only the more abundant or important isotopes are listed; others present include ^{65}Zn, ^{54}Mn, ^{55}Fe, ^{57}Co, ^{60}Co, ^{63}Ni, ^{95}Zr, ^{95}Nb, ^{103}Ru, ^{106}Ru, ^{125}Sb, ^{141}Ce, ^{144}Ce, ^{147}Pm, ^{155}Eu, ^{108m}Ag, ^{110m}Ag.

represented in organisms in the expected proportions to their stable counterparts. The representation of those artificial radionuclides which have no stable counterparts in aquatic organisms depends on: the extent and pattern of their release into the environment; their radioactive half-lives; the biological half-life of the nuclide in the organisms involved; and on certain other biological features of the organism involved.

A feature of the bioaccumulation of the heavier transuranic radioisotopes by animals is that these elements are strongly adsorbed by external tissues which are in direct contact with the ambient waters, such as the gills and exoskeleton (Phillips, 1980; Miramand *et al.*, 1989). Similarly, the accumulation of transuranic radioisotopes by algae is dominated by passive adsorption, as in the case of americium (Carvalho and Fowler, 1985).

As noted in Chapter 4, biomonitoring techniques were initially employed to monitor the abundances and distributions of radionuclides in aquatic environments, rather than stable isotopes of metals. These early studies included work on the discharge of nuclides such as ^{65}Zn from the Hanford reactors to the Columbia River in Washington, DC; monitoring along the coast of California; and investigations of fallout radionuclides in Lake Maggiore in Italy (e.g. see Watson *et al.*, 1961; Folsom *et al.*, 1963; Gaglione and Ravera, 1964; Osterberg *et al.*, 1964). By the 1970s, the technique was beginning to be employed to study stable trace elements also, as it had become apparent that the use of biomonitors provided advantages over the traditional methods of water or sediment analyses. This chapter reviews the further development and current state of the art of biomonitoring for both stable trace elements and radionuclides.

As discussed in Chapter 4, an ideal biomonitor of stable trace metals or radionuclides should fulfill several prerequisites (Phillips, 1980, 1990b; Bryan *et al.*, 1980):

- Biomonitors should be sessile or sedentary in order to be representative of the study area.
- Biomonitors should be abundant in study areas, easy to identify and sample at all times of year, and should be large enough to provide sufficient tissue for analysis of the contaminant of interest.
- Biomonitors should be hardy, tolerating wide ranges of contaminant concentration and of physicochemical variables such as salinity, thereby permitting the design of transplant experiments and laboratory studies of contaminant kinetics.
- Biomonitors should be strong accumulators of the relevant trace metal, with a simple correlation between the metal concentration found in the tissues of the biomonitor and the average ambient bioavailable metal concentration. This correlation should be the same at all study sites.

Many species which have been employed in aquatic biomonitoring surveys of trace metals fail to satisfy one or more of the above prerequisites. Most importantly, the accumulation strategy of a potential biomonitor for the particular trace metal should be known, which usually requires laboratory experimentation (Rainbow *et al.*, 1990). Accumulation strategies for stable trace metals and radionuclides depend on mechanisms for the uptake, excretion, and storage or sequestration of the individual elements, and these are reviewed in the next section.

B. METAL UPTAKE, EXCRETION AND STORAGE BY AQUATIC ORGANISMS

Trace metals are taken up by aquatic organisms by more than one route (Luoma, 1983). The most important routes of uptake are those from solution and from food. However, in some cases metals may also be taken up through the pinocytosis of metal-rich particles external to the alimentary tract (Bevelander and Nakahara, 1966; George *et al.*, 1976), or even by the mixing of the external medium directly with body fluids (Mangum, 1979; Depledge and Phillips, 1986). The uptake of metals from solution and from food is reviewed below, to provide a basis for estimating the usefulness of various species as biomonitors.

(i) Metal Uptake from Solution

Aquatic organisms are bathed in solutions of trace metals at dissolved concentrations ranging from nanograms per litre in the open ocean (e.g. Bruland, 1983), through levels approximating micrograms per litre in coastal seas, to even higher concentrations approaching or exceeding milligrams per litre in estuaries and acid metal-rich streams (Foster *et al.*, 1978; Bryan and Gibbs, 1983). By comparison with the trace element concentrations in the tissues of aquatic organisms, these ambient dissolved levels of metals are low. However, the uptake of many trace metals from solution into aquatic organisms at their permeable surfaces is generally considered to be a passive process not requiring the expenditure of energy (Williams, 1981a,b; Simkiss and Taylor, 1989a). This situation contrasts markedly with that pertaining to the major ions of alkali metals (e.g. sodium, potassium, calcium), which are taken up through active transport pumps. The key to this difference lies in the different chemistries of the two groups of metals (Nieboer and Richardson, 1980).

Strictly, trace metals or heavy metals (see Chapter 1 for definitions of these terms) should be defined by their chemical properties, particularly the relative binding affinities of their ions for specific ligands. Elements which are usually loosely classified as trace metals may be defined more precisely as those with metal ions falling into the Class B or Borderline categories of Nieboer and Richardson (1980). These metal ions have high affinities for ligands containing sulphur and nitrogen, and therefore bind

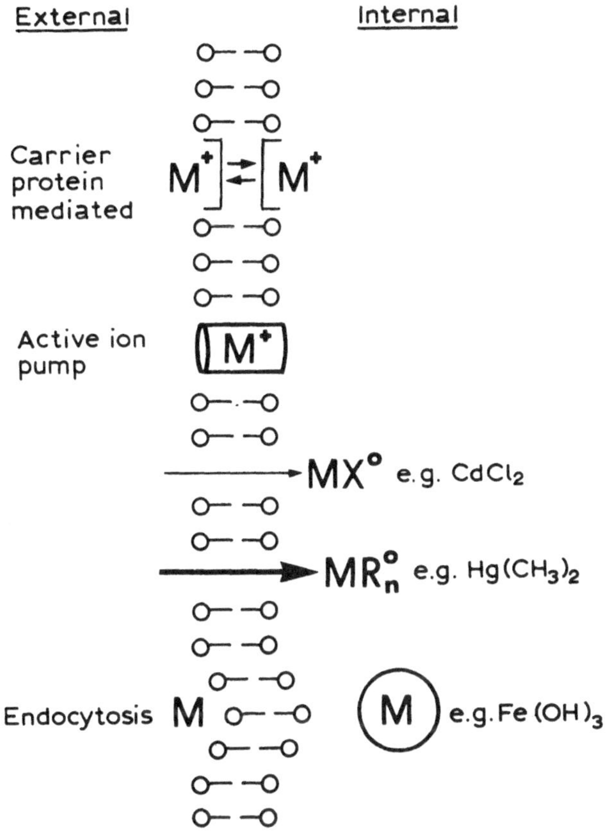

Fig. 28. Possible models of the uptake of trace metals from solution. After Simkiss and Taylor (1989a).

relatively easily to proteins and other cellular macromolecules (Nieboer and Richardson, 1980; Williams, 1981a,b).

The high affinity of such metal ions for proteins and other cellular constitutents provides the basis for one model of their passive uptake from solution by aquatic organisms (Fig. 28). This model proposes the binding of the dissolved metal ion in the external medium onto a transport protein in the membrane of the permeable surface of the aquatic organism. This initial binding is thought to be a passive process, given the high affinity of the metals involved for the ligands present in the transport protein. By the process of facilitated diffusion (which is also passive in nature), the element is transferred across the membrane and into the cell along a series of metal-binding ligands of increasing affinity. Thus, the metal is passed down a thermodynamic gradient into the cell, where it binds finally with the ligand of highest metal affinity. This will give rise to either storage of the metal or its transfer out of the cell, perhaps ultimately to ligands in circulating fluids or specific target organs. No significant reverse flow of elements occurs, because of the excess of non-diffusible metal binding sites of high affinity internally, both within and beyond the initial metal-receiving cell. Thus, metal uptake into the organism continues as a passive process, apparently against a concentration gradient.

Major ions, such as those of sodium, potassium and calcium (assigned to class A by Nieboer and Richardson, 1980) do not have a high affinity for organic ligands and do not bind to membrane transport proteins so readily for transport into the cell. Thus, active transport pumps are required for the movement of these "more ionic" elements against concentration gradients across the hydrophobic membrane.

This is also the case for other non-metallic ionic species, such as chloride and sulphate, and for a few of the trace elements which exhibit particular types of speciation in seawater. Thus, for example, the metals molybdenum and vanadium occur in seawater predominantly as the oxygenated anions molybdate (MoO_4^{2-}) and vanadate (HVO_4^{2-}) (Bruland, 1983), and they are therefore not available as the free metal ion to bind with membrane transport proteins to the same degree as elements such as cadmium or zinc. The uptake of these oxygenated metal ions from seawater is similar to that of anions such as sulphate (SO_4^{2-}); they are taken up by active transport pumps, either specific for that metal ion or by incorporation into pumps employed normally for sulphate, phosphate, or other anions.

It is inevitable that some trace metals will become incidentally incorporated into active transport pumps for the major metal ions. For example, the free metal ion of cadmium has a similar ionic radius to that of the calcium ion, and cadmium will therefore be taken up to some extent through calcium pumps. The relative significance of this route of entry as opposed to that of facilitated diffusion varies with the organism concerned and with environmental conditions, as discussed later in this chapter.

The binding of a metal to a membrane transport protein is commonly considered to involve the free (hydrated) metal ion. Thus, the uncomplexed metal ion is believed to be the most bioavailable chemical form in general (Sunda and Guillard, 1976; Sunda *et al.*, 1978; Engel and Fowler, 1979; Canterford and Canterford, 1980; Zamuda and Sunda, 1982;

Florence, 1983; Sanders and Jenkins, 1984; Jenkins and Sanders, 1986; Nugegoda and Rainbow, 1988a, 1989a; O'Brien *et al.*, 1990). Trace metal ions dissolved in aquatic media are partitioned in equilibria between complexing ligands, the latter being both organic and inorganic in nature.

For example, cadmium exists in seawater mostly as chloro-complexes (e.g. $CdCl^+$, $CdCl_2^0$, $CdCl_3^-$), with only about 2.5% of the total being present as the free (hydrated) cadmium ion Cd^{2+} (Zirino and Yamamoto, 1972; Bruland, 1983). For many trace elements in seawater, the free metal ion is present at a relatively low equilibrium percentage of the total dissolved metal (Bruland, 1983), this percentage varying with changes in physicochemical parameters such as salinity and pH. Mantoura *et al.* (1978) modelled the increase in concentration of Zn^{2+} at the expense of inorganically complexed species of Zn (e.g. chloro-complexes) down the salinity gradient of an estuary. This model proposed an increase of about 25% in the concentration of the Zn^{2+} species with a decrease in salinity from 32 parts per thousand (47% of Zn present as Zn^{2+}) to 15 parts per thousand.

In freshwater, the degree of inorganic complexation (especially by chloride) is reduced, and dissolved organic matter such humic and fulvic acids may be of increased significance as complexing agents. Away from the buffering capacity of seawater, pH changes in freshwater have greater potential to affect the complexing equilibria of dissolved metals, the low pH of acid streams often promoting the percentage contribution of the free metal ion, with important consequences for both metal uptake and toxicity.

Any physicochemical change that reduces the hydrophilic complexation of a dissolved metal enhances its bioavailability by increasing the availability of the free metal ion. This free ion may then enter an organism by binding with a membrane transport protein, or replacing a major ion in an active pump. The presence of the chelating agent EDTA (at constant salinity) reduces the absolute equilibrium concentration of many free metal ions, and will thus give rise to a decrease in the bioavailabilities of those metals. This is the case, for example, for zinc uptake by the decapod *Palaemon elegans* (Nugegoda and Rainbow, 1988a; O'Brien *et al.*, 1990; see Fig. 29(A)). Similarly, a decrease in salinity increases zinc uptake by *P. elegans* (Nugegoda and Rainbow, 1989a; Fig. 29(B)), as the absolute equilibrium concentration of the free metal ion increases under such conditions (Mantoura *et al.*, 1978).

Given that neutral complexes of metals (e.g. $CdCl_2^0$) are more lipid-soluble than ionic species, Simkiss (1983) has suggested that such neutral

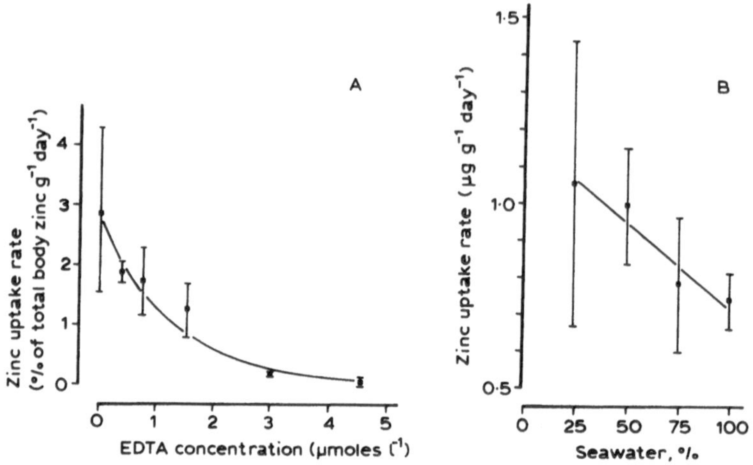

Fig. 29. The effects of changes in physicochemical parameters on the uptake of zinc by the decapod crustacean *Palaemon elegans* at 10°C. (A) Mean rates of zinc uptake (±1 sd) at 100 μg litre^{-1} labelled zinc, with increasing EDTA concentrations. After Nugegoda and Rainbow (1988a). (B) Mean rates of zinc uptake (±1 sd) at 56.2 μg litre^{-1} labelled zinc with change in salinity. After Nugegoda and Rainbow (1989a).

species may be the form transported (by direct diffusion) across hydrophobic cell membranes (Fig. 28). Whether such a diffusion process is more important than the binding of the free metal ion to a membrane transport ligand remains to be demonstrated. It is possible, for example, that the lipid solubility measured by Simkiss (1983) may have been affected by binding of the metal to an organic contaminant of the "model lipid" (olive oil) used in the studies (Simkiss and Taylor, 1989a).

Direct passage through the lipid membrane is probably the major route of uptake available to lipophilic organometallic compounds (Fig. 28), whether these are of natural occurrence (e.g. methylmercury, formed by microbial activity) or are anthropogenically derived (e.g. tributyltin compounds in anti-fouling paints). Similarly, the addition of a lipophilic complexing agent may increase metal bioavailability (see Ahsanullah and Florence, 1984) because the typical uptake route is bypassed.

The uptake of dissolved metal may take place all over the body surface of small and/or soft bodied organisms, as well as at particular sites of high permeability such as the gills. Metal uptake from solution will also take place in the alimentary tract, when any of the medium is swallowed during "drinking" or the ingestion of food. For example, hypo-osmoregulating crustaceans in littoral or salt lake environments will drink the medium to replace water lost by osmosis, and also excrete excess salts actively through the gills (Mantel and Farmer, 1983). Marine teleost fish are also hypo-osmotic regulators and drink seawater routinely, subsequently disposing of excess salts via the gills and faeces.

The uptake of metals from solution can also take place in unexpected fashions. Large marine gastropods such as *Busycon carica* take seawater into the blood sinuses of the foot on re-expansion after their retraction into the shell (Mangum, 1979). Since this seawater mixes freely with the blood, it is quite likely that dissolved metals could bind to the blood pigment haemocyanin or to other available ligands (see Depledge and Phillips, 1986). In some polychaetes, including *Neanthes arenaceodentata*, cilia lining the metanephridial canals cause a constant emission of fluid from the coelom through the nephridia to the external medium. These cilia may be inhibited by the narcotic effect of a xenobiotic, and the medium may then leak directly into the coelom without passage across the body wall or gut epithelium (Mason *et al.*, 1988).

Whether the usual uptake of trace metals from solution is by passive facilitated diffusion using membrane-associated transport proteins or by incorporation into active transport pumps, the rate of uptake of an element will be directly proportional to the external concentration of dissolved metal over typical environmental concentration ranges. Any enzyme-driven active uptake system has the potential to be saturated, and theoretically even internal ligands of high affinity for metals (which drive facilitated diffusion pathways) might become saturated under extreme conditions of external metal bioavailability. Under most circumstances, however, the rates of uptake of trace elements respond proportionally to increases in external dissolved concentrations (e.g. see White and Rainbow, 1984a; Rainbow and White, 1989, 1990). Such evidence does not contradict any model citing the free metal ion as the bioavailable form, as the absolute concentrations of free metal ions alter in direct proportion to total metal concentrations.

The relative importance of different routes for the uptake of trace metals from solution varies between different organisms and with environmental conditions. The passive facilitated diffusion model of metal uptake using a membrane-associated ligand can usually explain results for the uptake of trace elements by marine invertebrates. For example, the rate of uptake of zinc by the littoral decapod *Palaemon elegans* can be related to the concentration of the free zinc ion under changing conditions of salinity

(Nugegoda and Rainbow, 1989a; see Fig. 29(B)), or to the concentration of the hydrophilic chelating agent EDTA (Nugegoda and Rainbow, 1988a; O'Brien *et al.*, 1990). However, such data by themselves do not distinguish between the facilitated diffusion model and the random incorporation of free metal ions into active pumps for major ions.

Nugegoda and Rainbow (1989b) have shown that changes in salinity and osmolality have different effects on the uptake of zinc by *Palaemon elegans*. These results are compatible with the facilitated diffusion model of uptake, but not with the uptake of zinc through its incidental incorporation into a major ion pump. Salinity depends only on the inorganic content of seawater (i.e. predominantly its NaCl content), whereas osmolality depends ultimately on the total number of particles in solution, whether these are inorganic or organic in nature. Nugegoda and Rainbow (1989b) were therefore able to manipulate salinity levels (including chloride concentrations) independently of the total osmolality of the medium by the addition of an organic molecule, in this case sucrose. Zinc uptake by *P. elegans* was found to be dependent on changes in salinity but not osmolality, and this is compatible with the changed inorganic complexation of dissolved zinc affecting the available concentrations of the free zinc ion. If zinc were taken up in proportion to the activity of pumps, such as those for Na^+ or Ca^{2+}, the zinc uptake rate would be expected to be dependent on osmolality.

It may be concluded that in general, physicochemical changes in the medium control the uptake of metals by marine invertebrates, with no physiological control of uptake by the organism being evident. However, two exceptions exist to this general rule, both affecting estuarine organisms.

Firstly, some euryhaline crustaceans exhibit altered permeability to water and ions upon acclimation to low salinity (Mantel and Farmer, 1983); this is often referred to as the apparent water permeability (see Campbell and Jones, 1990). This physiological response has the potential to affect the uptake rates of trace metals. Indirect evidence for a physiological change in permeability affecting metal uptake is available for the littoral amphipod crustacean *Orchestia gammarellus* (Rainbow *et al.*, in press). The rates of uptake of both zinc and cadmium by this amphipod at decreasing salinities increase in proportion to the calculated concentrations of the free metal ion (see the example in Fig. 30). However, at salinities of 25 parts per thousand and below there is no further change in metal uptake rates, even though the absolute free metal ion concentrations are still increasing. In contrast to the case of *Palaemon elegans*, it appears

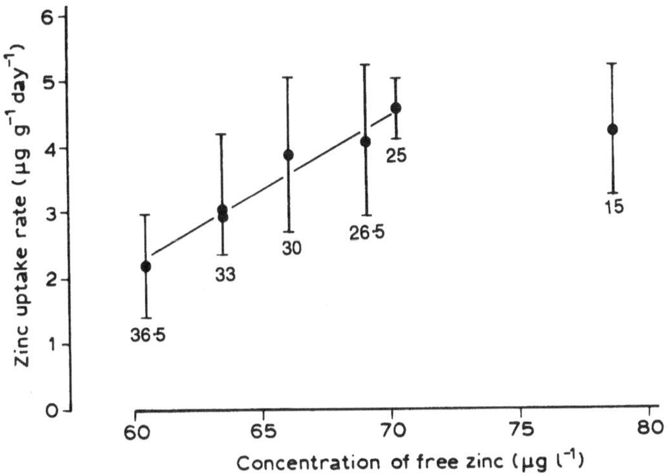

Fig. 30. The effects of changes in the concentration of free zinc ion on the mean rate of uptake of zinc (±1 standard deviation) by the littoral amphipod *Orchestia gammarellus* exposed to 100 μg litre^{-1} labelled zinc at 6 different salinities in NaCl solution at 10°C. Salinities are shown under each vertical bar, as %. After Rainbow *et al.* (in press).

that progressive physiological changes in the amphipod may offset any physico-chemical effects on metal uptake rates.

Secondly, molluscs and malacostracan crustaceans with a high physiological demand for calcium (whether for shell formation or calcification of the exoskeleton) may exhibit atypically high calcium pump activities, particularly in waters of low salinity. These pumps will incorporate cadmium, even to the extent that this becomes a predominant route for cadmium uptake. In low salinity, the shore crab *Carcinus maenas* takes up major ions including calcium actively in the gills (see Mantel and Farmer, 1983), and cadmium uptake similarly increases at low salinities (Wright, 1977a). This increased cadmium uptake appears to be sensitive to the external calcium concentration independently of the overall salinity effect (Wright, 1977b), perhaps reflecting that component of cadmium uptake entering via the calcium pump.

Similarly, unpublished work by Bjerregaard and Depledge, cited in Depledge (1990), has shown that cadmium uptake by the winkle *Littorina littorea* exhibits a dependency on the calcium concentration which is independent of any salinity effect. Such a gastropod mollusc with a thick calcareous shell has a large requirement for calcium; thus, the high activity of calcium pumps may have increased the significance of this pathway of cadmium uptake, to match if not overtake that of facilitated diffusion. It is also notable that the results of Jeffree (in press), showing a strong correlation between accumulated labelled cadmium and labelled calcium concentrations in the Australian freshwater mussel *Velesunio angasi*, can be completely explained by the uptake of both elements through the calcium pump.

(ii) Metal Uptake from Particulates

Particulates present in natural waters may be either inorganic (e.g. suspended sediments) or organic (e.g. phytoplankton) in nature, and metals may be taken up from each of these fractions, by various mechanisms.

The uptake of metals directly from particulates usually involves ingestion of the particles and subsequent uptake of the metal from the alimentary tract. Other direct routes not involving ingestion include the uptake of particulate iron by pinocytosis in the gills of mussels (George *et al.*, 1976), and the scavenging of metals from sediment particles by fronds of macrophytic plants in direct contact with the sediment (Luoma *et al.*, 1982).

Metal uptake from particles in the alimentary tract of animals usually occurs after digestive processes have released the elements from the particulate matrix. The bioavailabilities of metals ingested in solid food depend on the digestive processes of the consumer, and the concentration and chemical nature of the metal in the food. Digestive processes vary according to the pH of the gut and the availability of digestive enzymes in relation to the substrates present in the food.

It is notable here that one consequence of a particularly efficient transfer of metals through food could be the biomagnification of elements along food chains or through food webs. Biomagnification of metal concentrations has not been observed for most metals (Bryan, 1979; Luoma, 1983), although high concentrations of mercury occur in large predatory fish and in air-breathing aquatic mammals (Thompson, 1990). It appears that the long generation times of species of higher trophic levels (Bryan, 1979), and the specific characteristics of mercury (including perhaps organic transformations of the metal by bacteria in the gut of some species) may be involved in this exception (Luoma, 1983).

The relative uptake rates of metals in aquatic biota from the two major routes (i.e. from solution and from food) are difficult to define (Luoma, 1983), for the two are linked, especially in the case of deposit feeders. The relative significance of the two routes varies for each metal and each organism with ambient physicochemical conditions, physiological and developmental stage of the organism, food availability and type, and other factors. Microphagous filter feeders ingest many small organically-rich particles which are potentially metal-rich, but also pass large volumes of water containing dissolved metals over permeable surfaces (e.g. lamellibranch bivalve gills, barnacle cirri). The presence and concentration of suspended food often stimulate the production of feeding currents under such circumstances, affecting both routes of metal uptake. In certain species, the use of mucous sheets to trap small particles provides extra surface area for the adsorption of dissolved metals as part of the feeding process. While various experimental attempts have been made to quantify the significance of different metal uptake pathways, the estimates produced are indirect at best (Luoma, 1983).

However, the tissue distributions of elements within organisms often differ according to the route of exposure to metals in the laboratory, and this permits extrapolation to field conditions in attempts to define the predominant routes of metal uptake (Luoma, 1983). For example, Pentreath (1976a,b,c) exposed plaice (*Pleuronectes platessa*) in the laboratory to methylmercury in food. The fish exhibited tissue distributions of the element similar to those observed in samples collected from the field, indicating that food was the major route of mercury uptake in natural conditions. Similarly, Bryan and Uysal (1978) concluded that accumulated cadmium, cobalt, lead and zinc (but possibly not copper or silver) originated largely from the food of the deposit-feeding bivalve *Scrobicularia plana*.

Sediments constitute a concentrated pool of metals (Luoma, 1989, 1990) and are the food source of deposit-feeders. In recognition that the total metal concentration of a sediment is by no means a measure of bioavailable metal (see Chapter 4), sequential chemical extraction procedures have been used to measure the bioavailable fraction of sediment-bound trace metals (Luoma, 1983, 1989, 1990; Tessier and Campbell, 1987). However, such analytical procedures have certain major drawbacks. Firstly, chemical extraction procedures only model the possible bioavailable fractions to sediment-dwelling organisms, and these fractions are in any case likely to differ between organisms according to feeding mechanisms, digestive and assimilation efficiencies, and other factors. Secondly, any partitioning of metals in sediments occurs as a dynamic equilibrium; the extraction and removal of one portion of metal will thus alter the distribution of the same element in other chemical fractions. Thirdly, the state of oxidation of the sediments is often affected when sequential extraction procedures are undertaken, and the resulting data cannot therefore be considered typical of the situation *in situ*.

Nevertheless, sequential extraction techniques have provided some insights into factors controlling the bioavailability of metals in sediments. Thus, trace metal levels in deposit-feeding benthic organisms are best related not to total metal concentrations in adjacent sediments, but to concentrations of relatively easily extracted fractions (Diks and Allen, 1983; Tessier and Campbell, 1987; Luoma, 1989).

Several studies have been undertaken of trace element concentrations in surficial sediments and associated deposit-feeding invertebrates along gradients in metal abundance. These show that predictions of trace element bioavailabilities are often improved when extracted metal concentrations of the sediment are normalised with respect to iron and/or the organic content of sediments (Luoma and Bryan, 1978, 1982; Langston, 1980, 1982; Bryan, 1985; Luoma, 1989). Luoma and Bryan (1978) showed in a survey of English estuaries that 80% of the variance in lead concentrations in the tissues of *Scrobicularia plana* could be explained by the ratio of lead to iron in the sediments (Fig. 31). High concentrations of extractable iron in the sediment were associated with minimal enrichment of lead in the bivalves, even if lead levels were high in the sediments. By contrast, lead was accumulated extremely efficiently from sediments which exhibited low iron concentrations (Luoma, 1989).

The bioavailability of mercury is reduced in sediments which are rich in organic matter, as a result of strong associations between this element and organic ligands (Luoma, 1977a, 1989). This gives rise to a linear relation between the ratio of mercury to organic matter in sediments and the mercury concentrations of *Scrobicularia plana* and *Macoma balthica* in British estuaries (Langston, 1982).

The strong dependence of the metal concentrations of benthic organisms upon the metal levels (or ratios between these) in local sediments does not necessarily imply that the route of entry of metals is through ingestion of the sediment (Tessier and Campbell, 1987). For example, high concentrations of trace metals in the gills and mantle of the filter-feeding freshwater bivalve *Elliptio complanata* from lakes in Quebec correlated with low sediment metal concentrations (Tessier *et al.*, 1984). These two organs are remote from the digestive system and are in contact with large volumes of water necessary for feeding and respiration. Thus, the correlations recognised probably result from an interaction of dissolved trace metal concentrations with both sediment and bivalve (Tessier and Campbell, 1987). Similarly, anoxic estuarine sediments have the

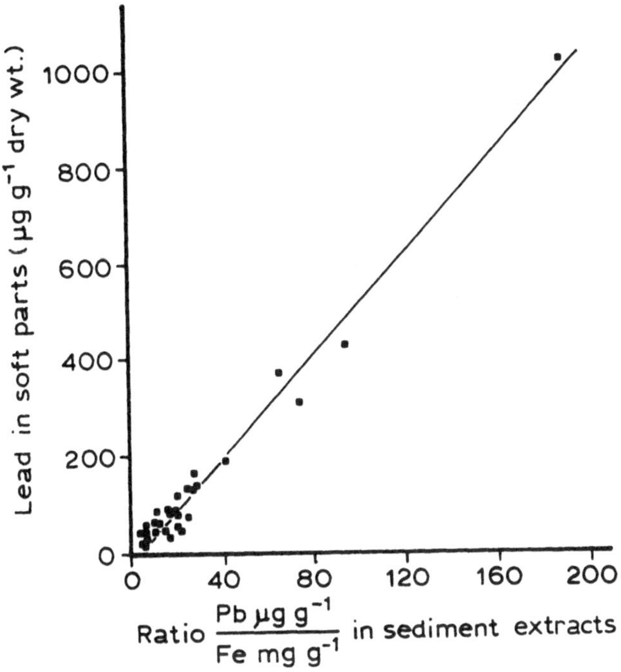

Fig. 31. Correlation between concentrations of lead in soft tissues of the deposit-feeding bivalve *Scrobicularia plana* and the ratio of extractable iron and lead concentrations (as defined by extracts employing 1 M HCl) in sediments. After Luoma and Bryan (1978).

potential to release soluble manganese interstitially and into the water column, to affect the accumulated manganese concentrations of estuarine seaweeds such as *Fucus vesiculosus* (Bryan and Hummerstone, 1973a).

Although the biomagnification of metals along aquatic food chains is not of general occurrence, certain organisms receive atypically high metal loads in their food as a result of the specific nature of their diet. For example, stegocephalid amphipods feed on cnidarians rich in iron. These amphipods have developed a storage mechanism to detoxify iron in the cells of the absorptive ventral caeca of the gut, preventing further incorporation of the element into other tissues, where toxic impacts could occur. The iron concentrations found in these species are atypically high for marine amphipods as a result of this adaptation (Moore and Rainbow, 1984).

Similarly, predators of barnacles such as the whelk *Nucella lapillus* or the nudibranch *Onchidoris bilamellata* on British shores are potentially exposed to a high dietary source of zinc, as barnacles are notoriously strong accumulators of this element (see Rainbow, 1987). Much of this zinc is in the detoxified form of zinc pyrophosphate granules (Pullen and Rainbow, 1991), and such granules appear in the faecal pellets of *Nucella lapillus* feeding on barnacles, indicating that the zinc is not in fact available to the consumer (Nott and Nicolaidou, 1990).

The predators of certain ascidian seasquirts may also receive unusually large dietary loads of vanadium (Carlisle, 1968), although the high vanadium concentrations present in ascidians may deter predators (Stoecker, 1980). Staggeringly high concentrations of copper in the branchiae of the ampharetid polychaete *Melinna palmata* (Gibbs *et al.*, 1981), and of arsenic in the cirratulid polychaete *Tharyx marioni* (Gibbs *et al.*, 1983) offer a severe dietary metal challenge to any predator, although again one possible function of at least the former is the deterrence of predators. The spindle shells *Hemifusus ternatanus* and *H. tuba* also accumulate large quantities of arsenic, possibly as a result of specific dietary preferences (Phillips and Depledge, 1986a,b). This arsenic is mostly organic in nature, and is almost certainly present as arsenobetaine (Phillips, 1990a), which is fortunate in view of the consumption of these species by the Chinese population in Hong Kong and elsewhere.

(iii) Metal Excretion

The accumulation of trace elements by an aquatic organism results from the net balance of the processes of metal uptake and excretion. All aquatic organisms take up metals in significant quantities, but for many species the excretion of accumulated metals may be insignificant.

Metals which are present in the gut contents of an organism or adsorbed to its surface (Fig. 32) may be lost through defaecation or desorption, and these are of little significance to the present discussion. The remainder of this section is thus restricted to consideration of the excretion of absorbed metal present in the body of organisms.

In the same way as many aquatic organisms have the potential to take up trace metals across any permeable surface, so might metals be excreted by this route. As discussed above, the net result of any passive exchange of metals across these surfaces is usually their accumulation internally, because of the availability of binding sites of high affinity for elements within the organism. Nevertheless, metals are sometimes excreted across permeable surfaces, although the mechanisms involved have not been

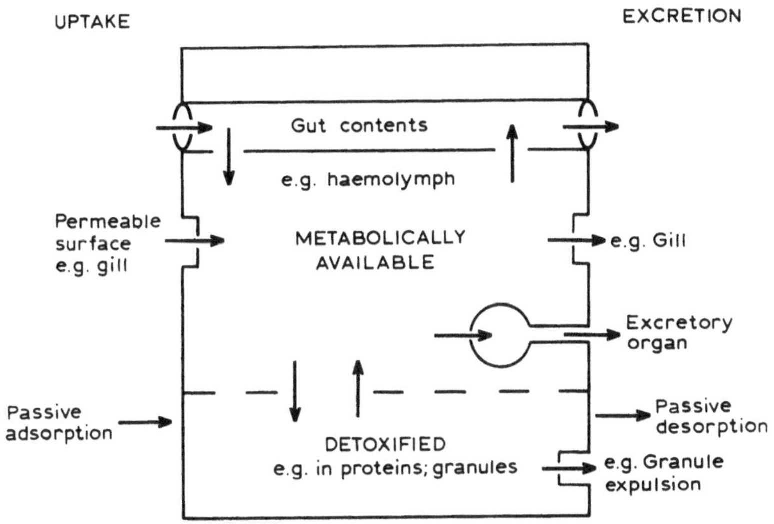

Fig. 32. A schematic representation of the uptake, sequestration and excretion of trace metals by an aquatic invertebrate. After Rainbow (1988).

clarified. For example, the crabs *Carcinus maenas* and *Cancer magister* excrete excess zinc across the gills (Bryan, 1966, 1968), as apparently does the prawn *Palaemon elegans* (White and Rainbow, 1984b). The lobster *Homarus gammarus* also excretes significant amounts of manganese across the body surface (Bryan and Ward, 1965).

The excretory tubules of animals fall into two major categories, nephridia and coelomoducts, the nomenclature of the two often being confused. These may both be present in certain species (e.g. some polychaetes), or in fact be combined in single tubules, as in some polychaetes and all crustaceans (Barrington, 1967). In terrestrial arthropods, nephridia are lacking and malpighian tubules (which are outgrowths of the alimentary tract) take over major excretory functions. Coelomoducts, if retained at all, play a limited excretory role in insects, being involved only in reproduction, to pass gametes to the outside.

Excretory tubules are also involved in ionic and osmotic regulation, essentially pumping out modified body fluids to the external medium. Such emissions are often formed by ultrafiltration of the blood, so that the primary urine contains all solutes of the blood except proteins of high molecular weight. The primary urine may be subsequently modified by selective resorption or secretion in the distal tubule, but the final urine

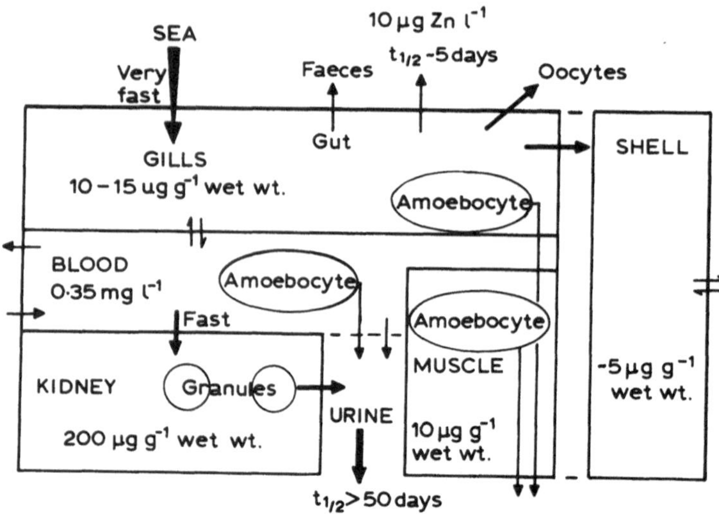

Fig. 33. A model for zinc dynamics in the mussel *Mytilus edulis*. After George and Pirie (1980).

which is excreted may contain trace metals. However, if the blood contains high concentrations of proteins (e.g. haemocyanin in malacostracan crustaceans or gastropod molluscs), most trace elements in the blood will be protein-bound and a limited percentage of the total metal content of the blood will thus be available for ultrafiltration and subsequent loss in the urine.

The excretory tubules of decapod crustaceans are the antennary glands, and the lobster *Homarus gammarus* excretes both zinc (Bryan, 1964, 1966; Bryan *et al.*, 1986) and manganese in the urine from these glands (Bryan and Ward, 1965). The paired kidneys of bivalve molluscs are derivatives of paired coelomoducts (Barrington, 1967) and are able to excrete metals. For example, the kidney of the mussel *Mytilus edulis* excretes zinc-rich granules into the urine (George and Pirie, 1980; Fig. 33). Similar metal-rich granules accumulate intracellularly and extracellularly in the kidney lumen of other bivalves, including scallops of the genus *Argopecten* (Carmichael *et al.*, 1979) and the quahog *Mercenaria mercenaria* (Sullivan *et al.*, 1988). Some excretion of metals is also possible in bivalves by the migration of amoebocytes containing metal-rich granules through the kidney and mantle (George and Pirie, 1980; George, 1990; Fig. 33).

Cells lining the alimentary tract (e.g. the midgut diverticula, caeca, or hepatopancreas) or more specialised derivatives (e.g. malpighian tubules) also have the potential to excrete trace elements. Metals present in granules in the hepatopancreas or caeca of malacostracan crustaceans are released into the lumen of the alimentary tract for expulsion via the faeces on completion of the epithelial cell cycle. Examples include iron-rich ferritin crystals in stegocephalid amphipods (Moore and Rainbow, 1984); granules containing lead in the crab *Carcinus maenas* (Hopkin and Nott, 1979); and copper-rich granules in the amphipod *Corophium volutator* (Icely and Nott, 1980). Manganese is excreted through the alimentary tract of the lobster *Homarus gammarus* (Bryan and Ward, 1965), as is zinc in the case of the freshwater crayfish *Austropotamobius pallipes* (Bryan, 1966). Any excretion of metals as metalliferous granules probably occurs as a series of episodic emissions, and if these are large enough, step-wise reductions in the metal content of an organism will result.

Another route of metal excretion involves the byssus of the mussels *Mytilus edulis* and *M. galloprovincialis*, as in the case of iron (George *et al.*, 1976), arsenic (Ünlü and Fowler, 1979), and certain radionuclides (Goldberg *et al.*, 1983). Aquatic arthropods can also lose both adsorbed and absorbed metals with the cast moult on ecdysis. However, both zinc and cadmium are resorbed from the cuticle into the underlying tissues of the decapod *Palaemon elegans* before moulting (White and Rainbow, 1984b, 1986). If the moulted arthropod consumes the cast moult, any metal excreted with the cast (or adsorbed thereto after moulting) might become available for uptake again.

A further route for the loss of metals from an organism is through the release of gametes, spores or larvae. Such reproductive products will inevitably contain essential metals such as copper and zinc, and this route of excretion may be significant in the overall metal budget of an aquatic organism.

(iv) The Sequestration of Metals

After metal uptake and excretion, the third factor determining the accumulation of trace elements by biota is the sequestration of the absorbed metal in the tissues (Phillips and Rainbow, 1989).

The sequestration of metals in marine organisms has been reviewed recently by Viarengo (1989) and George (1990). Several mechanisms exist, including the binding of metals to soluble metallothioneins and other metal-binding proteins (see also Cherian and Goyer, 1978; Roesijadi, 1980/81, 1992; Engel and Brouwer, 1989), and the sequestration of elements in metal-rich insoluble deposits or granules which may or may

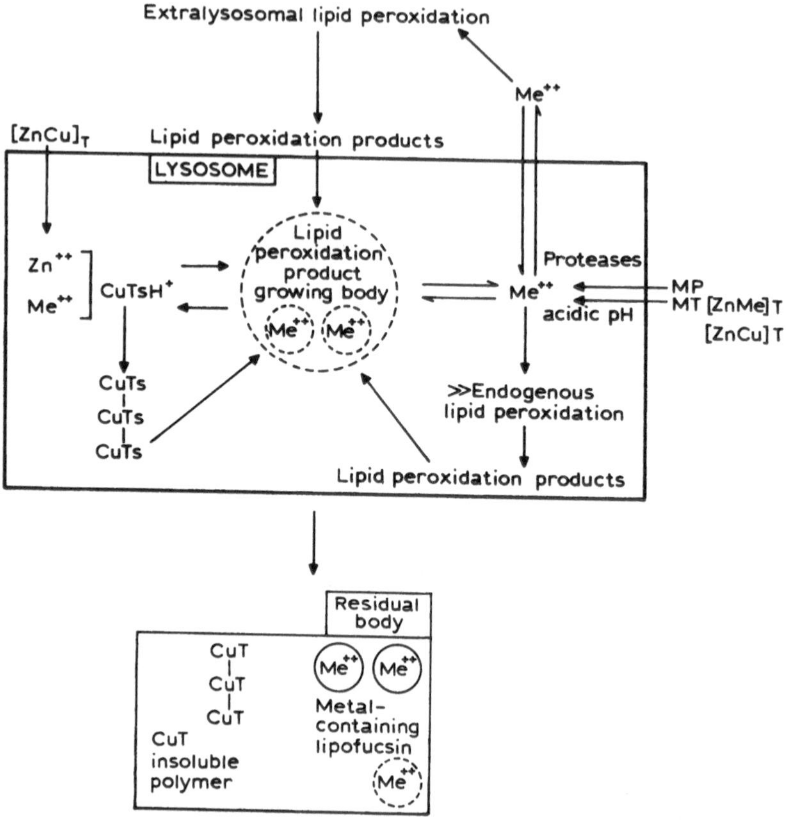

Fig. 34. Biochemical mechanisms involved in the compartmentation of trace metals in lysosomes. After Viarengo (1989).

not be associated with lysosomes (Simkiss and Taylor, 1989b). Metalliferous granules have been specifically reviewed by several authors (Mason and Nott, 1981; Simkiss, 1981; Brown, 1982; Simkiss *et al.*, 1982; Taylor and Simkiss, 1984; Hopkin, 1989).

Metallothioneins are soluble proteins found in the cytosol. They exhibit rapid turnover times in tissues, and may finally accumulate in large quantities in tertiary lysosomes (Fig. 34). These tertiary lysosomes are membrane-bound vesicles of variable half-life, and are recognised in the tissues as metal-rich granules (George, 1990). The sequestration of accumulated trace elements in such granular form in specific tissues (e.g. the hepatopancreas or kidney) is of major importance as a mechanism for final metal detoxification in animal tissues. The metal burdens of most invertebrates which exhibit significant accumulation of trace elements are generally attributable to metals sequestered in this fashion (George, 1990).

Many of the granules found in the tissues of aquatic biota appear to be based on calcium. Brown (1982) has suggested that these may be categorised separately from granules which contain sulphur and often copper, the latter being thought to have a metallothionein ancestry (e.g. Walker, 1977; Icely and Nott, 1980; Rainbow, 1987). However, more recent evidence suggests that metals such as cadmium and copper which are originally incorporated into metallothioneins might later become integrated into calcium-based granules. This is believed to occur, for example, in the winkle *Littorina littorea* (Nott and Langston, 1989) and the barnacle *Elminius modestus* (Simkiss and Taylor, 1989b; Pullen and Rainbow, 1991).

Calcium-containing granules may themselves be divided into two types (Simkiss, 1980, 1981; Mason and Nott, 1981). The first of these constitute relatively pure granules composed primarily of calcium carbonate, with some magnesium carbonate. They may be resolubilised readily, and probably act as a temporary physiological store for calcium (Simkiss, 1980, 1981). The second form of calcium-containing granule exhibits a significant proportion of pyrophosphate, and is able to bind many different metal cations to form very insoluble salts under cytosolic conditions. By contrast to the calcium carbonate-based granule discussed above, this pyrophosphate-containing granule apparently plays a detoxification role (Mason and Nott, 1981; Simkiss, 1981; Pullen and Rainbow, 1991).

Despite the need for an organism to detoxify metals in its tissues, some proportion of the accumulated body load of essential metals such as copper and zinc must be metabolically available in order to fulfill a metabolic role, for example in enzymes or respiratory pigments (Pequegnat *et al.*, 1969; White and Rainbow, 1985, 1987; Depledge, 1989a). This proportion will change according to the total amount of metal sequestered in an organism. Thus, decapod crustaceans such as *Palaemon elegans*, which regulate the body concentration of zinc (but see Depledge and Rainbow, 1990) will have a high percentage of their total zinc load in metabolically available form (Rainbow, 1988). By contrast, barnacles, which accumulate very large quantities of zinc in detoxified form (as zinc pyrophosphate granules) will have a very low body percentage of this as metabolically available zinc (Rainbow, 1987).

Table 15 Concentrations of zinc ($\mu g\ g^{-1}$ dry weight) accumulated by species of oysters and mussels (after Phillips, 1979b, 1985b; Eisler, 1981; Phillips and Yim, 1981)

Species	*Accumulated zinc*
Ostrea edulis	660–3280
Crassostrea virginica	322–12675
Saccostrea cucullata	430–8629
Mytilus edulis	14–500
Perna viridis	77–164
Septifer virgatus	74–116

Similarly, mussels contain a higher proportion of body zinc in metabolically available form than do oysters, as the latter exhibit far greater concentrations of the element than are present in mussels (Table 15). The high uptake of zinc and lack of any significant excretion of the element in barnacles and oysters are associated with the sequestration of accumulated zinc in a detoxified form. In both cases, the sequestered metal is present in zinc-rich granules (Walker *et al.*, 1975a,b; George *et al.*, 1978; Pirie *et al.*, 1984; Thomson *et al.*, 1985; Rainbow, 1987; Pullen and Rainbow, 1991).

By definition, essential metals are required by an organism in at least a minimum concentration, playing metabolic roles in enzymes and other macromolecules such as respiratory proteins, or even playing structural roles in some cases. White and Rainbow (1985) attempted to calculate the enzyme requirements of molluscs and crustaceans for copper and zinc, based on the earlier work of Pequegnat *et al.* (1969). Such calculations were later extended to iron and manganese (White and Rainbow, 1987). The requirements for enzymes in invertebrate tissues were calculated to approximate 26 $\mu g\ g^{-1}$ for copper; 27 $\mu g\ g^{-1}$ for iron; 4 $\mu g\ g^{-1}$ for manganese; and 34 $\mu g\ g^{-1}$ for zinc.

White and Rainbow (1985) extended their calculations to include copper in haemocyanin (which is present at high concentrations in many molluscs and crustaceans). They also considered the possibility that zinc bound to haemocyanin in the haemolymph of decapod crustaceans might be essential to the functioning of this respiratory pigment. However, recent work by Chan (1990) casts doubt on this possibility, as the stoichiometric ratios of zinc to haemocyanin vary widely in decapods, both intraspecifically and interspecifically.

These and other estimates (see Depledge, 1989a) provide guidelines to the minimum requirements of aquatic biota for essential metals, although it is important to differentiate between elements present in metabolically active soft tissues of organisms and those in relatively inert forms or tissues, such as the cuticles of crustaceans. It is relevant that these theoretical estimates agree closely with both the regulated levels of essential metals in decapod crustaceans, and the concentrations of such metals in soft tissues of invertebrates from habitats believed to be unpolluted.

As noted above, trace metals may also be involved in structural roles; in these instances, they may be present at very high concentrations in an organism. For example, zinc is a major structural component of the jaws of nereid polychaetes (Bryan and Gibbs, 1979, 1980), and copper has a similar role in the jaws of glycerid polychaetes (Gibbs and Bryan, 1980). Trace metals may also be involved in the deterrence of predators, for example in the cases of vanadium in sea squirts (Stoecker, 1980); copper in branchiae of the polychaete *Melinna palmata* (Gibbs *et al.*, 1981); and arsenic in the polychaete *Tharyx marioni* (Gibbs *et al.*, 1983).

C. ACCUMULATION STRATEGIES FOR METALS

The strategies employed by organisms for the accumulation of metals are of fundamental importance with respect to their usefulness as biomonitors of trace elements (Phillips and Rainbow, 1989). All metals are accumulated by aquatic organisms to body concentrations higher than those in an equivalent quantity of the ambient medium, measured for example in comparative wet weight terms. The degree of such accumulation, often expressed as a concentration factor, varies greatly between different species and metals. As noted above, the content of a trace metal in an organism (total metal load, in μg) results from the net balance between the processes of metal uptake and metal loss. The concentration of a trace metal in an organism (content per unit weight, typically in $\mu g\ g^{-1}$ dry weight) depends on changes in body weight and in the proportions of the various tissues, for example during growth, starvation, the development of gonads, gamete storage or expulsion, and the accumulation or depletion of energy reserves. The relationship between metal uptake and loss dictates the particular metal accumulation strategy of an organism.

Some further distinctions also require clarification. The total trace metal content of an invertebrate can be divided into three components (Rainbow, 1988; see Fig. 32). These are trace elements passively adsorbed onto the external surface of the organism; metals in the gut not assimilated into the body; and absorbed or accumulated metals accessible to physiological processes (and therefore potentially metabolically available to play either an essential physiological role, or to exert a toxic effect). The metal contents of aquatic plants clearly lack the ingested component, but the passively adsorbed portion includes trace elements entering the apparent free-space of plant cell walls, external to the cytoplasm of plant cells (Higgins and Mackey, 1987).

Passively-adsorbed metals are available for exchange (involving further adsorption or desorption) with elements in the aquatic medium. Such adsorbed metals represent a variable proportion of the total contents of trace elements in organisms, usually being more significant when the overall metal contents are low. The adsorbed component can be high in plants such as macrophytic kelps (Higgins and Mackey, 1987), or in small invertebrates with high surface area-to-volume ratios (e.g. see Rainbow and Moore, 1986). Passive adsorption may be most significant in the case of metals like iron and manganese, for these elements have a high propensity to come out of solution in seawater as hydroxides, even in the absence of available organically-rich surfaces with many potential metal-binding sites. Metals which are passively adsorbed are not metabolically available.

In most instances, the trace metal content of ingested food is a negligible proportion of the total content of an element in an invertebrate (e.g. see Weeks and Moore, 1991), although certain exceptions exist. In deposit-feeding invertebrates, the metal content in sediment in the gut may approach that of the accumulated metal content of the body, particularly if the metal is very strongly bound to the sediment and is unavailable for assimilation, absorption and subsequent accumulation by the organism over time. Thus, deposit-feeding species living in soft muds may exhibit different total body concentrations of iron and manganese before and after gut depuration, because of the contribution of the gut contents to the total body loads of elements. Certain other instances also exist in which metals in the gut of organisms have been found to constitute an appreciable proportion of the total body load of elements, and these are reviewed in section D. (viii) below. As for passively adsorbed metals, any trace elements in the gut lumen are not available to the metabolism of the organism, at least until they have been assimilated and entered the third body component — that of absorbed metals.

The concentrations of absorbed or accumulated metals in organisms (i.e. those of elements which have entered the tissues by whatever route of uptake, to become metabolically available) depend upon the metal accu-

NON REGULATORS	EXTREME UPTAKE
Barnacles	Zn in granules
Ascidians	V in vanadocytes
Bivalve molluscs	Cu, Fe, Mn, Pb, Zn in granules
Gastropod molluscs	Cu, Zn in granules
Isopods, amphipods	Cu, Fe, Pb, Zn in granules
Polychaetes	Cu in granules
Macroalgae	Most metals
Mussels/other bivalves	Most metals, metallothioneins
Polychaetes	Cd, Pb
Decapod crustaceans	Cd, Pb
Finfish	Cd, Pb
Bivalves (rare)	Cu, Zn
Macroalgae	Mn
Polychaetes	Cu, Zn
Decapod crustaceans	Cu, Mn, Zn
Finfish	Cu, Mn, Zn
REGULATORS	LOW UPTAKE

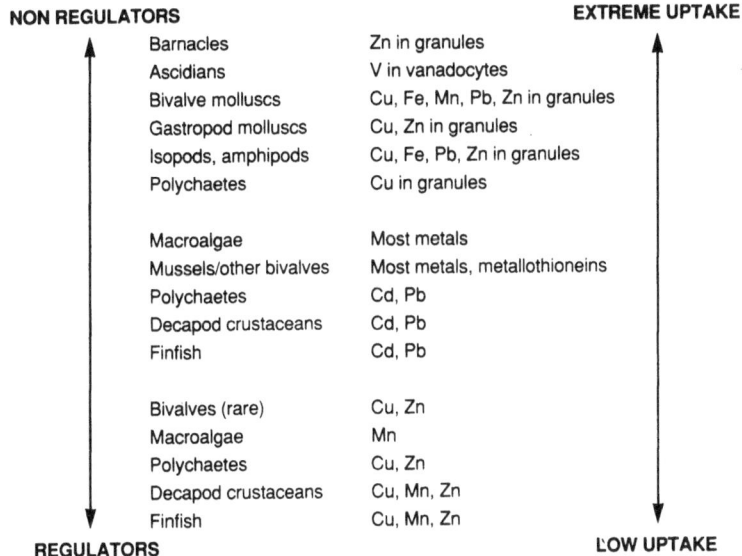

Fig. 35. A gradient of metal accumulation strategies in aquatic organisms. After Phillips and Rainbow (1989).

mulation strategy of the particular species concerned. The metal accumulation strategies exhibited by aquatic organisms fall along a spectrum or gradient (Phillips and Rainbow, 1989; Rainbow *et al.*, 1990; Fig. 35). At one extreme (the strategy of strong net accumulation), all the metal taken up into the body is accumulated, with no excretion at all. At the other extreme, particular species are able to regulate their total body concentrations of certain trace metals to approximately constant levels, over a wide range of dissolved metal availabilities in the ambient environment.

Net accumulators of trace metals do not match the rate of excretion of accumulated elements to that of uptake, thus inevitably increasing in metal content over time. The metal concentration in the whole body of the organism will therefore also rise with time, unless the rate of increase in body weight exceeds the rate of net metal uptake. The degree of such net accumulation of metals varies according to the relative rates of uptake and excretion of elements. If the uptake rate of a metal (through one or more routes) is very high and excretion is minimal, the organism will be a strong net accumulator of that element. This strategy is accompanied by the necessary physiological pathways for detoxification of accumulated metals, involving storage of the potentially toxic elements in non-toxic forms. For example, barnacles accumulate very considerable concentrations of zinc (Rainbow, 1987; Phillips and Rainbow, 1988; Fig. 35), as a result of high uptake rates and the lack of any significant excretion of this element (Rainbow and White, 1989). The accumulated metal is stored in the body in a detoxified form in granules, which contain high concentrations of zinc pyrophosphate (Walker *et al.*, 1975a,b; Pullen and Rainbow, 1991).

Other net accumulators may exhibit high metal uptake rates but also relatively high excretion rates, resulting in a weak net accumulation of trace elements. Alternatively, a weak net accumulator of metals may exhibit no significant excretion of accumulated elements, but a very low metal uptake rate (e.g. zinc in the littoral amphipod crustacean *Echinogammarus pirloti*; see Rainbow and White, 1989). A weak net accumulation strategy will approach the case of regulation under conditions where excretion matches uptake.

Such weak net accumulation may be referred to as the partial regulation of trace elements, as in the case of zinc in the tropical mussel *Perna viridis* (Phillips, 1985b; Chan, 1988; Phillips and Rainbow, 1989). Other partial regulators of zinc include the mussels *Mytilus edulis* (Amiard *et al.*, 1987), *Septifer virgatus* (Phillips and Yim, 1981) and *Trichomya hirsuta* (Klumpp and Burdon-Jones, 1982); the gastropods *Haliotis tuberculata* (Bryan *et al.*, 1977), *Littorina obtusata* (Young, 1975), *Littorina littorea* (Bryan *et al.*, 1983; Amiard–Triquet *et al.*, 1987) and *Gibbula umbilicalis* (Amiard–Triquet *et al.*, 1987); and the polychaete *Nereis (Hediste) diversicolor* (Bryan *et al.*, 1980; Bryan and Hummerstone, 1973b; Amiard *et al.*, 1987).

While mussels excrete considerable quantities of zinc as granules from the kidney (George and Pirie, 1980), oysters of the genera *Ostrea, Crassostrea* and *Saccostrea* do not excrete the element in significant amounts (George *et al.*, 1978). The result of this difference in excretion rates is that oysters exhibit a strong net accumulation strategy for zinc, storing the element in granules, whereas mussels exhibit weak net accumulation or partial regulation of the element (Table 15; Fig. 35).

The method of physiological sequestration of an absorbed metal is a key determinant of a metal accumulation strategy. An example is afforded by a comparative study of British populations of three marine molluscs: the gastropod *Littorina littorea*, the mussel *Mytilus edulis* and the clam *Macoma balthica* (Langston and Zhou, 1987; Langston *et al.*, 1989). *M. balthica* was found to exhibit a relatively slow rate of cadmium accumulation, and in contrast to the other two showed no binding of cadmium to metallothioneins after cadmium exposure. *L. littorea* bound newly incorporated cadmium (notably in the digestive gland) to previously-synthesised metallothioneins, while *M. edulis* sequestered such cadmium in newly-induced metallothioneins (Langston *et al.*, 1989). These interspecific variations in cytosolic metal-binding behaviour may explain the large observed differences in cadmium accumulation rates of the three species, both in the laboratory and the field. Laboratory-derived bioconcentration factors for cadmium varied from 3 for *M. balthica*, to 16 for *L. littorea*, and to 31 for *M. edulis* (Langston *et al.*, 1989).

Two reasons may exist for the low cadmium accumulation rate of this British population of *Macoma balthica*. Firstly, absolute cadmium uptake may be low as a result of the reduced amounts of metallothioneins present in this population (i.e. the low detoxification capacity for the element). Alternatively, the ability to synthesise metallothioneins may have been reduced to negligible proportions because cadmium accumulation is low in this population for another reason (see Langston and Zhou, 1987). Interestingly, the population concerned clearly differs fundamentally in its response to metal exposure from other populations of (apparently) the same species. Thus, Langston and Zhou (1987) have found that its inability to synthesise metallothioneins in any significant amount extends not only to cadmium exposure, but also to challenges with silver or zinc. By contrast, Johansson *et al.* (1986) showed a metal-tolerant population of *M. balthica* from San Francisco Bay to be capable of synthesising significant amounts of metallothionein-like proteins in response to exposure to copper, silver or zinc. These differences suggest the existence of either extreme intraspecific variability, or a problem of species identification (see below).

The littoral decapod *Palaemon elegans* provides an example of trace element regulation. This species regulates its total body content of zinc to an approximately constant level over a wide range of external bioavailabilities of the metal (Fig. 36). At extreme external concentrations of the element, however, regulation breaks down and net accumulation commences (Fig. 37). Regulation of zinc is achieved by an increase in the rate of excretion to match that of uptake, the latter being beyond the physiological control of the decapod (White and Rainbow, 1982, 1984a,b; Nugegoda and Rainbow 1987, 1988a, 1989a; Rainbow and White, 1989).

The concept of regulation of total body metal content is useful pragmatically, particularly with reference to biomonitors, but it is difficult to appreciate how a specific feedback regulatory control might act at the whole body level (Depledge and Rainbow, 1990). Whole body loads of

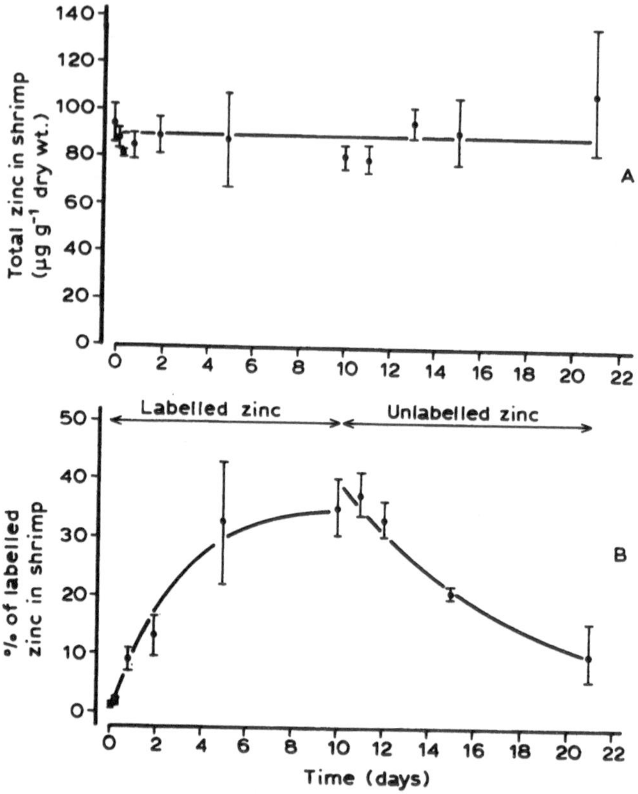

Fig. 36. Exposure of the decapod crustacean *Palaemon elegans* to 100 μg litre^{-1} zinc for 20 days at 10°C (days 0–10 in labelled Zn; thereafter in unlabelled zinc). (A) Total body concentrations of zinc. (B) Alterations in labelled zinc concentrations over time. After White and Rainbow (1984a).

elements reflect the summated loads of the constituent tissues, and the latter would appear likely to be the sites of separate feedback mechanisms (Depledge and Rainbow, 1990).

Evidence for the regulation of trace elements by biota may be derived from laboratory experiments (e.g. see White and Rainbow, 1982; Rainbow, 1985; Rainbow and White, 1989), or from comparative analyses of the metal concentrations of invertebrates from sites which exhibit different metal bioavailabilities (Bryan, 1968; Phillips and Yim, 1981). Strictly, radioactive tracers should be used to distinguish absolute uptake from net accumulation, in order to unequivocally establish the presence of

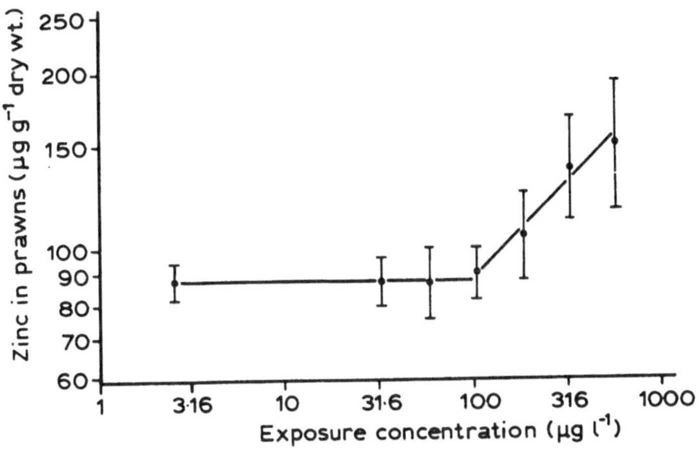

Fig. 37. The effect of increasing concentrations of zinc in solution on the mean body zinc concentrations ($\mu g\, g^{-1}$ dry weight ± 1 standard deviation) of *Palaemon elegans* at 15°C. After Nugegoda and Rainbow (1987).

regulation as an accumulation strategy (e.g. Rainbow and White, 1989). However, the lack of a suitable radiotracer (as in the case of copper, for example) renders this difficult in certain cases.

The accumulation of zinc by the shore crab *Carcinus maenas* illustrates this point well. Rainbow (1985) concluded from laboratory experiments that *C. maenas* regulates its concentration of zinc over a range of exposures to this element. He presumed, on the basis of similar data for the shrimp *Palaemon elegans*, that this was due to a capacity for significant excretion of zinc by *C. maenas*; however, it was noted that the data might also be explained by the existence of very low net accumulation of the metal. The latter is in fact the case. Chan (1990) has shown that the crab is actually a net accumulator of zinc at all ambient exposure levels, but that uptake (and subsequently accumulation) of the metal is so low as to be negligible at most environmentally realistic concentrations. The existence of a very low uptake rate for the element has produced a zinc accumulation strategy that approximates to regulation, although it is based in reality upon low net accumulation. A rapid growth rate under such circumstances might even produce a reduction in body zinc concentration (as opposed to zinc content) during growth.

This account has emphasised the interspecific variability of metal accumulation strategies, but such strategies may also vary intraspecifically to some extent (e.g. in the case cited previously for the clam *Macoma balthica*). Most specimens of *Palaemon elegans* are net accumulators of cadmium, with essentially no capacity for excretion of this metal. However, some individuals excrete significant amounts of cadmium, although they remain net accumulators of the element (White and Rainbow, 1986; Rainbow and White, 1989). In both decapods and amphipods, metal uptake rates may in fact vary by as much as 10-fold between individuals. In a study of zinc accumulation strategies (Rainbow and White, 1989), specimens of the amphipod *Echinogammarus pirloti* showed a wide range of net accumulation rates, and these correlated with the observed range in uptake rates for zinc between individuals.

It must be emphasised here that a species may exhibit a different accumulation strategy for each metal. Nevertheless, it is tempting to make generalisations with respect to the implications of general biology for metal accumulation strategies. Thus, for example, a low uptake rate for an element may be a preadaptation to the adoption of regulation as a metal accumulation strategy. Such a low uptake rate may result from morphological changes during evolution, e.g. the development of a tanned, calcified exoskeleton in large, benthic crustaceans which have adopted a walking habit (by contrast to the permeable cuticle of small, swimming or epibenthic crustaceans). Many decapods appear to regulate essential trace elements (including zinc, copper and manganese), but not non-essential metals such as cadmium and lead (Rainbow, 1988). Differences in regulated body concentrations of zinc, for example intraspecifically with temperature in the case of *Palaemon elegans* (see above and Fig. 38), or interspecifically between related species such as *Palaemon elegans*, *Palaemonetes varians* and *Pandalus montagui* (Fig. 39; Nugegoda and Rainbow, 1988b, 1989a), probably reflect small differences in metabolic requirements for zinc.

The feeding habits and the habitats of organisms are also important in defining their exposures to metals, and hence perhaps their adopted (or more strictly, the evolved) strategies for metal accumulation. Sedentary, suspension-feeding invertebrates with large permeable surface areas may be committed to high metal uptake rates that cannot be matched by excretion rates, ruling out regulation as a possible strategy (Rainbow *et al.*, 1990). By contrast, estuarine crustaceans which exhibit increased impermeability of their body surface and high urine production rates may be preadapted to low net accumulation rates for trace metals.

The significance of a metal accumulation strategy for biomonitoring is clear. No regulator of trace elements should be used as a biomonitor. Any

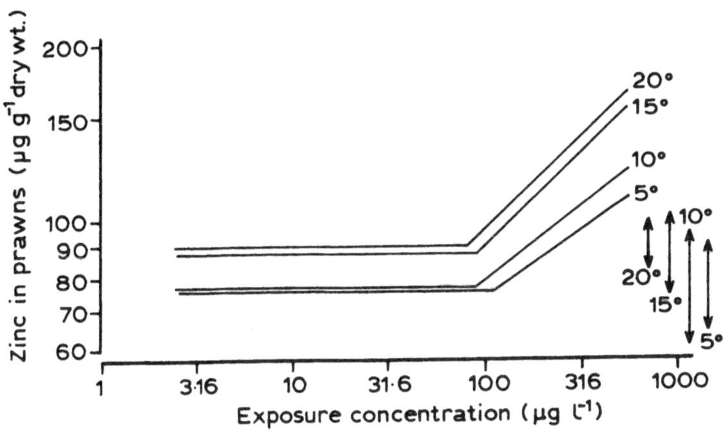

Fig. 38. Summary of the effect of a range of dissolved zinc exposures (μg litre^{-1}) on the zinc concentrations of *Palaemon elegans* (μg g^{-1} dry weight) at four temperatures. Vertical arrows indicate the different regulated ranges of body zinc concentrations at each temperature, equivalent to the 95% confidence limits of each mean. After Nugegoda and Rainbow (1987).

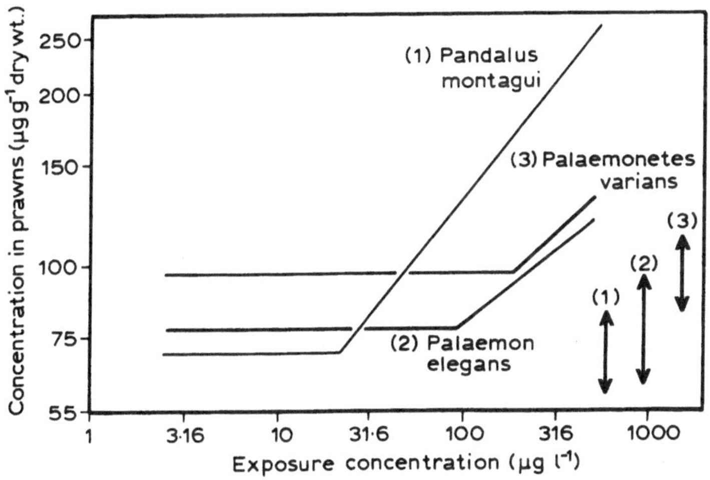

Fig. 39. Comparison of zinc regulation and regulation breakdown in three decapod crustaceans, *Pandalus montagui, Palaemon elegans* and *Palaemonetes varians,* under identical physicochemical conditions at 10°C. Vertical arrows represent regulated ranges of body zinc concentrations (μg g^{-1} dry weight) of each species, as in Fig. 38. After Nugegoda and Rainbow (1988b).

accumulator of metals is a potential biomonitor, so long as any change in body metal concentration brought about by a geographical or temporal change in the ambient bioavailability of a metal is measurable above the "background noise" of intraspecific variability. The latter is more likely to be realised through studies of strong net accumulators of metals, such as oysters or barnacles for zinc (e.g. Boyden and Phillips, 1981; Phillips and Yim, 1981; Phillips and Rainbow, 1989). However, it may also be possible using relatively weak net accumulators, as in the case of zinc in talitrid amphipods (Rainbow *et al.*, 1989; Weeks and Rainbow, 1991).

D. REVIEW OF BIOMONITORING

(i) Introduction

There have been several reviews published on the use of aquatic organisms as biomonitors of the availability of trace metals (Phillips, 1977, 1980, 1990b; Bryan *et al.*, 1980, 1985), with that of Phillips (1980) being exceptionally comprehensive, and that of Bryan *et al.* (1985) being particularly strong on methodology.

The first biomonitoring programmes, undertaken in the early 1960s, concentrated on radionuclides released into the Columbia River and those in coastal waters off California. Later in the 1960s and 1970s, biomonitoring programmes also investigated the availability of stable isotopes of trace metals (e.g. Portmann, 1971), and this was continued in large-scale geographical programmes, such as the Mussel Watch studies undertaken in the United States in the mid-1970s and in more recent years (Goldberg *et al.*, 1978, 1983; Farrington *et al.*, 1983; Lauenstein *et al.*, 1990).

The biomonitoring of trace metals provides information on the variation in time and space of the concentrations of metals which are available to a selected biomonitor. Thus, biomonitoring seeks to identify significant differences in metal bioavailabilities between samples from different sites, or between samples taken from the same site at different times. It is necessary in any such programme to identify significant changes in the accumulated metal concentrations of biomonitors that are caused by alterations in the ambient bioavailabilities of metals, as opposed to "background" or other variations which may be due to a number of biological and physicochemical factors. These factors include size, age, sex, season, contaminant interactions, salinity, and water temperature (see also Phillips, 1980); the effects of these are considered in more detail in the following sections. Even when all such factors are known and taken into account, some residual or "inherent" variability in trace metal levels exists in all biomonitors. Sampling programmes need to allow for this, in order to distinguish signals of raised metal bioavailabilities from "background" variation.

A crucial prerequisite in the design of biomonitoring programmes is the strict definition of the programme objectives (Phillips and Segar, 1986). The three most common objectives are: (i) the delineation of spatial variations in metal bioavailabilities; (ii) the elucidation of changes in metal bioavailabilities over time at one or more sites; and (iii) the identification of previously unknown contaminants in a given water body. The design of monitoring programmes should differ considerably for each of these objectives (Phillips and Segar, 1986). Further discussion on the design of biomonitoring programmes is provided in Chapter 8.

(ii) Species Identification

The reliable taxonomic identification of the species used is of vital importance to biomonitoring studies. The use of oysters in the genera *Crassostrea* and *Saccostrea* (Stenzel, 1971; Ahmed, 1975) provides a good example. While the Pacific oyster *Crassostrea gigas* is rarely mis-identified (for example, see studies by Ayling, 1974; Watling and Watling, 1976; Boyden and Phillips, 1981; Okazaki and Panietz, 1981; Thomson, 1982), greater problems exist with the tuberculated or denticulated oysters of the genus *Saccostrea*. The Sydney rock oyster has been variously termed *Saccostrea commercialis*, *Crassostrea commercialis* and *Saccostrea cucullata* (Mackay *et al.*, 1975; Ward, 1982; Talbot, 1985). The precise taxonomic relationships between this species and other oysters employed in biomonitoring programmes are uncertain in several cases (e.g. see reports on *Saccostrea echinata* in Australia by Denton and Burdon-Jones, 1981; *Saccostrea glomerata* in New Zealand by Nielsen and Nathan, 1975, and in Hong Kong by Phillips, 1979b; and *Saccostrea cucullata* in Thailand by Brown and Holley, 1982).

Similarly in the case of mussels, *Septifer virgatus* was originally mis-identified as *Septifer bilocularis* in another Hong Kong study (Phillips and Yim, 1981). Such taxonomic problems are almost inevitable in biomonitoring investigations from tropical and subtropical waters with poorly characterised fauna and flora, but similar problems arise even in the North Atlantic. Lobel *et al.* (1990) have pointed out that mussels identified as *Mytilus edulis* in North Atlantic studies were not always clearly distinguished from the similar species *M. galloprovincialis* and *M. trossolus*. Such mis-identifications may well affect the conclusions drawn from biomonitoring investigations. For example, *M. trossolus* exhibited higher concentrations than *M. edulis* from the same site in Newfoundland for 24 out of 25 elements, and these differences were statistically significant in the case of 11 of the elements involved (Lobel *et al.*, 1990).

Some types of organisms proposed as potential biomonitors of trace metals have inherent problems of identification that can be overcome only by specialists; nemertean worms provide one such example (McEvoy, 1988). For comparative work, potential biomonitors should preferably be geographically widespread or cosmopolitan, such as the mussel *Perna viridis* and the barnacles *Balanus amphitrite* and *Tetraclita squamosa* found throughout the Indo-Pacific (see Phillips, 1984; Phillips and Rainbow, 1988). Such comparative work will inevitably entail the identification of organisms by different (probably non-expert) workers, and the reliability of such identifications should be ensured wherever possible.

(iii) The Effects of Size, Weight and Age

The size, weight or age of biomonitors may affect their accumulated concentrations of trace elements, and this may also interfere with their use as biomonitors (Boyden, 1974, 1977; Phillips, 1980).

In metal-accumulating organisms increasing in size and weight with time, two components of metal concentration are changing: (i) the metal content of the organism; and (ii) the weight of the individuals. However, these changes do not necessarily occur in tandem. Thus, if growth is slow relative to the rate of accumulation of a metal, the concentration of that element will increase with age and weight. Alternatively, if growth is rapid compared to metal accumulation, the observed concentration of the element will decrease with age and weight, even though the overall metal content may be increasing. Boyden (1974, 1977) discussed various relationships between body weight and metal content and concentration in molluscs, concluding that some generalisations might be possible. However, more recent data argue against this, and differences between samples in size, weight or age are now considered important potential interferences in biomonitoring studies (Phillips, 1980).

For example, Rainbow and Moore (1986) showed that littoral amphipod crustaceans typically exhibit a decrease in metal concentration with dry weight, modelled by the power relationship $y = ax^b$. Metal concentrations were particularly high at body dry weights below 2 mg, interpreted to be a probable result of the high contribution of adsorbed metal to the total body burden in small animals with high surface area-to-volume ratios. Allowance for the impacts of body weight could be made by

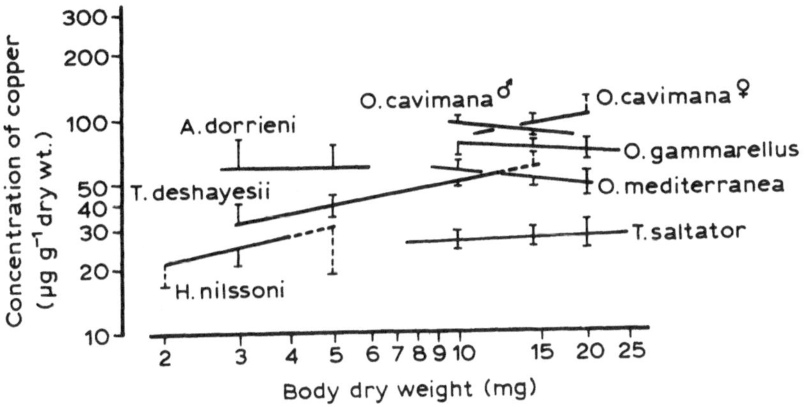

Fig. 40. Best-fit regression lines of copper concentration ($\mu g\,g^{-1}$ dry weight) against body dry weight for whole bodies of six species of talitrid amphipods. Species names in full are *Hyale nilssoni*, *Talitrus saltator*, *Talorchestia deshayesii*, *Arcitalitrus dorrieni*, *Orchestia gammarellus*, *O. mediterranea* and *O. cavimana*. Data for *O. cavimana* are shown for males and females separately. After Moore and Rainbow (1987).

using only amphipods of greater than 2 mg dry weight in analyses, or more effectively by logarithmic transformation of the data and comparison of the weight/concentration regression lines (Rainbow and Moore, 1986; Moore and Rainbow, 1987; Fig. 40). Such regression lines may be compared by Analysis of Covariance (ANCOVA), which may also be employed to establish the lack of significant differences between regression coefficients (slopes) of the relationships compared. Rainbow *et al.* (1989) have employed ANCOVA techniques in a biomonitoring study for copper and zinc using the talitrid amphipod *Orchestia gammarellus*.

ANCOVA techniques have also been used to accommodate size effects in studies of metal concentrations in deep sea crustaceans (Rainbow, 1989; Ridout *et al.*, 1989) and in barnacles (Phillips and Rainbow, 1988; Fig. 41). In the latter case involving a biomonitoring study of barnacles and mussels in Hong Kong, ANCOVA revealed the existence of significant intersite differences in body metal concentrations in the barnacles *Capitulum mitella*, *Tetraclita squamosa* and *Balanus amphitrite*, after allowance for size effects on metal concentrations in each species. Rank orders of the metal concentrations in barnacles were then compared against those for the same elements in soft tissues of the green-lipped mussel *Perna viridis*. For the mussel, size allowance was effected by pooling the tissues of

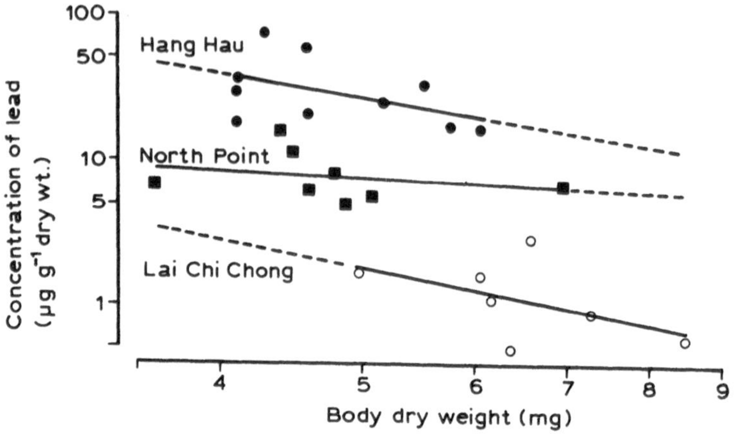

Fig. 41. Best-fit regression lines of body concentrations of lead ($\mu g\, g^{-1}$ dry weight) against body dry weight for whole bodies of the barnacle *Balanus amphitrite* collected from three sites in Hong Kong in April 1986. From unpublished data of P.S. Rainbow.

mussels of similar size from each site (Phillips and Rainbow, 1988). In another example, Bryan *et al.* (1980, 1985) catered for size variation in biomonitoring studies with the bivalve *Scrobicularia plana* by using replicate pooled samples of animals of a standard shell length. The various options available to cater for the possible impacts of organism size (weight; age) on contaminant concentrations in biomonitors are addressed further in Chapter 8.

Physiological and ecological characteristics may also change with age in biomonitors, with important consequences for accumulated metal concentrations. As already noted, surface area-to-volume ratios decrease with size, and this affects the relative contribution of adsorbed metal content to the total body burden of trace elements. Changes in diet or habitat may also occur through the life stages of a biomonitor, and these may affect their exposure to metals. For example, the veliger larvae of benthic gastropods are often suspension feeders of phytoplankton, moving relatively large volumes of water across permeable surfaces and feeding microphagously on potentially metal-rich particles. Such larvae have very different metal uptake routes and rates from adults of the same species, which may be highly selective carnivores, or specialist or generalist browsers. A further example is provided by adult decapod crustaceans living in hyposaline media, which migrate to more saline waters for breeding and subsequent larval development. Larvae and adults of these species thus occupy habitats which exhibit different physicochemistries, with fundamental consequences for their rates of metal uptake and accumulation.

In the case of macrophytic algae, growth gives rise to changes in metal concentrations of different parts of the thallus. Thus, for example, metal concentrations are low in the growing tips of *Fucus vesiculosus* and increase towards the older part of the thallus (Bryan and Hummerstone, 1973a). Analysis of the younger tissue at the tips generates information on more recent pollution conditions, whereas the study of the older portions of the thallus provides an integrated picture of the extent of contamination, over several months in some cases (Bryan *et al.*, 1985). In a biomonitoring survey using macroalgae, it is therefore important to use a standard region of the thallus for all samples analysed (Barnett and Ashcroft, 1985).

The process of growth itself has effects on accumulated metal concentrations of aquatic arthropods, which increase in size through a series of step-wise moults. Moulting is associated with changes in the internal tissue distributions of metals (see Depledge and Bjerregaard, 1989), and elements may also be lost in the cast moult. Moreover the cuticle of recently moulted arthropods is relatively permeable, and a period of rapid uptake of metals may occur (White and Rainbow, 1984a,b) before the more usual impermeability is restored by tanning and calcification of the new cuticle.

(iv) The Effects of Season

Seasonal changes in trace element concentrations may be an important source of interference in biomonitoring studies. This generally arises where biological changes in the organisms employed in biomonitoring programmes serve to alter the concentrations (or sometimes, contents) of metals present. The great majority of such biological changes are associated with the processes of reproduction and growth (e.g. see Bryan, 1973).

Metal concentrations vary between tissues of animals (e.g. Depledge and Bjerregaard, 1989; Depledge and Rainbow, 1990); thus, changes in the percentage contribution of a tissue to the whole body can affect metal concentrations in the whole organism. Such changes can be caused by: the storage or depletion of energy reserves; the development of gonads on sexual maturity; and the subsequent differential storage and emission of gametes. Where growth itself is seasonal (as in macroalgae; see above and Bryan *et al.*, 1985), care must be taken to sample at the same time each year in biomonitoring studies of temporal changes in the bioavailabilities of metals.

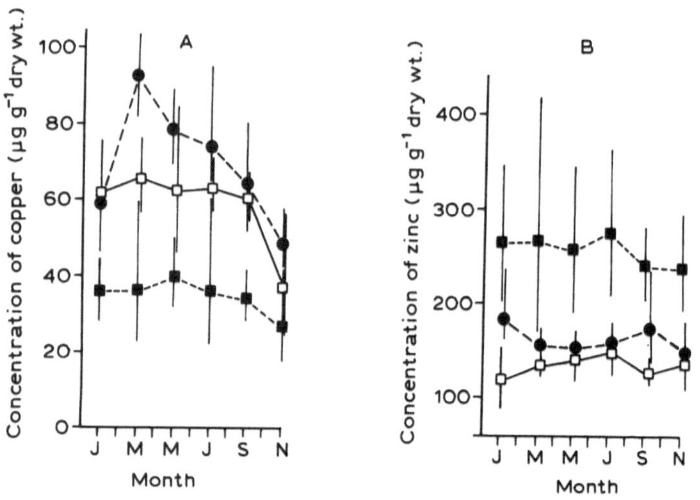

Fig. 42. Seasonal variation in copper and zinc concentrations ($\mu g\, g^{-1}$ dry weight) in the amphipods *Orchestia gammarellus* (●), *O. mediterranea* (□) and *Talitrus saltator* (■) collected from the Isle of Cumbrae, Scotland, in 1987. Data are for amphipods of 10 mg standard dry body weight, as derived from log-log weight-concentration regressions. After Rainbow and Moore (1990).

In such studies of temporal changes in the bioavailabilities of trace elements, it is necessary to assess the significance of any observed seasonal changes in metal concentrations. Rainbow and Moore (1990) found no significant seasonal variation in body zinc concentrations in the talitrid amphipod *Orchestia gammarellus* in Scottish waters, although copper concentrations of this amphipod were elevated in March and lowered in November (Fig. 42). Given the constancy of body zinc concentrations, it is unlikely that changes in tissue weights would have caused the observed seasonal changes in copper concentrations, which may be associated with variation in the levels of the copper-bearing respiratory pigment haemocyanin.

Nor is it possible to produce generalisations on the occurrence of seasonal changes in metal concentrations of particular biomonitors. Thus, Goldberg *et al.* (1978) and Boalch *et al.* (1981), working in the USA and southern England respectively, reported a lack of seasonal alterations in soft tissue concentrations of trace metals in the mussel *Mytilus edulis*. However, Gault *et al.* (1983) detected significant seasonal changes in soft tissue metal concentrations of *M. edulis* in Northern Ireland, as did Amiard *et al.* (1986) for the same species from the Bay of Bourgneuf in

France. Other authors have also noted significant seasonal changes in the concentrations of metals in this species (Phillips, 1976a; Simpson, 1979; Popham and D'Auria, 1982), and Fowler and Oregioni (1976) reported similar effects in the related mussel *M. galloprovincialis* from the north-west Mediterranean Sea.

Seasonal variations in trace metal concentrations appear to be more marked in oysters than in mussels, perhaps reflecting the more variable estuarine habitat of some oyster populations. Frazier (1975) noted considerable seasonal variation in the concentrations of cadmium, copper and zinc in *Crassostrea virginica* from Rhode River, USA, as did Wright *et al.* (1985) for copper and zinc in the same oyster from Chesapeake Bay in the USA. Similarly, Harrison (1978) found zinc concentrations in *C. gigas* from the Humboldt Estuary in the USA to vary widely with season. Talbot (1986) showed that minimum and maximum copper and zinc concentrations in the tuberculated oyster *Saccostrea cucullata* occurred in summer and spring respectively in Australia, probably in reflection of the reproductive status of these populations.

Seasonal changes in weight, for example with variation in reproductive status or energy stores, certainly affect the accumulated metal concentrations of a biomonitor (e.g. see Phillips, 1976a; Simpson, 1979). However, it is often difficult to identify these physiological variables as being of importance in determining metal levels, against the background of other seasonal changes. The latter may include temporal fluctuations in the input of metals into an estuarine habitat, or in hydrological processes, which may vary seasonally to affect both total dissolved metal concentrations and their bioavailabilities. Thus, Luoma *et al.* (1985) could not separate these factors in terms of their importance in defining the temporal variation of copper and silver in the clam *Macoma balthica* in south San Francisco Bay (Fig. 43). However, in a later study, Cain and Luoma (1990) concluded that the most marked temporal variations in metal concentrations were associated with seasonal changes in soft tissue weights, while the smaller fluctuations in metal levels of *M. balthica* of standard shell length were correlated with changes in copper and silver concentrations in sediments (Fig. 44).

In a study of the effect of sexual maturation as a source of variation in tissue cadmium concentrations in the mussel *Mytilus edulis*, Cossa *et al.* (1979) compared regressions of cadmium concentration against tissue dry weight in mature and immature mussels from the Gulf of St Lawrence, Canada (Fig. 45). Regression coefficients were considerably more variable after sexual maturity, probably as a result of physiological and biochemi-

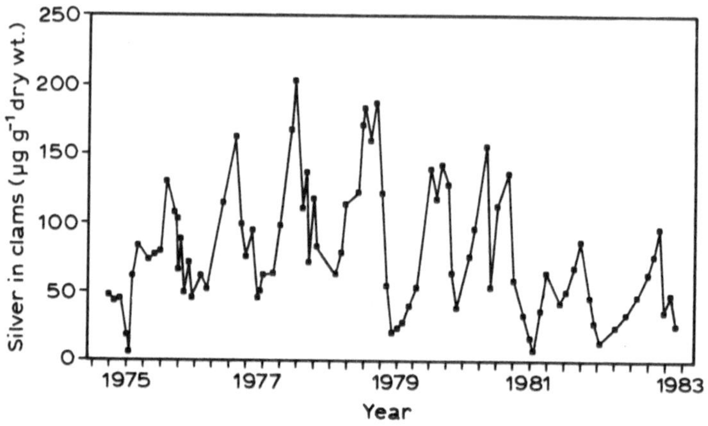

Fig. 43. Temporal variations in the concentrations of silver ($\mu g\ g^{-1}$ dry weight) in whole soft tissues of clams (*Macoma balthica*) from a station in south San Francisco Bay. After Luoma *et al.* (1985).

cal changes associated with the reproductive cycle. Cossa *et al.* (1979) thus recommended the use of immature mussels in biomonitoring studies, as a method of reducing or eliminating the impacts of size-related changes in metal concentrations. This has not been followed up by other authors, but in some instances would appear to be worthy of further consideration.

Sexual maturation may also lead to a further complication in biomonitoring surveys, as in some species, male and female individuals are known to vary in their accumulated concentrations of trace metals. Latouche and Mix (1982) found tissue concentrations of manganese and zinc (but not of cadmium, copper or nickel) to differ between the gonads of male and female *Mytilus edulis* from Yaquina Bay, Oregon. By contrast, metal concentrations of somatic tissues did not differ between the sexes for metals other than cadmium. Orren *et al.* (1980) found that the females of a population of mussels (*Choromytilus meridionalis*) collected prior to spawning exhibited higher soft tissue concentrations of copper, iron, manganese and zinc than did males. These differences disappeared after spawning, suggesting that they were due to variations in the amounts of metals present in the gametes of the two sexes. Such sex-based differences in the bioaccumulation of metals may be important in certain instances in biomonitoring surveys. Where they exist, samples should be stratified for sex of the individuals collected.

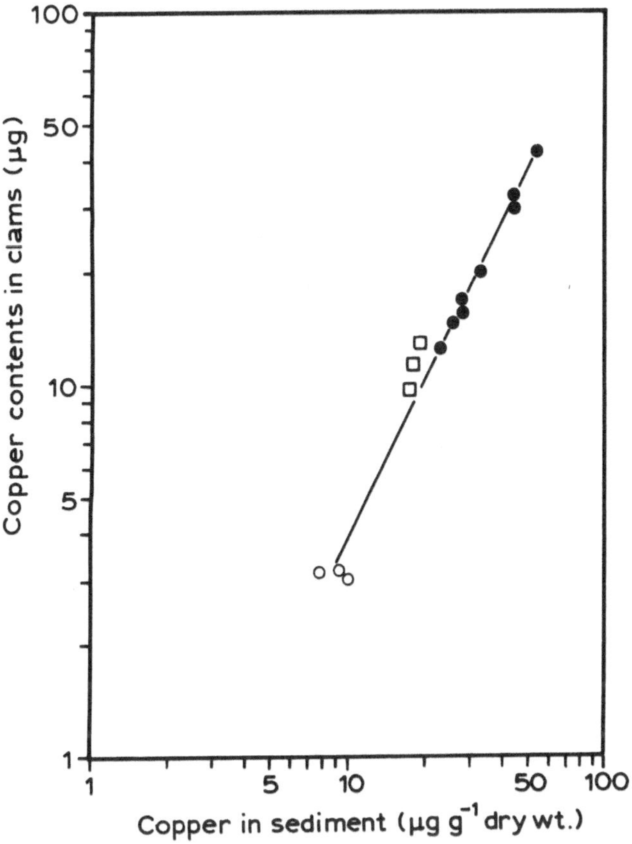

Fig. 44. Relationship between copper concentrations in sediments ($\mu g\ g^{-1}$ dry weight) and copper contents (μg; yearly means) of clams (*Macoma balthica*) of age four years from three stations in San Francisco Bay. After Cain and Luoma (1990).

(v) Interactions Between Contaminants

The discussion above on the uptake of trace metals from solution emphasises the importance of physicochemical parameters in controlling the availability of the free metal ion, and hence the bioavailability of a dissolved trace metal. The physicochemical parameters discussed included salinity and the presence of water-soluble metal chelating agents.

It follows that other physicochemical parameters may also affect the uptake of trace elements, and these include the relative concentrations of

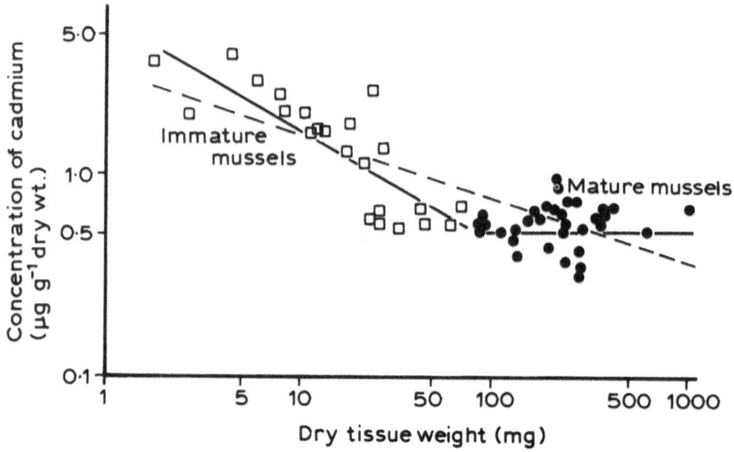

Fig. 45. Relationships between cadmium concentrations ($\mu g\ g^{-1}$ dry weight) and mussel soft tissue weights in mature and immature *Mytilus edulis* from Pointe Métis, St. Lawrence, Canada. After Cossa *et al.* (1979).

other metals, and changes in pH and redox potential. Dissolved metals may act indirectly, affecting the complexation equilibria of an element by competing for particular organic or inorganic ligands, and thereby changing the concentration of the free metal ion available for uptake and accumulation by biota. Such indirect interactions clearly depend on the relative concentrations of the elements in question. Thus, a metal present in solution at concentrations of nanograms per litre will have a negligible effect on the bioavailability of a second element present at levels of tens of micrograms per litre, even if the former metal has higher affinity for every ligand present.

Other trace metals may have more direct effects on the accumulated concentrations of a metal by competing at the site of uptake. For example, cadmium concentrations in the biota of Restronguet Creek, England, are not unusually elevated even though the ambient dissolved levels of this element are extremely high. The unexpectedly low cadmium concentrations accumulated by these organisms probably result from the competition with cadmium for uptake sites by the enormously high concentrations of other metals, particularly zinc (Bryan and Gibbs, 1983; Bryan *et al.*, 1985). Phillips (1976a) has shown that the accumulation of copper by the mussel *Mytilus edulis* is affected by high concentrations of cadmium and zinc, but not of lead. By contrast, the uptake of cadmium, lead or zinc by *M. edulis* appears unaffected by the presence of the other metals. In a further example, Bjerregaard (1982) noted that selenium promoted the accumulation of cadmium in the gills of the shore crab *Carcinus maenas*, and that cadmium promoted selenium accumulation also. The precise mechanisms involved in such interactions are often obscure, but they pose an intractable problem for the use of certain biomonitors to define general metal bioavailabilities, as noted by Phillips (1977, 1980).

Stauber and Florence (1985) proposed that the amelioration by manganese of copper toxicity to the marine diatom *Nitzschia closterium* was mediated through the adsorption of copper by manganese (III) hydroxide at the membrane surface, thereby limiting copper penetration into the cell. Iron behaves similarly to manganese in this respect, but is only active at much higher concentrations, probably as a result of differences in the relative affinities of iron and manganese hydroxides for copper (Stauber and Florence, 1985).

Changes in pH and redox potential affect physicochemical equilibria, which control the complexation and even the solubility of trace metals. Iron and manganese precipitate out of solution as oxides and hydroxides when acidic freshwaters meet alkaline seawater in estuaries. Most of the iron remains in the sediment, but manganese may return into solution if the redox potential of the sediments promotes the reduction of Mn(III) to Mn(II) (Burton and Young, 1980; Förstner and Solomons, 1983). During their initial precipitation, iron and manganese hydroxides often act as sites for the adsorption of other metals (Förstner and Solomons, 1983), thereby affecting total metal concentrations in solution and also the complexation equilibria for elements. Changes in pH and redox potential (which may affect manganese and iron availability) may occur on an extremely local scale in the vicinity of phytoplankton or aquatic macrophytes, which acquire inorganic carbon from the equilibrated pool of CO_3^{2-}, HCO_3^- and CO_2, a pH-dependent process. Such local changes have consequential effects on the availability of other trace metals.

The interstitial solutions of sediments are particularly prone to changes in redox potential, with important consequences for the bioavailabilities of trace elements. Dissolved metals in interstitial waters can be taken up by burrowing infauna and by the roots of aquatic angiosperms (Luoma, 1983, 1989). The elements will redistribute between the sediment and the interstitial waters (and between different binding components of both these compartments) as changes in sediment chemistry occur. Organically-rich sediments contain many metal binding sites, and such sediments are also likely to be anoxic. The reducing conditions favour the reduction of manganese (and iron), with the possible release of other metals previously adsorbed onto their hydrous oxides (Luoma, 1990). It is clear from this that metal interactions external to the organism are commonplace, and these may have important consequences for the bioavailabilities of trace elements in aquatic ecosystems, and hence for the use of biomonitors.

(vi) The Effects of Salinity and Temperature

As noted previously, changes in salinity exert a direct physicochemical impact on the inorganic complexation of dissolved trace metals, with consequent effects on metal uptake rates by biota. Such changes in uptake rates will give rise to alterations in the accumulated concentrations of trace elements in organisms exhibiting a net accumulation strategy. These effects of salinity are real reflections of changes in the bioavailabilities of metals, and no allowance for this need be made in any comparison of accumulated concentrations of metals in biomonitors.

Changes in salinity may, however, cause physiological responses in an organism, thereby affecting its accumulation of trace elements. Such physiologically-based changes in metal accumulation are not indicative of alterations in external metal bioavailabilities, and therefore constitute a source of potential error in biomonitoring surveys.

Mention has already been made of certain physiological processes which may affect the rates of metal uptake by biota (e.g. changes in apparent water permeability, and in the rates of activity of pumps employed for the uptake of major ions). Such physiological changes may stimulate excretion rates, for example through the incorporation of trace metals into active pumps expelling major ions in marine teleosts, which drink seawater to replace water lost osmotically from the tissues. It is also possible that the atypically high rates of production of isosmotic urine by decapod crustaceans maintained in hypo-osmotic media may provide an exit route for accumulated metals (see Nugegoda and Rainbow, 1989a).

Exposure to salinity fluctuations may cause changes in behaviour (including feeding), or in rates of gill ventilation or blood circulation, and these may affect the rates of metal uptake and excretion (see Depledge, 1990). For example, mussels and barnacles exhibit closure responses, with consequent cessation of feeding and respiratory currents, on exposure to fluctuating salinities. This response has been shown to affect the uptake of copper by the mussel *Mytilus edulis* (Davenport, 1977; Davenport and Manley, 1978; Davenport and Redpath, 1984; Redpath, 1985). The differential abilities of organisms to acclimatise to new physicochemical conditions (including changes in salinity) affect their relative metal accumulation rates (Depledge, 1990), and such abilities vary both intraspecifically and interspecifically.

Changes in water temperature also influence metal accumulation rates, with consequences for biomonitors. An elevated temperature might be expected to increase thermodynamically the rate of any physicochemical reaction in the aquatic medium, including the rate of binding of a free metal ion to a membrane transport ligand, leading to a consequent increase in metal uptake. Temperature changes may also cause alterations in the rates of physiological processes within an organism, such as metabolism, ventilation, and heart beat (e.g. see Newell, 1976), and in certain cases these may affect the uptake of trace elements.

Increases in water temperature generally give rise to elevated rates of metal uptake (and hence toxicity; see McLusky *et al.*, 1986), and this may be translated into a higher net accumulation of elements. Elevated temperatures cause an increase in the uptake rate of zinc in *Palaemon elegans* (White and Rainbow, 1984a), and also lead to regulation of zinc concentrations in the whole body of this decapod to a higher absolute level (Nugegoda and Rainbow, 1987). Cadmium accumulation by the fiddler crab *Uca pugilator* (O'Hara, 1973), the blue crab *Callinectes sapidus* (Hutcheson, 1974), the mussel *Mytilus edulis* (Phillips, 1976a) and the oyster *Saccostrea echinata* (Denton and Burdon-Jones, 1981) is increased with elevated temperatures, as is zinc accumulation in *S. echinata* (Denton and Burdon-Jones, 1981). However, temperature increases have little effect on lead accumulation by *S. echinata* (Denton and Burdon-Jones, 1981), nor on zinc accumulation by *M. edulis* (Phillips, 1976a).

(vii) Inherent Variability

Boyden and Phillips (1981) coined the term "inherent variability" to describe the residual variation in trace metal concentrations found in biomonitors, after all known sources of such variation in a given population have been accounted for. A striking feature of metal accumulation in aquatic organisms is the presence of significant inherent variability in tissue and body metal concentrations within given populations of a species. This is important in biomonitoring investigations, in that inherent variability in metal levels creates "background noise", against which any "signal" (of altered metal bioavailability) must be measured.

Lobel (1987a,b) discussed the high inherent variability of zinc concentrations in a population of the mussel *Mytilus edulis* from Newfoundland. He concluded that this variability in the whole soft tissue levels of the element resulted almost entirely from the large variations in zinc con-

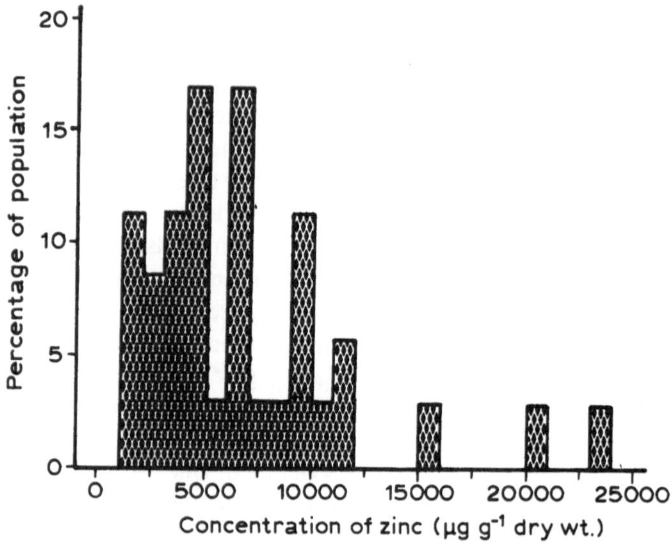

Fig. 46. Frequency distribution of zinc concentrations ($\mu g\ g^{-1}$ dry weight) in kidneys of mussels (*Mytilus edulis*) collected from a contaminated area in Hawke's Bay, Newfoundland. After Lobel (1987b).

centrations of the kidney (Fig. 46). Interestingly, the variability of zinc concentrations in the kidney was not a result of differences between individuals in the quantity of zinc-rich granules stored in that tissue (Lobel, 1987c), but appeared to result from variations in short-term processes concerned with the uptake or excretion of the element (Lobel, 1987a).

Lobel *et al.* (1989) have proposed a method for quantifying and comparing the inherent variabilities of tissue metal concentrations, again using *M. edulis* as a model. A multiple regression equation is constructed, with metal concentration as the dependent variable and all factors known to affect this parameter as independent variables. The variance of the residuals about each regression line may then be calculated, and may be compared to indicate differences in the handling of elements between individual organisms. It is assumed that such differences are genetically-based (Lobel *et al.*, 1989), as other factors which may affect the net accumulation of metals have been accounted for in this treatment.

Biomonitoring programmes should recognize the presence of such inherent variability in trace element concentrations in organisms, and be designed in such a manner as to allow its statistical accommodation.

Various statistical approaches have been used to this end. Gordon *et al.* (1980) derived estimates of means and variances in metal concentrations for several populations of mussels (*Mytilus californianus*), in order to calculate the sample sizes needed to produce the desired resolution in biomonitoring surveys. Realistically, there must be a compromise for cost reasons between the number of samples analysed and the resolution desired in any given programme (Gordon *et al.*, 1980; see Chapter 8).

Wright *et al.* (1985) addressed the problem of individual variability in studies of trace elements in the oyster *Crassostrea virginica* in Chesapeake Bay, USA. The presence of outliers ("superaccumulators") in these oyster populations had important implications for the requisite size of representative samples, causing disproportionate effects on mean trace metal concentrations and producing non-normal distributions of data. These authors computed a cumulative mean for a sample by incorporating individual data sequentially in random order, and thereby estimated the number of individual data points required for the cumulative mean to converge within 10% of the actual mean for a given population. The required sample number was 15 individuals at one site, but 20 individuals at a second site where both tissue metal concentrations and inherent variability were increased (Wright *et al.*, 1985). Reciprocal square root transformation techniques were used to convert skewed metal distributions to normality, before applying parametric statistics (Lobel and Wright, 1983; Wright *et al.*, 1985).

Popham and D'Auria (1983) have proposed a statistical approach to assist in decisions on whether mussels (*Mytilus edulis*) have been collected from a water body that can be classified as metal-polluted. Principal component analysis was applied to a correlation matrix of variables (tissue concentrations of manganese, iron, copper, lead and zinc; dry and wet tissue weights; shell length; and day of collection) measured in mussels over a 12-month period from a relatively unpolluted site in British Columbia. The principal component scores determined for these samples were used as a reference to indicate whether other mussel populations were derived from "metal-polluted" waters (Popham and D'Auria, 1983). In essence, this technique attempts to resolve a "pollution signal" from "background noise", and to indicate whether or not the signal is significantly different from the values obtained in a reference area.

Phillips and Segar (1986) discussed various principles to be applied in the design of biomonitoring programmes, including statistical considerations. These authors emphasised the usefulness of time-bulking and space-bulking techniques in reducing data variance. While such methods may in certain instances be useful to overcome the problems raised by inherent variability of metals in biomonitoring populations, their principal advantages involve other aspects. They are therefore covered elsewhere in this text (see Chapter 8).

(viii) Laboratory Handling of Collected Samples

Samples collected as part of a biomonitoring programme are likely to be associated with sediment particles. These may be adherent to the body surface, be trapped in the gills, and/or be present in the gut. Such particles may contain trace metals in both their matrix and adsorbed to their surfaces, and this may interfere significantly with attempts to measure elements incorporated into the tissues of the sampled organism. The presence of sediment particles in the gut of biomonitors tends to be of greatest importance for deposit feeding organisms, but grazing invertebrates such as limpets or winkles may also contain significant amounts of such materials, as may certain fish species (e.g. Flegal and Martin, 1977; Lobel *et al.*, 1991).

Sediment particles adhering to the surface of an organism can be removed by routine rinsing of all collected biota in a standardised fashion. Thus, for example, Bryan *et al.* (1985) washed samples of seaweed in deionised water, and employed a soft brush to remove any remaining particles adhering to the surfaces of each sample.

With respect to gut contents, many authors have employed a laboratory-based depuration period to permit the defaecation of sediment and any undigested food material prior to sample digestion and analysis. NAS (1980) recommended that biomonitoring samples of bivalve molluscs which may contain significant amounts of ingested inorganic particulates should be subjected to depuration for 36–48 h prior to analysis for trace elements. However, the depuration periods employed tend to vary between authors and between species. Bryan *et al.* (1980, 1985) stated that depuration should be undertaken for all studies of trace elements in biota (particularly for metals such as iron and aluminium), and devised various protocols for different species. Thus, they maintained *Scrobicularia plana* for 7 days in regularly changed 50% seawater in the laboratory to remove ingested sediment; *Mytilus edulis* were exposed to clean seawater for about 24 h; and *Cerastoderma edule* were depurated in settled seawater for several days. Specimens of the polychaete worm *Nereis (Hediste) diversicolor* were allowed to depurate in acid-washed sand covered with 50% seawater for 2–6 days, before transfer for 1 day to 50% seawater only (Bryan *et al.*, 1980, 1985).

In practice, the gut contents of some potential biomonitors (particularly strong net accumulators of metals) do not contribute significantly to the total body load of an element (e.g. see Weeks and Moore, 1991). Where it has been confirmed experimentally that there is no significant difference between the metal concentrations of an organism before and after gut depuration, it is acceptable to analyse samples without recourse to depuration (e.g. Phillips and Rainbow, 1988; Rainbow *et al.*, 1989). This may on some occasions be the preferred course of action, given the possibility of extraneous contamination of samples during the depuration period (NAS, 1980). It is notable here that depuration is not recommended for samples destined for hydrocarbon or organochlorine analysis, given the short biological half-lives of these contaminants (see Chapter 6).

E. THE BIOMONITORING OF RADIONUCLIDES

As noted at the beginning of this chapter, in most instances the techniques employed for the biomonitoring of radionuclides are similar or identical to those used to monitor stable isotopes of trace metals. For the sake of completeness, however, brief discussion is provided here of particular programmes employing biomonitors to measure the abundances and bioavailabilities of radionuclides in aquatic ecosystems.

The contamination of coasts of the USA by artificial radionuclides (including ^{238}Pu, ^{239}Pu + ^{240}Pu, and ^{241}Am) was monitored in the Mussel Watch Program of the mid-1970s, employing a total of four species of mussels and oysters (Goldberg *et al.*, 1978, 1983; Farrington *et al.*, 1982, 1983). It was concluded that samples taken on the east and Gulf coasts of the USA generally exhibited levels of transuranic radionuclides and ^{137}Cs which could be attributed to the fallout from nuclear weapons testing. However, one anomalous result in a sample originating in 1976 from Plymouth, Massachusetts, led to the discovery of local contamination from a nearby nuclear power reactor (Farrington *et al.*, 1983). Bivalve samples from the west coast of the USA exhibited generally higher concentrations of plutonium and much higher ratios of americium to plutonium nuclides than were found in samples from the east or Gulf coasts, but the source of these transuranics was unknown (Goldberg *et al.*, 1978).

All bivalve tissues analysed exhibited ratios of caesium to plutonium which were much lower than those reported from the analysis of coastal waters. This was considered to reflect differential bioavailabilities of the

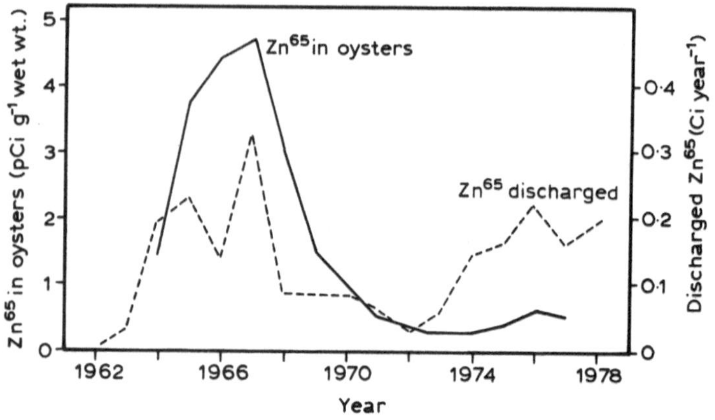

Fig. 47. Relationship over time between discharges of ^{65}Zn from the Bradwell Power station (dashed line) and the accumulated concentrations of the same radionuclide in oyster tissue (*Ostrea edulis*; solid line) in the Blackwater Estuary, England. After Woodhead (1984).

radionuclides, and/or different accumulation strategies of the bivalves for caesium and plutonium. It appeared that the bivalves analysed obtained much of their radionuclide burden from suspended particles rather than from solution (but see Bjerregaard *et al.*, 1985, discussed below). The byssal threads of mussels showed much greater enrichment of plutonium and often of americium than did the soft parts (Goldberg *et al.*, 1978).

Discharges of ^{65}Zn from the Bradwell Power station on the Blackwater Estuary in England have been monitored in the native oyster *Ostrea edulis* (Woodhead, 1984). Figure 47 shows the variation of ^{65}Zn in soft tissues of *O. edulis* sampled 0.5 km downstream of the power station outfall, and the annual discharges of the same isotope. Oysters are strong accumulators of zinc, and were found to provide an accurate time-integrated measure of total ^{65}Zn released prior to 1972. However, after this date, the relationship between the accumulation of ^{65}Zn by oysters and the discharge of the radionuclide apparently changed, for no clear reason (Woodhead, 1984). The reduction in concentrations of ^{65}Zn, ^{60}Co and ^{137}Cs in oysters at four stations up to 10 km distant from the outfall are shown in Fig. 48. ^{137}Cs apparently behaves conservatively in the Blackwater Estuary. By contrast, concentrations of ^{65}Zn and ^{60}Co reduced more rapidly over distance, suggesting the presence of a local sink for these two nuclides. Woodhead (1984) suggested that the nuclides might be scavenged by silt particles, which are transported differentially along the two sides of the estuary.

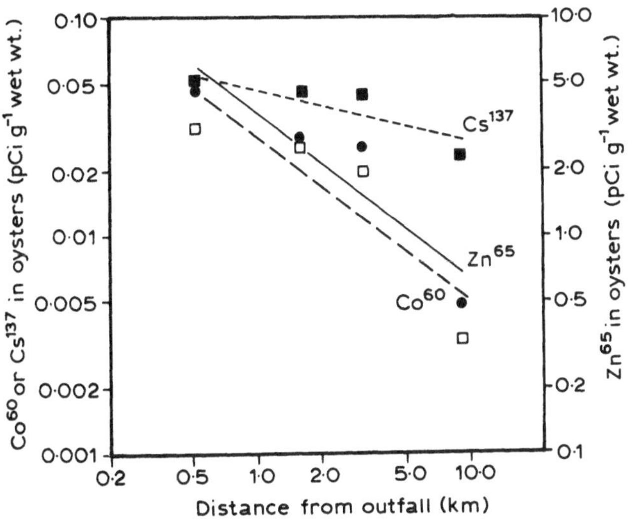

Fig. 48. Concentrations of ^{65}Zn (●), ^{60}Co (□) and ^{137}Cs (■) in oyster tissue (*Ostrea edulis*) from the Blackwater Estuary, England, as a function of distance from the outfall of the Bradwell Power Station. After Woodhead (1984).

Measurements of the accumulated concentrations of ^{106}Ru in the seaweed *Porphyra umbilicalis* have been employed to delineate the geographical extent of the contamination from the Sellafield (Windscale) nuclear reprocessing plant in north-east England. Data for 1961 for ^{106}Ru in *P. umbilicalis* (Fig. 49) showed significant contamination by this isotope at distances up to 160 km from the source. At more distant locations, the levels of the isotope became indistinguishable from the background concentrations, the latter resulting from fallout (Woodhead, 1984). A reduction in ^{106}Ru availability to the macroalga over distance reflected the combined effects of dilution and dispersion, the removal of the nuclide to sediment, and radioactive decay. The abrupt change in the rate of reduction in the levels found between 10 and 20 km (Fig. 49) may reflect a boundary between advective transport along the coast in tidal currents and more widespread dispersion by turbulent diffusion (Woodhead, 1984).

Hamilton and Clifton (1980) reported data on the distribution of isotopes of plutonium and americium discharged from Sellafield, using the mussel *Mytilus edulis* and the seaweed *Fucus vesiculosus* as biomonitors. Mussel tissues associated with the uptake of radionuclides from food (e.g. the digestive gland) exhibited enrichment of plutonium relative to ameri-

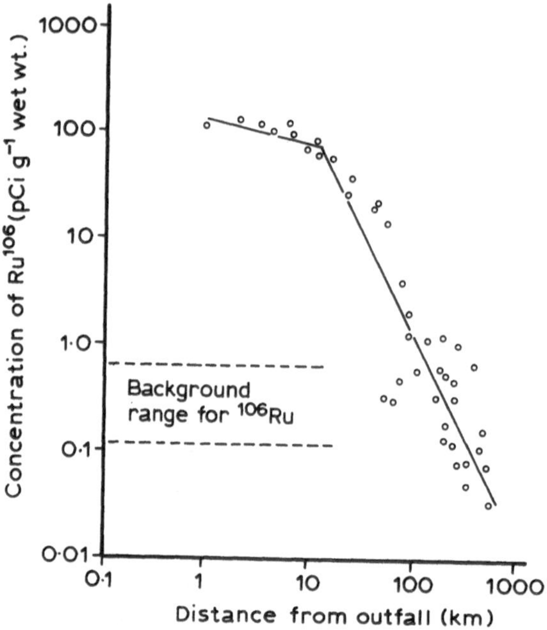

Fig. 49. Accumulated concentrations of ^{106}Ru in the macroalga *Porphyra umbilicalis* as a function of distance from the Sellafield nuclear reprocessing plant in north-west England. Data relate to collections in 1961. After Woodhead (1984).

cium. The opposite enrichment profile was found in tissues (e.g. the mantle) where radionuclides were predominantly taken up from solution. The byssal threads of mussels exhibited higher concentrations of americium than any soft tissues (see discussion of the data of Goldberg *et al.*, 1978, above), and plutonium concentrations in the byssal threads were similar to the high concentrations found in the digestive gland of the mussel. It was considered that both nuclides were taken up directly into the byssus from the ambient seawater. Samples of *F. vesiculosus* exhibited a similar ratio of americium to plutonium to that found in the byssal threads of mussels, supporting this hypothesis (Hamilton and Clifton, 1980).

Koide *et al.* (1981) proposed on the basis of studies of americium/plutonium ratios in seawater, particulates and mussels, that the ingestion of contaminated particles represents the principal means of plutonium and americium accumulation by the soft parts of mussels. Bjerregaard *et al.*

(1985) investigated the kinetics of americium and plutonium in *Mytilus edulis*, attempting to cast light on variations in the ratios of these nuclides reported for mussels worldwide. Both water and ingested food acted as sources of the two isotopes to *M. edulis*, the mussels retaining 1 to 15% of the radionuclide ingested with labelled diatoms. Americium was absorbed from food more efficiently than plutonium, and dissolved americium was more biologically available than plutonium.

Bjerregaard *et al.* (1985) extrapolated their experimental data, to speculate that the different concentrations of americium and plutonium observed in oysters and mussels in the Mussel Watch Program in the USA (see above and Goldberg *et al.*, 1978, 1983) simply reflected differential levels of these elements in the coastal waters of the USA. Furthermore, the high ratio of plutonium or americium concentrations in the shells of mussels to those in the soft tissues (Koide *et al.*, 1982) agreed with experimental data on radionuclide accumulation by these tissues from water. This suggests that most of the radionuclides in the Mussel Watch samples were accumulated from the dissolved phase (Bjerregaard *et al.*, 1985).

Data for uranium kinetics in mussels (Hamilton, 1980) are similar to those for plutonium and americium (Hamilton and Clifton, 1980). High concentrations of uranium are associated with byssal threads, in reflection of their high content of tanned proteins with SS—SH and disulphide groups (Hamilton, 1980). Accumulated uranium concentrations in *Mytilus edulis* from a variety of sites in the United Kingdom revealed minimal impacts from nuclear effluent discharges (Hamilton, 1980).

F. CONCLUSIONS

Much progress has been made over the last three decades in the use of biomonitoring techniques to identify spatial or temporal changes in the bioavailabilities of trace metals or radionuclides in aquatic ecosystems. Our understanding of the kinetics of uptake and excretion of these contaminants in organisms has improved markedly through a combination of laboratory-based and field investigations, and this has provided a more balanced view of the most appropriate species for use in monitoring programmes. Clearly, the most efficient biomonitors are strong net accumulators of trace elements (Phillips and Rainbow, 1989), although weak net accumulators may also be of use under certain circumstances. Regulators of metals should not be employed as biomonitors, as they do not respond to differences in the abundance of trace elements in the external medium.

One result of this improvement in understanding has been the increasing use of biomonitoring techniques in local, national and international programmes to monitor the abundances and bioavailabilities of trace metals and radionuclides in aquatic ecosystems. There is no doubt that such techniques have considerable advantages over the more traditional methods of water or sediment analysis (see Chapter 4).

However, the use of biota also involves certain disadvantages, and these merit careful consideration in the design phase of all monitoring programmes (Phillips and Segar, 1986). The correct taxonomic identification of the species used is of fundamental importance, as even closely-related species are known to exhibit considerable differences in their accumulation strategies for trace elements (e.g. see Eisler, 1981).

In addition, several variables exist which may interfere with the efficient use of biomonitors to measure the comparative bioavailabilities of trace elements over space and/or time, and these should be addressed in all investigations. The most important variables involve the size (weight or age) of the individuals sampled, and the impacts of seasonal changes on biomonitoring data (the latter depending on both biological events and changes in the external environment). Under certain circumstances, the effects of these may give rise to spurious conclusions in monitoring surveys, and the minimisation of such effects is thus of considerable importance. Several methods exist to this end, and these will be addressed further in Chapter 8.

Additional variables which merit consideration in biomonitoring studies include: animal sex; interactions between contaminants; and the impacts of salinity and water temperature. Sampling strategies have been proposed to minimise the effects of such variables on the results of biomonitoring surveys (e.g. see Phillips, 1976a; Bryan *et al.*, 1980, 1985), and these should be designed into all programmes as required.

Even when the impacts of all such variables have been minimised or eliminated, however, different individuals within a discrete population of biomonitors will exhibit differences in their accumulated concentrations of trace elements. This so-called inherent variability is probably due mostly to genetic differences between individuals, and it provides "background noise" against which the desired signal (of altered metal bioavailability with space and/or time) must be measured. The techniques employed in the design of monitoring programmes to maximise their efficiency in this regard will again be further discussed in Chapter 8.

Chapter 6

The Biomonitoring of Organochlorines and Hydrocarbons

A. SOURCES AND TRANSPORT

Organochlorines and hydrocarbons are addressed together in this chapter, as their cycling through aquatic ecosystems is broadly similar and many of the issues related to the use of biomonitors to quantify these contaminants are shared, at least to some degree (see Connell, 1988). The following text emphasises data for organochlorines and biomonitoring, as rather more information exists for these compounds than for hydrocarbons. Where relevant, however, discussion is provided on the distinctions between the two classes of contaminants and their quantification through the analysis of biomonitors.

As a preliminary to discussion of the biomonitoring of organochlorines and hydrocarbons in aquatic environments, it is useful to consider briefly the sources of these compounds and their modes of global transportation. These differ significantly from those of trace metals or radionuclides discussed previously, and an understanding of the general cycling of organochlorines and hydrocarbons through the ecosphere is important if biomonitoring programmes are to be designed appropriately.

(i) Organochlorines

The organochlorines discussed here include the chlorinated hydrocarbon pesticides, polychlorinated biphenyls (PCBs), dioxins and dibenzofurans. The pesticides are employed in both agricultural and industrial applications, and all the compounds involved are synthetic in nature. Their release to the environment is frequently deliberate and systematic (as in agricultural applications of pesticides, for example). By contrast, PCBs are employed solely by industry and their release to the external environment is generally accidental (see reviews by Peakall and Lincer, 1970; Hutzinger *et al.*, 1974; NAS, 1979; Richardson and Waid, 1982; Waid, 1986). Dioxins and dibenzofurans are produced by pyrolysis, but the former may also be produced by the chlorination of certain types of aqueous effluents (such as those from pulp mills; see Waldichuk, 1990).

It is important to note that the total amounts of these types of contaminants entering the ecosphere remains significant to the present, despite restrictions placed on the utilisation of many of them over the last two decades in western nations. Goldberg (1975) was the first to draw attention to the so-called "southward tilt" in pesticide usage, noting that the global production and use of DDT has remained relatively constant following the bans placed on its use in the developed nations of the temperate zones. This was due to an increasing utilisation of DDT for anti-malarial, agricultural and industrial applications in the developing nations, most of which are situated in the subtropics and tropics (Latin America, Africa, the Far East). The rapidly increasing human populations of such areas are fuelling a massive expansion in intensive agricultural practices and industrial development, and the "southward tilt" in the global utilisation and abundance of DDT is certainly accompanied by a similar geographical shift for many of the other persistent organochlorines (Phillips and Tanabe, 1989).

The situation with respect to PCBs is slightly different, as it is thought that the world production of PCBs peaked in the 1970s and is now diminishing (Tanabe, 1988). Despite this, however, PCBs remain a significant threat to aquatic ecosystems, either through the recycling of residues already released and present mainly in coastal sediments or open ocean waters, or by the mobilisation of PCBs which remain in industrial use or have been placed in insecure (i.e. not environmentally isolated) landfills (Table 16). Thus, Tanabe (1988) estimated that 31% of the accumulated net tonnage of PCBs has already reached open terrestrial or aquatic environments; only 4% has been destroyed through high-temperature incineration or by metabolic breakdown; and the remaining 65% of the

Table 16 Estimated loads of PCBs (tonnes) in the global environment, and their percentage contributions to the total (after Tanabe, 1988)

Environment	*PCB load (t)*	*Percentage of PCB load*	*Percentage of world production*
Terrestrial and coastal			
Air	500	0.13	
River and lake water	3500	0.94	
Seawater	2400	0.64	
Soil	2400	0.64	
Sediment	130000	35	
Biota	4300	1.1	
Total (A)	143000	39	
Open ocean			
Air	790	0.21	
Seawater	230000	61	
Sediment	110	0.03	
Biota	270	0.07	
Total (B)	231000	61	
Total load in the environment (A + B)	374000	100	31
Degraded and incinerated	43000		4
Land-stocked[a]	783000		65
World production	1200000		100

[a] Still in use in electrical equipment and other products, and deposited in landfills and dumps.

cumulative tonnage produced to date is either in use currently (mostly in ageing electrical equipment) or has been landfilled with no real guarantee of future containment. Given the rapid industrial expansion and the inadequacies of environmental controls in most developing nations, it is probable that global PCB abundance will also shortly exhibit a "southward tilt", just as is observed (or at least predicted) for most organochlorine pesticides.

While this geographical shift in the utilisation and abundance of chlorinated hydrocarbons is important in defining the locations of "hot-spots" of particular contamination, it is to some extent of academic interest in relation to overall global contamination. Thus, it has been known for several decades that these compounds are subject to co-distillation processes and are therefore primarily transported through the atmosphere. As a

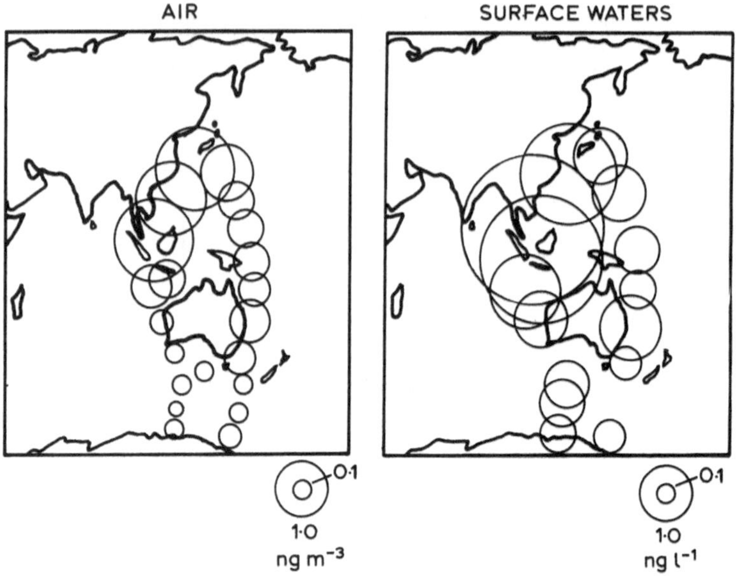

Fig. 50. Distributions of ΣHCH concentrations in air masses and surface waters between Japan and the Antarctic. After Tanabe *et al.* (1982a).

result, they may be found in measurable quantities even in remote pristine locations such as the Antarctic (e.g. Tatton and Ruzicka, 1967; Harrison *et al.*, 1970; Woodwell *et al.*, 1971; Tanabe *et al.*, 1982a, 1983a). It is now accepted that the aerial route of transport of organochlorines predominates in the global cycling of these compounds, despite their relatively low vapour pressures.

This does not, however, imply that the organochlorines are distributed homogeneously over the surface of the earth, nor that the routes of entry for organochlorines to coastal or offshore waters are restricted to the rain-out or dry deposition of these compounds from the atmosphere. There is ample evidence of increased concentrations of organochlorines in both the atmosphere and surface waters close to areas of known utilisation of these compounds; an example of this is provided in Fig. 50, for the distribution of hexachlorocyclohexane (HCH) isomers in the Indo-Pacific. Thus, aerial concentrations of organochlorines are greatest close to regions where these compounds are heavily utilised, and reduce significantly with distance from such areas, due to dispersion processes.

Routes of entry (additional to the atmospheric route) for organochlorines to marine environments include agricultural and urban runoff, wastewater discharges, rivers, and direct entry from sewage sludge or other sources, such as PCBs from anti-fouling paints (Phillips, 1980, 1986). The superimposition of loads from these various non-aerial sources onto the organochlorines entering aquatic environments from the atmosphere gives rise to a complex mosaic of organochlorine distributions and concentration gradients, particularly in coastal waters. The very considerable toxicities and bioaccumulation tendencies of such compounds thus create a demand for information on their distribution and abundance, i.e. a need to monitor them in aquatic ecosystems. The use of biomonitors for this purpose is discussed below, subsequent to a brief review of hydrocarbon sources and transportation mechanisms.

(ii) Hydrocarbons

The discussion provided here centres around petroleum-derived hydrocarbons, and does not include biogenic hydrocarbons of recent origin (mostly alkanes and isoprenoids, with smaller amounts of alkenes and aromatics). The global production of crude oil presently amounts to greater than 3 billion (10^9) tonnes per annum. The majority of this (about 80%) is refined, mostly for fuel use, while about an additional 12% is employed in non-fuel products (e.g. naphthas, bitumen, lubricants and waxes).

Crude petroleum is made up of complex mixtures of hydrocarbons, together with trace metals and various organic compounds containing nitrogen and sulphur. The hydrocarbons may be classified as aliphatics (comprising various forms of alkanes), and aromatic compounds (Fig. 51). Certain of these classes, such as the polynuclear aromatic hydrocarbons (PAHs, which are formed by the pyrolysis of organic materials or are naturally present in fossil fuels) have received greater study in aquatic environments than have other classes of compounds, such as the monoaromatics. This is a function both of the availability and simplicity of analytical methods and of the estimated toxicological importance of the contaminants concerned (e.g. many of the PAHs are known to be carcinogenic).

Certain differences exist between hydrocarbons and organochlorines with respect to their sources and modes of transport or entry into aquatic ecosystems, although such distinctions are largely quantitative rather than qualitative in nature. Useful reviews of the sources, transport, fate and effects of hydrocarbons in marine waters have been provided by several authors (NAS, 1975; Miller and Connell, 1982; Whittle *et al.*, 1982; IUCN,

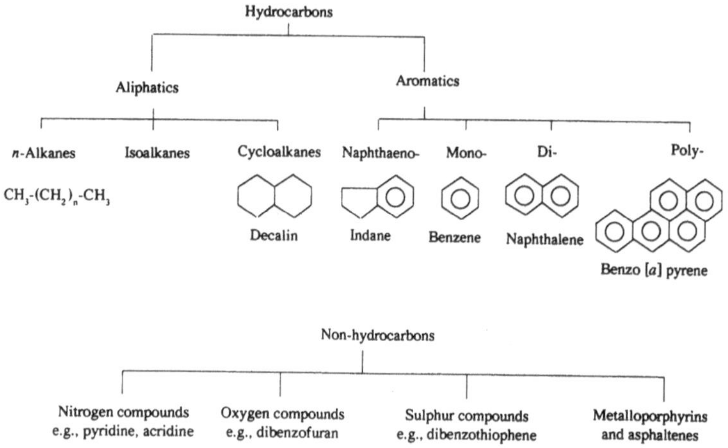

Fig. 51. Examples of the chemical structures of certain of the more common components of crude petroleum. After Connell and Miller (1984).

1983; Connell and Miller, 1984; Samiullah, 1985), and these data will be covered only briefly here.

Various estimates have been provided by these authors for the gross inputs of petroleum-derived hydrocarbons to the marine environment (Table 17). The earlier prediction provided by NAS (1975) of greater than 6.1 million tonnes per annum (about 2% of the annual global production of crude oil) has since been revised downwards, and there is some evidence that the recent introduction of stricter controls on tanker operations and on aerial discharges may be reducing the total loads of petroleum-derived contaminants reaching marine waters.

In any event, it is clear from Table 17 that atmospheric inputs, while significant, do not predominate as a route of transfer of hydrocarbons to marine waters in the same fashion as seen for organochlorines. Tanker-related operations are a significant source of entry for hydrocarbons to the marine ecosphere (see also Chapter 2), although accidents leading to massive spillages account for only about 10% of total estimated inputs over time. Chronic contamination (especially of coastal waters) through bilge discharges and minor spillages from tankers constitutes a major problem, which contributes significantly to the contamination of tanker routes and coastal environments. Sewage and stormwater discharges are

Table 17 Estimates of the inputs of petroleum hydrocarbons (million tonnes *per annum*) into the marine environment (after NAS, 1975; Miller and Connell, 1982; IUCN, 1983)

	1973[a]	*1980*[b]	*1983*[c]
Terrestrial discharges			
Coastal refineries	0.2	0.02	—
Runoff, including: municipal/industrial waste, sewage discharge/other	2.5	2.40	1.40
Atmospheric discharges			
Rainout/fallout	0.6	0.60	0.30
Marine discharges			
Natural seeps	0.6	0.60	0.30
Tanker operation	1.08	0.50	0.71
Dry docking	0.25	0.15	0.03
Bilge discharge	0.5	—	0.32
Tanker accidents	0.2	0.15	0.39
Non-tanker accidents	0.1	—	0.02
Oil terminal operation	0.003	—	0.02
Offshore production	0.08	0.20	0.05
Totals	6.113	4.62	3.54

[a] NAS (1975).
[b] Miller and Connell (1982).
[c] IUCN (1983).

also important sources of these compounds, the latter providing an efficient delivery of residues from urbanised areas in particular. In coastal waters, there is ample evidence of such chronic contamination (e.g. see Connell, 1981; Burns and Smith, 1982; Burns *et al.*, 1982; Brown *et al.*, 1985; Cocchieri *et al.*, 1990; Emara, 1990), deriving from tanker traffic, industrial discharges (often entering marine waters through rivers or sewers), or from stormwaters contaminated by crankcase and other oils.

As a result of this complex mix of sources and transport mechanisms, petroleum hydrocarbons tend to be present at greatest concentrations in coastal waters adjacent to areas of high industrialisation and urban development, and in regions close to tanker routes. Contamination due to chronic inputs is overlaid by episodic spillages of a more or less massive nature due to accidents (see Chapter 2), giving rise to rapid temporal fluctuations in the abundance of these hydrocarbons in most coastal areas. This offers a considerable challenge in monitoring terms, and biomonitoring studies are only now beginning to address the complexities involved in monitoring transient or fluctuating levels of petroleum-derived compounds in natural waters, and differentiating these from biogenic compounds (e.g. see Theobald, 1989).

B. BIOMONITORING TECHNIQUES

(i) The Rationale

The drive to use biomonitors to measure organochlorine abundance in aquatic ecosystems was essentially provided by the difficulties in monitoring pesticides and PCBs in natural waters in the 1960s (e.g. see reviews by Portmann, 1975; Holden, 1981). This was primarily a function of the low concentrations of organochlorines in most waters, and the poor detection limits of the analytical methods which were then in common usage.

While some improvements in the latter respect had been realised by the 1970s and the analysis of PCBs and some of the pesticides in marine waters was becoming more widespread, analytical uncertainties remained. Thus, for example, Harvey *et al.* (1973a, 1974a) reported that PCB concentrations in the surface waters of the North Atlantic declined 40-fold between 1972 and 1974, and they ascribed this to the restrictions on PCB utilisation in industrialised nations of the northern hemisphere in the early part of that decade. More recently, however, these data have been criticised, and it appears likely that improvements in analytical accuracy were at least partly responsible for the apparent decline in PCB levels noted (see Schulz *et al.*, 1988; Fowler, 1990). The difficulties in eliminating the inadvertent contamination of samples have not been completely solved to date (Schulz *et al.*, 1988), and the accuracy of reported data remains uncertain in many instances.

It is revealing that no such criticisms have been levelled at the data produced concurrently by Harvey *et al.* (1973b, 1974b) for pesticides and PCBs in organisms from the North Atlantic. Evidently, the analytical methods then available were sufficiently robust to define the relatively high concentrations of organochlorines present in biota, but were probably inadequate for accurately quantifying the extremely low levels of these contaminants in most natural waters. This situation is reminiscent of the problems encountered until the late 1970s in the accurate analysis of stable trace metals or radionuclides in open marine waters (see Chapter 5), and in each of these cases, a trend towards the use of biomonitors emerged to overcome the difficulties experienced.

Similar problems have been encountered in attempts to analyse natural waters for petroleum hydrocarbons. Most of the individual compounds present in petroleum are of a lipophilic nature (as are most pesticides), and these tend therefore to be scavenged from solution by particulates, depositing eventually in sediments. In addition, both photochemical and microbially-mediated processes exist which degrade many hydrocarbons at significant rates, further reducing the concentrations of these compounds in natural waters (e.g. see Connell and Miller, 1984). The trace levels of hydrocarbons which remain in solution are difficult to quantify accurately, not least because of the analytical challenge involved in separating the individual contaminants which are present. While the analysis of samples by ultraviolet fluorescence spectroscopy (UVF) may yield results which are reasonably reproducible (e.g. Knap *et al.*, 1986; Law *et al.*, 1987), the individual compounds present are not identified by this technique. Indeed, more recent work using high performance liquid chromatography has concluded that many of the compounds quantified by UVF are in fact polar compounds which are not derived from petroleum (Theobald, 1989). As a result of these continuing analytical difficulties in monitoring hydrocarbons in natural waters, many authors have preferred to analyse either sediments or biota to provide an indication of the degree of contamination of aquatic environments by these compounds.

A further problem bedevilling the use of water analysis to monitor organochlorines or hydrocarbons has been considered by Phillips (1978, 1980, 1986) as of equal importance in the drive towards the use of biomonitors, and this concerns the temporal variability of such contaminants in many natural waters. In the case of agricultural pesticides, the seasonal patterns of application of pesticides to crops, and of rainfall (and consequent runoff, delivering the contaminant to inland and coastal waters) are the primary factors which define this temporal variability.

An example of the impacts of these factors is afforded by data on HCH abundance in the rivers of Japan. Several authors have shown that HCH isomers are present at considerably higher concentrations in these inland waters in summer than in winter, and this pattern holds true for rivers in distinct regions of the country (e.g. see Suzuki *et al.*, 1974; Ochiai and Hanya, 1976; Yamato *et al.*, 1980). Prior to the prohibition of HCH use in Japan in late 1970, such seasonal profiles were a function of both the timing of HCH application to local crops (mostly rice) and the annual changes in rainfall and patterns of irrigation of the ricefields (see also Ramesh *et al.*, 1989, for a more recent example relevant to India). After the ban on HCH use in Japan, however, the same seasonal fluctuations continued (Fig. 52), as historically-applied residues were washed out from the agricultural areas in solution and suspension (adsorbed to soil parti-

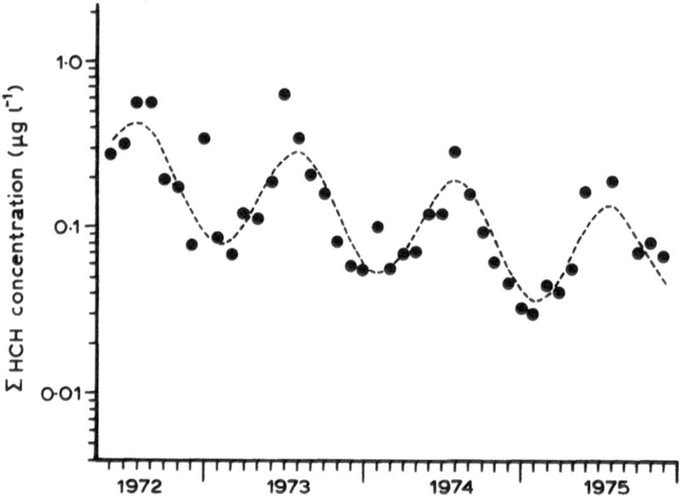

Fig. 52. Seasonal fluctuations in the concentrations of total HCH between mid-1972 and late 1975 in the Onga River, Japan. After Yamato *et al.* (1980).

cles) during the summer period of flooding the fields and harvesting of the crops. It is interesting to note the relatively slow decline in residue levels in the receiving freshwaters with time; clearly, HCH isomers are of considerable persistence in soils.

Even greater temporal fluctuations in organochlorine abundance may be encountered in rivers draining catchments where the industrial utilisation of such compounds predominates over other uses. Thus, for example, Boryslawskyj *et al.* (1985) reported marked temporal changes over short time periods in the concentrations of dieldrin in the River Holme in northern England (Fig. 53). The industrial use of dieldrin existed in this area for the mothproofing of textiles, and the batch discharge of wash waters from this process gave rise to slug-loading of the pesticide to sewage treatment works. While some attenuation of these peaks in dieldrin levels occurred through the sewage works, the discharges therefrom nevertheless contained concentrations of dieldrin which were highly variable over time.

Such fluctuations in pesticide levels are typical of the waters of industrial catchments (e.g. see Foehrenbach, 1972; Duinker and Hillebrand, 1979; Duinker *et al.*, 1980; Couillard, 1982), and may be exacerbated by concurrent changes in river flow rates, the loads of suspended particulates, estuarine mixing, and other factors. The considerable impacts of suspended

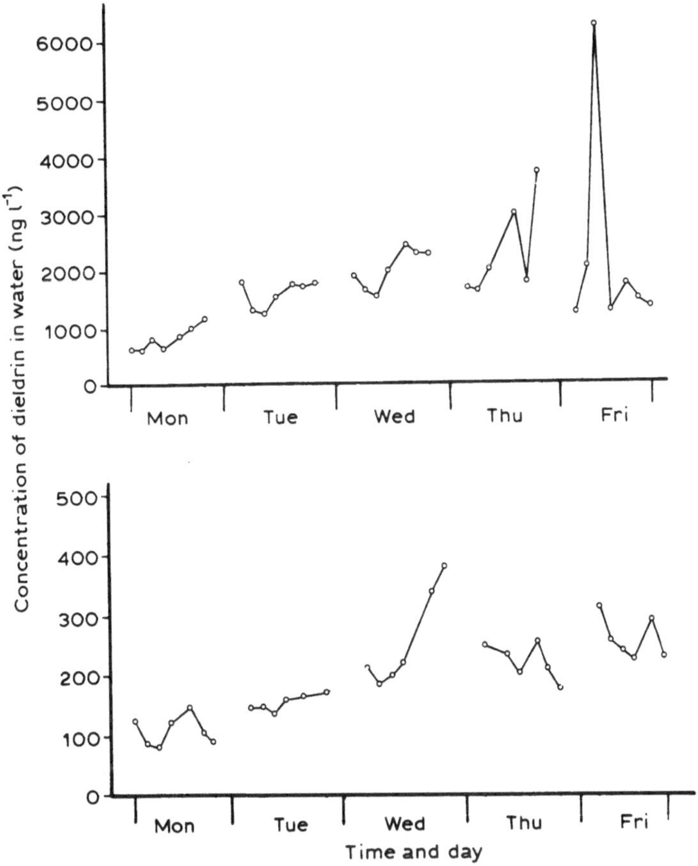

Fig. 53. Short-term temporal variations in the concentrations of dieldrin at two sites in the River Holme, northern England. Note the differences in the vertical scales. The monitored period extended from 09.30 to 16.00 hours each day. After Boryslawskyj *et al.* (1985).

particulate loads on the mass transport of pesticides by rivers (and on the total concentrations of these compounds in unfiltered waters) have been noted by several authors (e.g. Kellogg and Bulkley, 1976; Duinker *et al.*, 1980). Although this varies to some extent between individual compounds (being greater for pesticides of a more hydrophobic nature, which exhibit higher octanol:water partition coefficients), it is significant for all pesticides (Cox, 1971; Wilson, 1976; Gschwend and Wu, 1985).

While fewer data are available on the concentrations of petroleum hydrocarbons encountered in natural waters, it may be anticipated that these will fluctuate at least as widely over time as do those of organochlorines, particularly in view of the complexity of the sources of hydrocarbons and the importance of episodic spills as a route of their entry to inland and coastal waters. The effect of such temporal fluctuations is to frustrate attempts at routine monitoring of the abundance of such contaminants in the natural waters themselves, as representative samples cannot be taken in any simple manner. Clearly, there is a need for some method of time-averaging of contaminant concentrations, and this too has been cited as a rationale for the use of biomonitors (Phillips, 1978, 1980, 1986; Waldichuk, 1985).

(ii) The Emergence of Biomonitoring Programmes

The monitoring programme headed by Butler (1966, 1969a,b, 1971, 1973), concerned with the elucidation of organochlorine abundance in estuaries of the USA, provides an appropriate introduction to this section. In many senses, this programme (which pre-dated the US Mussel Watch project by almost a decade) was a forerunner to later biomonitoring efforts, and it therefore merits considerable discussion here.

The development of the monitoring programme commenced in 1958, with studies by the Bureau of Commercial Fisheries Biological Laboratory in Gulf Breeze, Florida, on pesticide toxicities. It was concluded from these investigations that pesticides differed substantially from each other in toxicity to aquatic biota, and that certain species tended to be more sensitive to organochlorines than others. This generated concern for the potential impacts of pesticides on commercial fisheries, and Butler (1966) noted that estuaries were particularly at risk, being important nursery areas for many coastal fish species and also receiving episodic inputs of trace organic contaminants from both agricultural and industrial activities in their upstream catchments. The establishment of a regular surveillance or monitoring programme for pesticides in estuaries of the USA was thus considered to be urgently required, as a fundamental component of fisheries protection measures as a whole in the nation (Butler, 1966).

Coincidentally, the Gulf Breeze Laboratory had also been undertaking extensive investigations of the biology of the eastern oyster, *Crassostrea virginica*, and the preliminary studies carried out on pesticides included work on both the toxicity of DDT to this species (impacts being greatest on shell growth) and the bioaccumulation of DDT by *C. virginica* from solution. The bioaccumulation studies were later extended to eight other

Table 18 The accumulation and excretion of pesticides by the clams (a) *Mya arenaria* and (b) *Mercenaria mercenaria* (after Butler, 1971)

Pesticide	*5-day exposure concentration* (*μg litre*$^{-1}$)	*Concentration in clams* (*ng g*$^{-1}$ *wet weight*)					
		Initial tissue residue		*Residue after 7-day flush*		*Residue after 15-day flush*	
		a	*b*	*a*	*b*	*a*	*b*
Aldrin	0.5	2300	190	290	120	22	70
DDT	0.1	880	126	200	36	47	18
Dieldrin	0.5	870	380	20	40	10	12
Endrin	0.5	620	240	10	42	n.d.[a]	25
Heptachlor	0.5	1300	110	90	26	10	n.d.
Lindane	5.0	200	63	10	n.d.	n.d.	n.d.
Methoxychlor	1.0	1500	470	n.d.	43	n.d.	n.d.

[a] n.d. = Not detected.

species of bivalve: three additional oysters (*Crassostrea gigas*, *Ostrea edulis* and *O. lurida*); three clams (*Mya arenaria*, *Mercenaria mercenaria*, and the brackish water species *Rangia cuneata*); and two mussels (*Mytilus edulis* and a species of the genus *Brachydontes*). Rates of uptake and excretion of DDT and other pesticides were studied in these bivalves, and it was noted that these differed considerably between species and with the pesticide involved. An example of these data is provided in Table 18, showing that the hard clam *Mercenaria mercenaria* accumulates pesticides from solution at a much slower rate than does the softshell clam *Mya arenaria*, and that depuration rates are also slower in the former species than in the softshell clam. Such data were valuable in defining the potential usefulness of the various bivalve species in monitoring organochlorines in field situations.

Pursuant to this initial work, an extensive field monitoring programme was commenced in 1966, the principal objective of which was to determine the extent of the pollution threat to commercial fisheries posed by pesticides in estuaries (Butler, 1971). The early years of the monitoring programme involved the monthly sampling of a total of seven species of bivalve molluscs from wild populations at 170 sites on the Atlantic, Pacific and Gulf coasts of the USA, and the concentrations of 10 organochlorine pesticides were determined in these samples (Butler, 1969b, 1971). Later studies extended the suite of contaminants analysed, most notably to include PCBs (Butler, 1971, 1973).

This monitoring gave rise to a very considerable improvement in the knowledge then available on the abundance of organochlorines in estuaries. DDT was found to be by far the most commonly encountered pesticide in the USA at that time, followed in decreasing order of occurrence by dieldrin, endrin, and chlordane. In addition, occasional samples exhibited detectable concentrations of toxaphene, mirex and PCBs. In several instances, contamination was noted by residues which were not previously known to be present in the particular locality involved (Butler, 1971, 1973).

In general, both the overall patterns of organochlorine abundance and their seasonal trends in these bivalves were interpretable on the basis of information on the utilisation of the various compounds in the catchments of the estuaries studied. For example, with respect to overall abundance, DDT was found to be present most frequently and in highest concentrations in bivalves from estuaries draining agricultural catchments (e.g. in California, Florida and southern Texas). However, urbanised areas also constituted significant sources of this and other organochlorines in certain cases, due to the use of pesticides in industrial applications. By contrast, estuaries with small catchment areas or those draining principally forested regions (e.g. in the States of Maine and Washington) were not significantly contaminated by pesticides.

The seasonal trends in organochlorine residues present in the bivalves monitored reflected the application rates of pesticides and the timing of rainfall and consequent runoff in the various drainage basins, and this differed significantly from area to area (Butler, 1971, 1973; Butler *et al.*, 1972). In many cases, concentrations of DDT and other pesticides in the bivalves collected exhibited a major peak in spring or summer (the seasons of major crop growth, and hence of most pesticide applications), with low or undetectable pesticide levels being present in samples taken in autumn or winter. However, certain estuaries drained catchments where multiple harvesting (and more than one pesticide application period) occurred annually, and this was reflected in both the bivalves studied and certain fish species taken for analysis (Fig. 54).

Such relatively simple seasonal profiles of pesticide abundance were not found for all the estuaries studied in the programme. Thus, for example, Modin (1968, 1969), reporting on the Californian component of the project, noted that the concentrations of DDT and its metabolites in Pacific oysters (*Crassostrea gigas*) from sites in San Francisco Bay varied quite erratically between months. This is hardly surprising, given the size and hydrological complexity of the catchment of San Francisco Bay

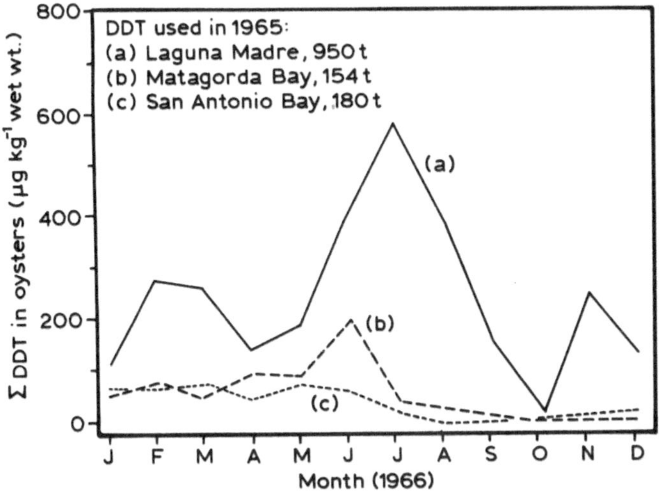

Fig. 54. Seasonal variations in concentrations of ΣDDT ($\mu g\,kg^{-1}$ wet weight) in the whole soft parts of oysters, *Crassostrea virginica*, from three estuaries in Texas in 1966. Catchment (a) involved multiple harvesting and DDT was applied three times annually; catchments (b) and (c) received only one application of DDT each year. The tonnages applied in the previous year are also shown. After Butler *et al.* (1972).

(Phillips and Spies, 1988; Wright and Phillips, 1988), and is indicative of multiple applications of DDT which occurred at that time in the various sub-catchments of the system. Butler (1969b, 1971) noted that the seasonal profiles of organochlorines in bivalves from estuaries with urbanised drainage basins were also often erratic, and ascribed this to temporal variations in industrial discharges and in river flows.

The emphasis of this early monitoring programme on the use of bivalve molluscs as biomonitors is important, as many of the efforts elsewhere did not reflect this trend. Butler (1971) justified this decision on two grounds. Firstly, he claimed that accumulation and excretion of organochlorines were more rapid in bivalves than in fish, and that bivalves therefore provided a more sensitive biomonitoring system. No actual data were cited to support this contention, however, and more recent information on the kinetics of organochlorines in a range of species suggests that such a generalisation is not always valid (Phillips, 1980).

However, Butler (1971) also noted that the variations in organochlorine levels between different individuals in a population were considerably greater for fish species than for bivalves, and this point certainly has merit.

Thus, the concentrations of contaminants in non-sedentary species (such as most fish) are a complex function of the exposure of each individual over space and time (Phillips, 1980; Phillips and Segar, 1986). If different patterns of movement are exhibited by the various individuals in any given population, one consequence would be an increase in intra-population variability in contaminant accumulation, and this tends to interfere with the ability of the researcher to define consistent differences in contaminant abundance between organisms taken from distinct locations (or from the same location at different times; see Chapter 8).

This important consideration was not always acknowledged in the studies which followed the estuarine monitoring programme in the USA, discussed above. In certain cases, the course followed was legitimate, as the authors of the new work were interested primarily in organochlorine cycling through food webs, and in establishing whether residues were biomagnified in aquatic ecosystems (e.g. see Robinson *et al.*, 1967; Woodwell *et al.*, 1967). In other instances, finfish were abundant in the waters of interest and were of importance in commercial or recreational fisheries; thus, the potential impacts of the organochlorine contamination of such species on human health was of relevance to the investigations undertaken (e.g. Duffy and O'Connell, 1968; Godsil and Johnson, 1968; Modin, 1969; Henderson *et al.*, 1969, 1971; Hansen and Wilson, 1970; Morris and Johnson, 1971; Sprague and Duffy, 1971; Zitko, 1971; Stout *et al.*, 1972; Butler and Schutzmann, 1979). However, other studies were also conducted which were intended primarily to monitor the environmental abundance of organochlorines, and many of these would have benefited from the use of sedentary or sessile species rather than their reliance on studies of fish (Phillips and Segar, 1986).

Investigations of organochlorines in European waters in the late 1960s and early 1970s employed several species, including bivalve molluscs, fish and seals. Holden (1970) and Holden and Portmann (1970) described the evolution of these studies, noting that several objectives existed, from the desire to delineate organochlorine abundance in various regions (mostly satisfied through analyses of contaminants in shellfish and seals), to the needs to improve analytical techniques and protect both commercial fisheries and public health. As a result of these mixed objectives, both the national monitoring programme in the United Kingdom and the international investigations undertaken by the member countries of the OECD (Organization for Economic Cooperation and Development) tended to be dominated by the analysis of fish species. While useful data were undoubtedly generated through such studies, the existence of high intra-population

Table 19 Concentrations of ΣDDT ($\mu g\,g^{-1}$ by wet and lipid weights) in six tissues of a single porpoise (*Phocaena phocaena*) from the coast of Scotland (after Holden and Marsden, 1967)

Parameter	*Blubber*	*Liver*	*Brain*	*Muscle*	*Spleen*	*Kidney*
ΣDDT (wet weight)	3.8	0.58	0.02	0.56	0.12	0.04
Percentage lipid	67.0	13.2	8.3	6.1	5.1	1.4
ΣDDT (lipid weight)	5.6	4.8	0.27	9.2	2.4	2.9

variations in organochlorine levels in fish (which is due principally to their non-sedentary nature) gave rise to a less robust database than would have been generated by the study of sessile organisms such as bivalve molluscs.

The early investigations in Europe and elsewhere of organochlorines in seals and porpoises merit particular mention here. These were initiated due to concerns relating to the biomagnification of organochlorines through food chains; it was theorised that marine mammals could be at significant risk from the toxic impacts of these compounds, due to their position at high trophic levels of aquatic ecosystems. Holden and Marsden (1967) reported preliminary data on seals and porpoises from the coasts of Scotland and Canada. Few consistent differences were found between the organochlorine concentrations accumulated by the various seal species at any one site, but significant variations existed with geographical location, and the organochlorine levels present in seal pups were somewhat lower than those found in adults from the same site.

Holden and Marsden (1967) also reported that the tissue distribution of total DDT (denoted generally as ΣDDT, this being the sum of the parent compound and its identified metabolites) in marine mammals depended primarily on the amounts of lipids in each tissue (Table 19). Thus, when concentrations of ΣDDT were expressed on the basis of wet tissue weights, considerable variability between the different organs was found. However, most of this variation was eliminated when ΣDDT levels were recalculated on a lipid weight basis, although pesticide amounts in the brain (which exhibits a distinct lipid chemistry from that of the other tissues analysed) remained low. This phenomenon had been previously reported for DDT in the tissues of brown trout (*Salmo trutta*) by Holden (1962), and the lipid contents of tissues and whole organisms are now known to be of general importance in defining organochlorine accumulation by biota (Phillips, 1978, 1980, 1986); this will be discussed in greater detail later in the present chapter.

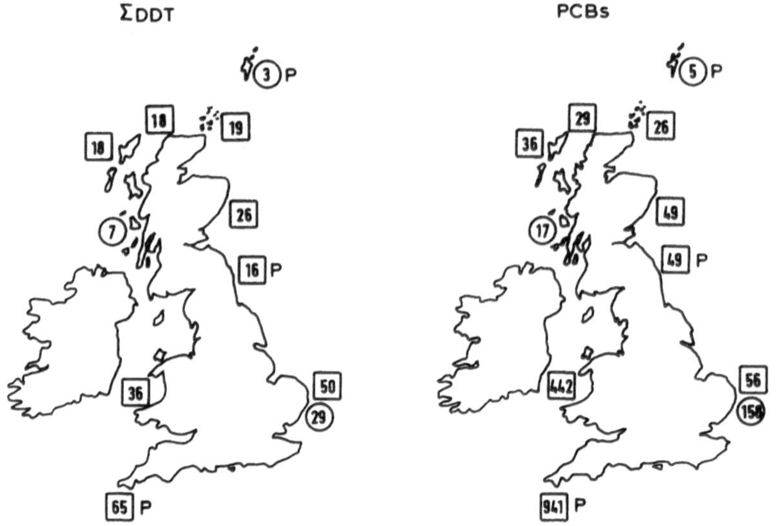

Fig. 55. Mean concentrations of ΣDDT and PCBs ($\mu g\, g^{-1}$ lipid weight) in the blubber of seals from the coasts of the United Kingdom. Figures in circles denote residues in common seals (*Phoca vitulina*); those in squares refer to grey seals (*Halichoerus grypus*). P denotes that the sample was of pups, rather than adult seals. Two data points are omitted due to inadequate information for the calculation of lipid-based concentrations (the Wash) or because the original data were cited for a mixture of species (the Clyde). After Holden (1972).

A further report by Holden (1972) provided additional data on the variations in pesticide levels in seals with location around coasts of the United Kingdom. These data are shown in Fig. 55, and it is clear that the variations found were considerable for many of the pesticides studied. At first sight, this appears unlikely, given the known tendencies of seals to migrate substantial distances. However, the populations studied were thought to differ considerably in their patterns of migration, and the profiles of contamination represented in Fig. 55 were considered to accurately reflect regional differences in organochlorine abundance at the seal feeding grounds around the United Kingdom.

Additional early information on organochlorine levels in marine mammals was provided by Anas and Wilson (1970a,b) and Gaskin *et al.* (1971). The former authors studied residues in northern fur seals (*Callorhinus ursinus*) from the Pribiloff Islands in Alaska. PCBs were not detected in these samples, and dieldrin was found in only trace amounts. However,

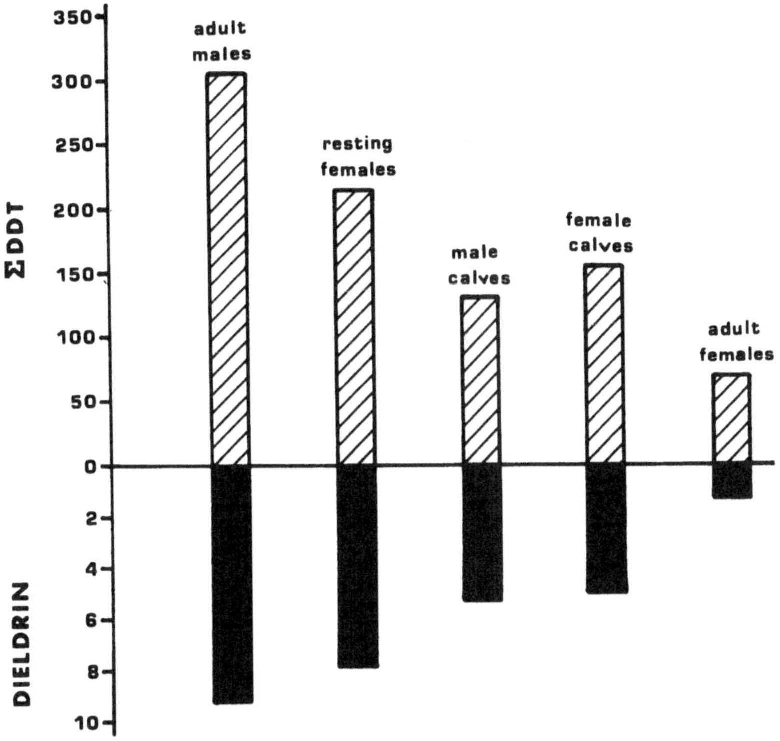

Fig. 56. Variations with age and sex in concentrations of ΣDDT and dieldrin ($\mu g\ g^{-1}$ in extractable fat) in the blubber of porpoises (*Phocaena phocaena*) from the Bay of Fundy, Canada. After tabulated data in Gaskin *et al.* (1971).

all the adults analysed exhibited significant accumulation of DDT and its metabolites, DDE being the most abundant compound present. Fur seal foetuses were shown to contain considerably less ΣDDT than their mothers, but the transplacental transfer of contaminants nevertheless occurred. Residue levels in fur seal pups (after birth) rose markedly with age, due to the further transfer of DDT and metabolites from the mother in milk (Anas and Wilson, 1970a,b).

The impacts of age on organochlorine levels in marine mammals and the loss of residues from mature females during pregnancy and lactation were confirmed by the work of Gaskin *et al.* (1971) on harbour porpoises (*Phocaena phocaena*) from the Bay of Fundy in the western North Atlantic (Fig. 56). Mature male porpoises were found to contain the highest concentrations of both ΣDDT and dieldrin, with immature or resting females exhibiting about 70% of these levels. There was no significant difference between the contamination of unweaned male and female sucklings, these containing about half the concentrations of pesticides present in the mature males. By contrast, the levels of ΣDDT and dieldrin found in pregnant or lactating female porpoises were only 15–20% of those present in the mature males sampled, demonstrating the very considerable transfer of residues from females to their offspring, this occurring both transplacentally before birth, and through lactation thereafter.

These early studies of organochlorines in bivalves, fish and marine mammals pre-dated any attempt to employ biota to monitor petroleum-derived hydrocarbons in aquatic ecosystems. They nevertheless laid a foundation for the later development of biomonitoring techniques for both classes of compounds, and many of the conclusions reached in the early investigations have been confirmed and extended by the more recent studies, which are discussed below.

(iii) Review of Biomonitoring to Date

Studies prior to 1980 which employed biomonitors to delineate the abundance of organochlorines and hydrocarbons in aquatic ecosystems have been reviewed by NAS (1980) and Phillips (1978, 1980, 1986). The present text provides a synthesis of these reviews and updates them, to include the more recent work. For presentational purposes, data on invertebrates are considered first, followed by studies on fish. Discussions of the more recent investigations of organochlorines and hydrocarbons in marine mammals conclude the present chapter.

The success of the early monitoring programme for organochlorines in the USA described above and by Butler (1966, 1969a,b, 1971, 1973) led to the use of bivalve molluscs and certain other invertebrate species in later work also, and this was eventually extended to include the study of hydrocarbons in addition to organochlorines. Invertebrates take up both classes of compounds readily, predominantly through a simple lipid:water partitioning mechanism (Hamelink *et al.*, 1971). Laboratory studies of the kinetics of uptake of these contaminants have added measurably to our overall understanding, and have assisted in the interpretation of field data on biomonitors.

Laboratory-based studies of the uptake of organochlorines by phytoplankton have revealed rapid rates of accumulation from solution, final concentration factors at equilibrium generally being of the order of 10^3 or 10^4 (e.g. Södergren, 1968; Cox, 1970; Reinert, 1972; Canton *et al.*, 1977;

Wang *et al.*, 1982). The uptake rates exhibited by zooplankton tend to be rather lower than those seen in phytoplankton, but equilibration is nevertheless approached within 24 h in most instances (Crosby and Tucker, 1971; Johnson *et al.*, 1971; Clayton *et al.*, 1977). Several authors have proposed that organisms of these lower trophic levels are at significant toxicological risk from the release of compounds such as PCBs to aquatic environments (e.g. Fisher *et al.*, 1973). Neither phytoplankton nor zooplankton can be considered appropriate as biomonitors, however, principally because of the interspecific variations noted in the accumulation of organochlorines and the fact that field collections always contain a mixture of species (Phillips, 1980; Harding, 1986; Fowler, 1990).

The accumulation of organochlorines by bivalve molluscs has also been shown to be a rapid process, and residues may be taken up from solution or suspension (e.g. Brodtmann, 1970; Petrocelli *et al.*, 1973; Roberts, 1975; Bahner *et al.*, 1977; Langston, 1978a,b; Morales-Alamo and Haven, 1983). Examples involving the kinetics of PCBs may be cited here to illustrate several points of importance to the use of bivalves (and, by extension, other species) as biomonitors.

Phillips (1986) noted that individual PCB homologues or congeners differ markedly in their solubility in water, and hence in their tendencies to be accumulated by biota. In general, PCB congeners of lower chlorination exhibit greater water solubilities (and therefore, lower bioaccumulation rates and extents) compared to those of higher chlorination. However, within groups of PCBs possessing the same degree of chlorination, significant differences in solubility exist (Table 20), these being dependent on the relative positions of the chlorine atoms on the biphenyl rings (see Tanabe, 1988).

In addition to this physicochemical effect on PCB uptake, there is evidence for considerable biological selectivity. For example, Vreeland (1974) found that the uptake of PCBs by the eastern oyster (*Crassostrea virginica*) varied with the extent of chlorination, being greatest for congeners of higher chlorination and least for lower-chlorinated PCBs. By contrast, the clam *Mercenaria mercenaria* apparently selectively accumulates lower-chlorinated PCBs (Courtney and Denton, 1976; Deubert *et al.*, 1981), and both the clam *Macoma balthica* and the cockle *Cerastoderma edule* exhibit a preference for pentachlorobiphenyls over PCBs with either higher or lower degrees of chlorination (Langston, 1978a,b).

Langston (1978b) also noted that the positions of chlorine substitution on the biphenyl ring affected excretion rates for individual PCBs; congeners with the most *ortho*-substituted chlorine atoms were of least

Table 20 Comparative aqueous solubilities of isomers and congeners of Aroclor 1242. Note the differences in rate of dissolution of different isomers and congeners as equilibrium is approached between water and Aroclor 1242. (after Lee *et al.*, 1979)

Peak no.	*No. of chlorines*	*Identification*	*Concentrations (µg litre^{-1}) attained in days shown*					
			7	*20*	*30*	*50*	*149*	*251*
1	1	2					124.60	121.59
2	1	3						
3	1	4					7.50	15.07
4	2				1.00	1.69	5.83	9.06
5	2	2,2′			3.18	1.78	20.80	21.18
6	2					2.93	23.24	23.76
7	2	2,3′			9.10	14.07	19.63	30.60
8	2	2,4′	2.90	27.30	71.18	78.45	141.85	138.93
9	3	2,6,2′			1.40	1.72	8.13	8.33
10	3	2,5,2′	5.52	22.97	41.05	41.20	62.48	61.35
11	3							2.15
12	3	2,4,2′						0.48
13	3					1.71	10.00	4.50
14	2	3,4′						2.98
15	2&3	4,4′;2,3,2′		8.83	16.25	17.19	26.15	35.30
16	3	2,4,3′					29.83	27.47
17	3	2,4,4′	2.55		4.08	4.63	9.83	9.87
18	3	2,5,4′	20.22	35.15	43.28	48.20	63.76	65.07
19	3	3,4,2′					2.59	2.59
20	4	2,5,2′,5′	1.58	6.67	8.94	8.78	24.05	22.25
21	4					0.64	2.66	3.59
22	4				0.55	0.49	3.78	4.09
23	3&4	2,4,2′,5′				2.11	17.95	19.51
24	4	2,4,2′,4′	11.12	7.39	8.93	9.12	19.85	15.88
25	4						4.15	3.37
26	4						1.11	1.09
27	4	2,3,2′,5′				0.69	12.75	12.83
28	4					0.74	4.10	3.13
29	4					1.42		2.50
30	4	2,5,3′,4′					9.15	7.21
31	3					0.59	4.53	4.70
32	4						4.10	2.70
33	4						5.43	2.63
34	3						0.73	0.82
35	4	2,4,3′,4′	1.96			2.18	13.83	11.62
36	5							
37	5						2.53	0.99
38	5							
39	5		0.71				4.79	2.47
40	5							2.84
Total			46.56	108.31	208.94	240.33	671.86	702.67

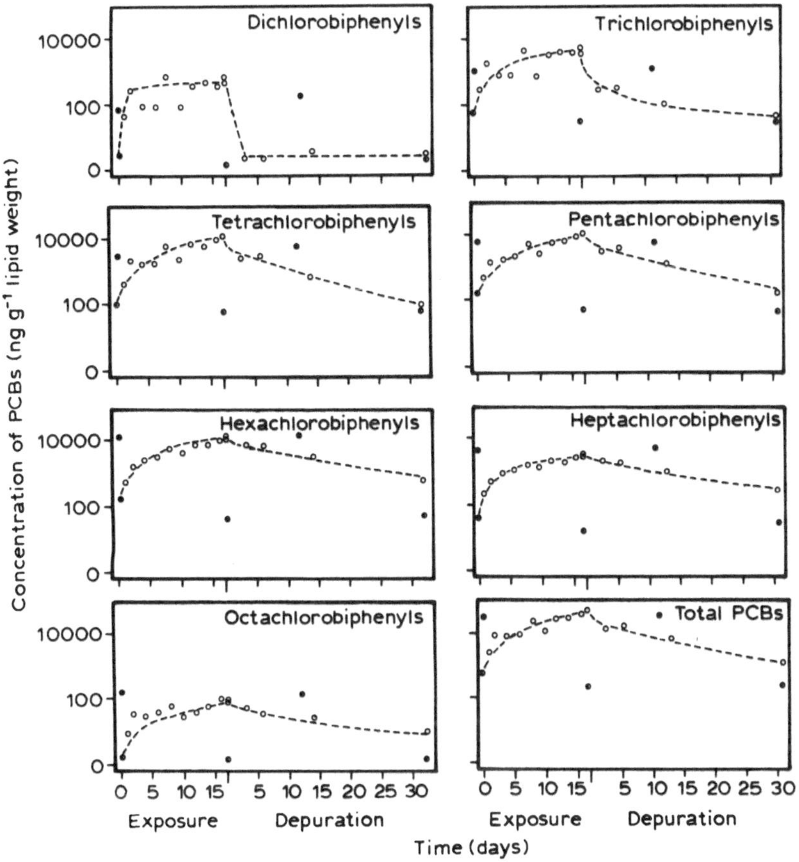

Fig. 57. Uptake and depuration under field conditions of PCBs of differing degrees of chlorination by green-lipped mussels (*Perna viridis*): ○, concentrations in transplanted and back-transplanted individuals; ◑, concentrations in native samples at the uncontaminated site; ●, concentrations in native samples of the contaminated site. After Tanabe *et al.* (1987b).

persistence. This observation has been significantly extended recently by studies on the uptake and excretion of PCBs in field situations by the green-lipped mussel *Perna viridis*. Tanabe *et al.* (1987b) showed that both the net uptake and excretion of PCBs by *P. viridis* were generally more rapid for lower-chlorinated isomers than for PCBs of greater chlorination (Fig. 57), but that the impacts of the position of chlorine substitution were substantial within any of the groups of PCBs of the same degree of chlorination (Table 21). The most persistent PCBs were those which exhibited a coplanar structure, with no *ortho* chlorine substitutions (Kannan *et al.*, 1989). Notable amongst these were 3,3′,4,4′-tetrachlorobiphenyl, 3,3′,4,4′,5-pentachlorobiphenyl and 3,3′,4,4′,5,5′-hexachlorobiphenyl; it is these non-*ortho* chlorine substituted coplanar PCBs which are now believed to be of the greatest toxicological importance in terrestrial and aquatic environments (Tanabe, 1988).

These data for PCB uptake by bivalve molluscs have been confirmed and extended by similar investigations on organisms of other phyla. Thus, for example, biological selectivity in PCB uptake has been noted for various species of zooplankton, crustaceans, polychaetes and fish (e.g. Sanders and Chandler, 1972; Clayton *et al.*, 1977; McLeese *et al.*, 1980) and the effects of the position of chlorine substitution on PCB bioaccumulation by fish have also attracted study (e.g. Bruggeman *et al.*, 1981; Tanabe *et al.*, 1982b).

While much less information exists on the kinetics of hydrocarbons in aquatic biota, the data available suggest that uptake and excretion rates of these compounds are broadly comparable to those of organochlorines. Connell and Miller (1984) compiled data on the reported half-lives of individual PAHs and other hydrocarbons in a range of organisms (Table 22), showing that half-lives vary from less than a day to several weeks, depending on the precise compound involved.

It should be noted here that hydrocarbons differ quantitatively from organochlorines in one important respect: their susceptibility to metabolic transformation. While most aquatic organisms are capable of metabolising hydrocarbons, it is known that in general, vertebrates exhibit the highest propensity for such transformations. The site of greatest activity is the liver, and mixed function oxidase (MFO) systems mediated by cytochrome P-450 are involved (see Chapter 9). Microsomal enzymes (which are inducible in nature) transform the parent compounds and their derivatives to less complex conjugates, which generally exhibit a more hydrophilic nature and are thus less biologically persistent (e.g. see Fig. 58). By comparison, organochlorine pesticides and PCBs are degraded very slowly by biota.

The above examples are important for two reasons with respect to the use of biota to monitor trace organic contaminants. Firstly, they demonstrate that the kinetics of any given contaminant in an organism are unique, i.e. that generalisations cannot be made concerning the bioaccumulation kinetics of any one contaminant in different organisms, or of different contaminants in any given species. Thus, bioaccumulation

Table 21 Data for clearance rate (k_2), biological half-life (BHL), and days to 90% uptake equilibrium (t) for PCB isomers and congeners in green-lipped mussels (*Perna viridis*) exposed to PCBs under field conditions (after Tanabe *et al.*, 1987b)

Peak no.	k_2 (day^{-1})	r[a]	BHL (day)	t (day)	Peak no.	k_2 (day^{-1})	r[a]	BHL (day)	t (day)
21	>1.31		<0.5	<1.8	58	0.140	−0.98	5.0	16.4
22	>0.92		<0.8	<2.5	59	0.102	−0.97	6.8	22.6
23	>1.18		<0.6	<1.9	510	0.122	−0.98	5.7	18.9
24	>1.80		<0.4	<1.3	511	0.117	−0.95	5.9	19.7
25	>0.28		<2.5	<8.2	512	0.173	−0.94	4.0	13.3
26	>1.50		<0.5	<1.5	513	0.172	−0.94	4.0	13.4
					514	0.102	−0.96	6.8	22.6
31	>1.29		<0.5	<1.8	516	0.109	−0.96	6.4	21.1
32	0.108	−0.74	6.4	21.3					
34	>1.41		<0.5	<1.6	61	0.119	−0.99	5.8	19.3
35	0.131	−0.87	5.3	17.6	62	0.108	−0.99	6.4	21.3
36	>1.52		<0.5	<1.5	63, 64	0.098	−0.99	7.1	23.5
38	0.106	−0.79	6.5	21.7	65	0.097	−0.99	7.1	23.7
39	0.127	−0.87	5.5	18.1	66	0.111	−0.99	6.2	20.7
310	0.118	−0.86	5.9	19.5	68	0.080	−0.99	8.7	28.8
					69	0.106	−0.96	6.5	21.7
41	0.182	−0.98	3.8	12.6	610	0.079	−0.99	8.8	29.1
42	0.185	−0.99	3.7	12.4	611	0.176	−0.95	3.9	13.1
43	0.139	−0.88	5.0	16.6	612, 613	0.110	−0.98	6.3	20.9
44	0.142	−0.90	4.9	16.2	614	0.084	−0.99	8.3	27.4
45	0.125	−0.94	5.5	18.4	615	0.076	−0.99	9.1	30.3
46	0.149	−0.98	4.7	15.4	617	0.092	−0.99	7.5	25.0
47	0.147	−0.99	4.7	15.7	618	0.109	−0.99	6.4	21.1
48	0.148	−0.99	4.7	15.5	619	0.105	−0.98	6.6	21.9
49	0.119	−0.88	5.8	19.3					
410	0.123	−0.92	5.6	18.7	71	0.081	−0.99	8.6	28.4
411	0.131	−0.95	5.3	17.6	72	0.074	−0.97	9.4	31.1
412	0.159	−0.98	4.4	14.5	73	0.084	−0.99	8.3	27.4
416	0.105	−0.86	6.6	21.9	74	0.067	−0.99	10.3	34.3
417	0.115	−0.91	6.0	20.0	75	0.067	−0.99	10.3	34.3
418	0.119	−0.95	5.8	19.3	78	0.064	−0.99	10.8	35.9
420	0.118	−0.93	5.9	19.5	79	0.079	−0.99	8.8	29.1
					711	0.104	−0.99	6.7	22.1
51	0.141	−0.98	4.9	16.3	713	0.102	−0.98	6.8	22.6
52	0.125	−0.99	5.5	18.4	714	0.119	−0.98	5.8	19.3
53	0.124	−0.98	5.6	18.6					
54	0.118	−0.98	5.9	19.5	81	0.059	−0.98	11.7	39.0
55	0.112	−0.98	6.2	20.5	82	0.064	−0.95	10.8	35.9
56	0.084	−0.93	8.3	27.4					
57	0.130	−0.98	5.3	17.7	Total PCBs	0.102	−0.98	6.8	22.6

[a] Correlation coefficient in the equation used to obtain k_2.

Table 22 Reported half-lives of certain hydrocarbons in aquatic organisms (after Connell and Miller, 1984)

Organism	*Compound*	*Half-life (days)*
Oysters (*Crassostrea virginica*)	Naphthalenes	2
	Anthracene	3
	Fluoranthene	5
	Benzo[*a*]anthracene	9
	B(*a*)P	18
Mussels (*Mytilus edulis*)	B(*a*)P	16
	Aliphatics and aromatics	2
	Aliphatics	4
	Aromatics	48–60
Copepod (*Calanus helgolandicus*)	Naphthalene	1.5
Water flea (*Daphnia pulex*)	Anthracene	0.4
Rainbow trout (*Salmo gairdneri*)	Naphthalene Methyl naphthalene	Depending on exposure time, 0.3–38
Sea mullet (*Mugil cephalus*)	*n*-Alkanes	18 (max)

kinetics depend not only on the physicochemical characteristics of the contaminant involved, but also on various attributes of the chosen organism (e.g. lipid content and chemical type; capacity to accumulate contaminants from suspension or food in addition to solution; ability to metabolise the contaminant after its uptake). Secondly, the examples provided emphasise the relatively rapid rates of uptake and excretion of most trace organic contaminants in biota. Even the more highly chlorinated PCBs (which are thought to be among the most persistent organic contaminants of concern in aquatic ecosystems) exhibit half-lives in mussels ranging from only about 5 to 26 days (Tanabe *et al.*, 1987b). By comparison to the persistence of many trace metals in biota (see Chapter 5), these rates of uptake and excretion may be considered to be rapid.

The implications of these findings have not been fully recognised to date by many authors employing biota to monitor trace organic contaminants in natural waters. In particular, the importance of the time-integration capacity of an organism for any given contaminant has been paid insufficient attention. Thus, if accumulated contaminants are excreted relatively rapidly (i.e. over days or weeks) by a biomonitor, and if they are present only episodically in aquatic environments, their quantification will be

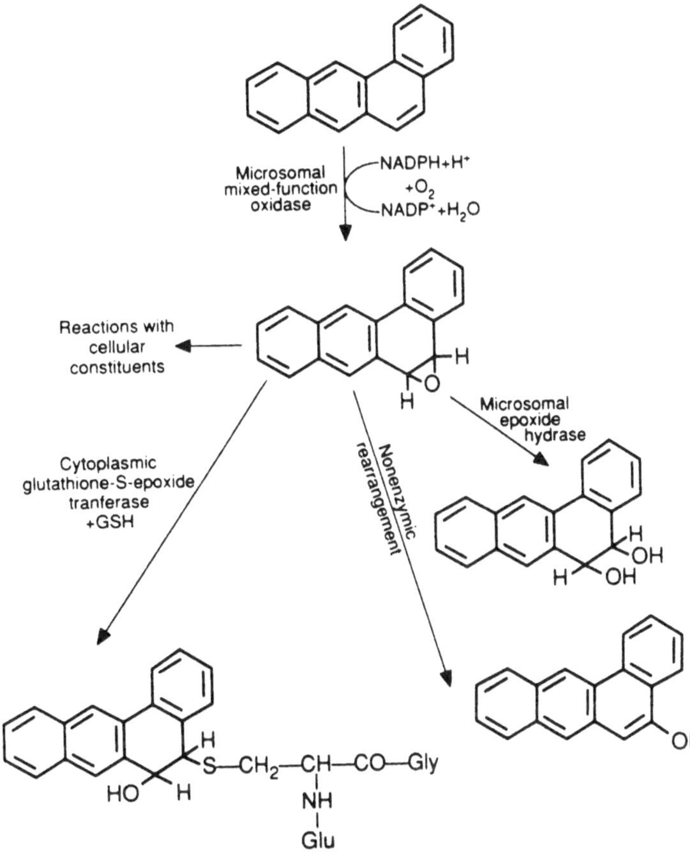

Fig. 58. An example of the metabolism of hydrocarbons by aquatic biota: the metabolic transformation of benzo(*a*)anthracene. After Connell and Miller (1984).

extremely challenging. Phillips and Segar (1986) discussed this problem, and recommended the use of so-called "time-bulking" of biomonitoring samples. This involves the collection of several samples of the selected biomonitor over time, from each location studied. The inter-sample period may be adjusted to match the kinetics of the contaminant(s) of concern, and the tissues may be composited in any desired fashion for bulk analysis. This method essentially provides an artificial extension of the time-integration capacity offered by the biomonitor itself for the contaminants analysed. It has been only rarely employed to date, but offers considerable advantages when studies are to be undertaken for contaminants of relatively short half-life in aquatic biota, particularly if such contaminants are likely to be discharged episodically or at fluctuating levels with time.

In addition to the studies discussed above of contaminant kinetics in biota, many field investigations of biomonitors have been undertaken since the early work discussed previously in this Chapter. Bivalve molluscs have been a frequent choice for such monitoring programmes, but species of other phyla have also been employed. In more recent years, the work on organochlorines has also been extended to petroleum-derived hydrocarbons.

As noted previously, neither phytoplankton nor zooplankton are suitable for use as biomonitors of organic contaminants in the field, because of the species-dependent nature of the bioaccumulation of these contaminants. Thus, plankton samples invariably contain many species, and no two locations will provide identical assemblages. In these circumstances, no meaningful conclusions may be reached concerning the relative contamination of different sites, as species differences between samples (and the consequent effects of these on contaminant bioaccumulation) confound any attempt at interpretation.

Macroalgae have been rarely employed to monitor trace organic contaminants, and this contrasts markedly with their popularity as biomonitors of trace elements and radionuclides (see Chapter 5). The low lipid contents of macroalgae may constitute one reason for the paucity of such investigations, as this mitigates against the accumulation of high concentrations of organochlorines or hydrocarbons. Preliminary surveys have been undertaken in certain locations (e.g. Parker and Wilson, 1975), but these are not thought adequate as a basis to gauge the true potential of macroalgae as biomonitors of organochlorines or hydrocarbons.

However, the surveys of Amico *et al.* (1979a,b) on organochlorines in 12 species of macroalgae and in mussels (*Mytilus galloprovincialis*) from the east coast of Sicily merit brief mention here. While precise comparisons between the two sets of data cannot be made due to differences in study locations, it appears from these investigations that the macroalgae contained significantly lower concentrations of ΣDDT, PCBs and HCH compared to mussels, and that the ratios of DDT:DDE and lower chlorinated:higher chlorinated PCBs were greater in the primary producers than in the bivalves. The distinction in overall bioaccumulation cannot be entirely explained by the lower total lipid contents of the macroalgae than the mussels, and differences in lipid chemistries may be involved; this could be a fruitful area for further study. It is possible also that macroalgae exhibit unusually low capacities for metabolising accumulated organochlorines, thus explaining the differences in ratios between the DDT-related and PCB compounds in the various species studied.

It may be noted here in addition that Coates *et al.* (1986) employed the filamentous green alga *Chlorodesmis fastigiata* in investigations of hydrocarbons in the Great Barrier Reef area of Australia. The concentrations of total hydrocarbons found correlated to the lipid contents of the algae from the various locations studied, and were very high compared to those of the other species analysed. The composition of these hydrocarbons was considered to reflect a biogenic rather than an anthropogenic origin, being dominated by C_{17} and C_{19} mono-alkenes and revealing little relationship to petroleum-derived residues. The high levels of biogenic hydrocarbons in primary producers of aquatic environments mitigate against their use to monitor anthropogenically-derived hydrocarbons, as noted by Murray *et al.* (1977) and other authors.

Crustaceans have been quite frequently employed to monitor organochlorines in aquatic ecosystems. Nimmo *et al.* (1970) reported that the rates of uptake of DDT by pink and white shrimp (*Penaeus duorarum* and *P. setiferus*) varied somewhat between species, but were quite slow in either organism. Bioaccumulated DDT was found to be present mostly in the hepatopancreas, which is a lipid-rich organ.

The first extensive survey of organochlorine abundance utilising a crustacean was that of Burnett (1971), who employed the surf-zone crab *Emerita analoga* to study the distribution of DDT and its metabolites off Californian coasts. No interference due to animal size was noted, but the sexual condition of the female crabs used (and hence, season of sampling) was found to affect the data. Thus, intra-population differences in the concentrations of ΣDDT were considerable, and these were ascribed to the transfer of DDT and its metabolites from the tissues to the lipid-rich eggs, and the eventual loss of this ΣDDT from the female crabs at spawning. Despite this, however, inter-population differences in ΣDDT levels were highly significant (Fig. 59), and the peaks in ΣDDT abundance occurred in areas of known sources of DDT contamination. These were at San Francisco and Monterey Bays (due to agricultural runoff from their respective catchments) and at the Los Angeles County sewer outfall, which was heavily contaminated by residues from the Montrose Chemical Corporation plant which manufactured DDT at that time. Burnett (1971) calculated a sediment-bound load of 102 tonnes of DDT and its metabolites in the Palos Verdes region, and later studies confirmed the high ΣDDT and PCB contamination of this region. While residues have

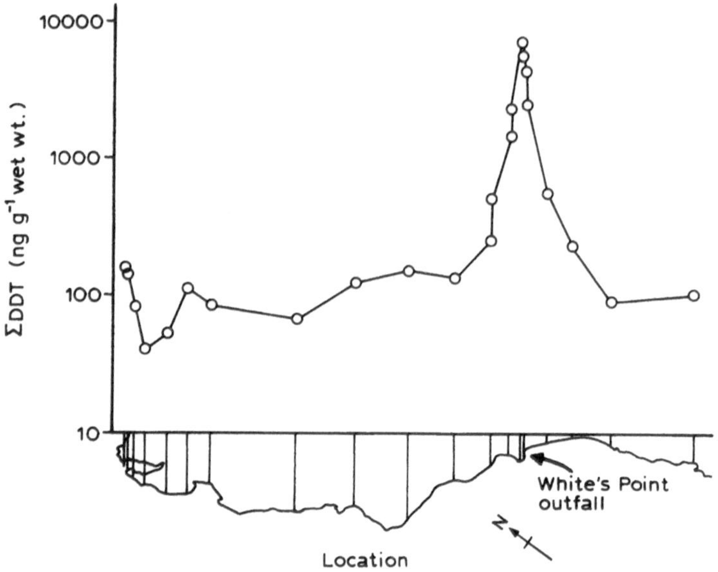

Fig. 59. Concentrations of ΣDDT (ng g^{-1} wet weight) in female surf-zone crabs (*Emerita analoga*) from sites as shown on the coast of California in 1970–71. After Burnett (1971).

declined in this area since the cessation of DDT production and the imposition of controls on the use of PCBs in the USA (e.g. see Young *et al.*, 1976; Martin, 1985; Green *et al.*, 1986), concerns continue with respect to the toxicological impacts of the remaining residues. Thus, for example, Hose *et al.* (1989) reported that white croaker (*Genyonemus lineatus*) from the San Pedro Bay area close to the Los Angeles outfalls exhibited reproductive impairment, which correlated closely to elevated DDT concentrations. A similar problem is believed to exist further north in California, involving PCB impacts on the reproduction of starry flounder (*Platichthys stellatus*) in San Francsico Bay (Spies and Rice, 1988).

Seasonal changes in organochlorine concentrations of crustaceans have been noted by other authors also. Södergren *et al.* (1972) found that the concentrations of ΣDDT and PCBs in amphipods (*Gammarus pulex*) from streams in southern Sweden varied significantly with month of sampling. These fluctuations were due both to temporal changes in the lipid contents of samples and the increased delivery of organochlorines to inland waters during times of snow-melt in spring. Temporal variations in organochlorine emissions from local sources and in lipid contents of the organisms studied are the most frequently cited causes of seasonal fluctuations in these contaminants in crustaceans (e.g. Amico *et al.*, 1979a; van den Broek, 1979; Satsmadjis and Gabrielides, 1983).

Other factors may also interfere with the use of crustaceans as biomonitors of organochlorines, however. Certain species exhibit significant variations in bioaccumulated concentrations of organochlorines with size, although this is by no means universal. Ravid *et al.* (1985) noted that PCB concentrations decreased with size in the prawn *Parapenaeus longirostris*, although ΣDDT levels did not vary significantly with size in this species. Negative correlations between organochlorine concentrations and size (age and weight) are rare in biota (Phillips, 1980), but have occasionally been reported for finfish (see below).

Organism sex may also affect organochlorine accumulation, as was noted previously in discussions of data on the surf-zone crab *Emerita analoga* (Burnett, 1971). Roberts (1981) found that the muscle of male blue crabs (*Callinectes sapidus*) from the James River and lower Chesapeake Bay contained higher concentrations of Kepone than the same tissue of female crabs from this area. It appeared that this difference was due both to the loss of Kepone residues to eggs by female crabs and to differences between the sexes in the tissue distribution of Kepone; both of these effects may depend in part at least on the covariation of lipids and Kepone levels in the crabs studied.

While significant differences in bioaccumulation exist between the various species of crustaceans used as biomonitors of organochlorines (e.g. Ravid *et al.*, 1985; Villeneuve *et al.*, 1987), most authors agree that both bivalves and finfish tend to accumulate greater amounts of these contaminants than do crustaceans (e.g. Wharfe and van den Broek, 1978; Amico *et al.*, 1979a; Contardi *et al.*, 1979; Bastürk *et al.*, 1980). As a result, rather more use has been made of bivalves and finfish as biomonitors of trace organic contaminants, including studies of both organochlorines and hydrocarbons.

Marchand *et al.* (1976) used mussels (*Mytilus galloprovincialis*) as biomonitors of ΣDDT and PCBs in the north-western Mediterranean Sea. While significant differences were found in organochlorine levels with location, the authors noted that lipid contents of the samples taken varied widely due to the non-synchronous spawning of mussel populations in the study area; this was thought to hamper interpretation of the inter-location variations in organochlorine abundance. Risebrough *et al.* (1976) voiced similar concerns, noting that apparent concentration factors for DDE or

PCBs in mussels varied widely between different locations. It is probable, however, that the water samples taken for organochlorine analysis in this study were not representative of the time-integrated organochlorine levels to which the mussels had been exposed in the field.

Data from the US Mussel Watch Program were reported by Goldberg *et al.* (1978) and Farrington *et al.* (1982, 1983). This programme developed from the earlier work of Butler discussed previously, and covered three years (1976–1978). Trace metals, radionuclides, organochlorines and hydrocarbons were included, and several species of mussels and oysters were employed as biomonitors. The resulting data provided a useful baseline for contaminant levels on the Pacific, Gulf and Atlantic coasts of the USA, and several previously unknown "hot-spots" of enhanced bioavailability of contaminants were identified.

Amongst these were PCB hot-spots in south San Francisco Bay, San Pedro Harbor and San Diego Harbor in California (see also Martin, 1985), and similar areas of elevated PCB abundance on the east coast of the USA (New Bedford Harbor in particular; see also Pruell *et al.*, 1990). Several samples exhibited evidence of significant petroleum-derived contamination, although hydrocarbons of biogenic origin were also found in many of the bivalves collected. Phenanthrene, alkyl phenanthrenes and alkyl dibenzothiophenes were the most commonly reported anthropogenic hydrocarbons in the samples analysed.

It should also be noted here, however, that the monthly samples which were taken at only two sites (*Mytilus edulis* from Narragansett Bay, Rhode Island, and *M. californianus* from Bodega Bay, California) exhibited very considerable seasonal fluctuations in hydrocarbon contents (Fig. 60). Phillips and Segar (1986) drew particular attention to this fact and noted that the sampling strategy (which called for single annual collections of bivalves at each study site apart from Narragansett Bay and Bodega Bay) did not permit the accurate quantification of episodically discharged contaminants of rapid kinetics in bivalves; as noted above, most hydrocarbons are included in this category.

Seasonal fluctuations of organochlorine and hydrocarbon levels in bivalves have also been reported in other studies (e.g. Clegg, 1974; Wharfe and van den Broek, 1978; Cajal-Medrano and Gutierrez-Galindo, 1981; Cowan, 1981; Satsmadjis and Galrielides, 1983; Folke and Birklund, 1986; Sericano *et al.*, 1990), although the degree to which they occur varies substantially from site to site. Phillips (1980, 1986) noted that this is a function of both the changes in lipid contents of samples with time (tied principally to the gametogenesis/spawning cycle) and the fluctuations in

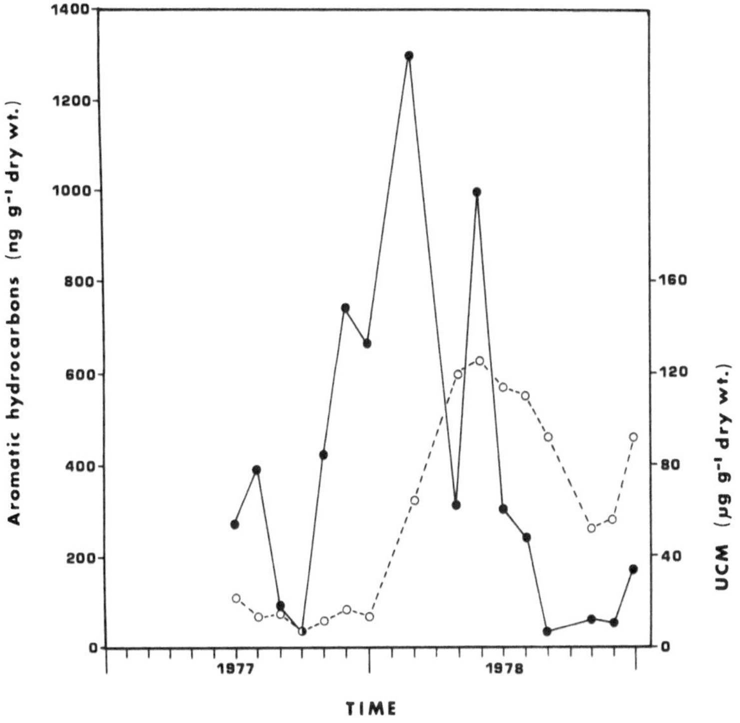

Fig. 60. Temporal changes in the concentrations of aromatic hydrocarbons (continuous lines; left axis) and hydrocarbons represented in the Unresolved Complex Mixture (UCM; dashes, right axis), in mussels (*Mytilus edulis*) sampled at monthly intervals during 1977 and 1978 in Narragansett Bay, USA. After Farrington *et al.* (1983).

ambient abundance of the contaminants concerned; however, in certain cases, other factors such as water temperature and oxygenation may also be relevant (Lunsford and Blem, 1982).

Data similar to those from the US Mussel Watch Program have been reported for certain other areas also. Cowan (1981) analysed mussels (*Mytilus edulis*) from Scottish coasts for DDT and its metabolites, dieldrin, hexachlorobenzene (HCB), isomers of HCH, and PCBs. Significant variation in concentrations with location was found, especially for dieldrin and PCBs (Fig. 61). The highest levels of dieldrin were present in mussels from the Firth of Clyde, the sample from Irvine being particularly contaminated; the source was believed to be the industrial use of dieldrin

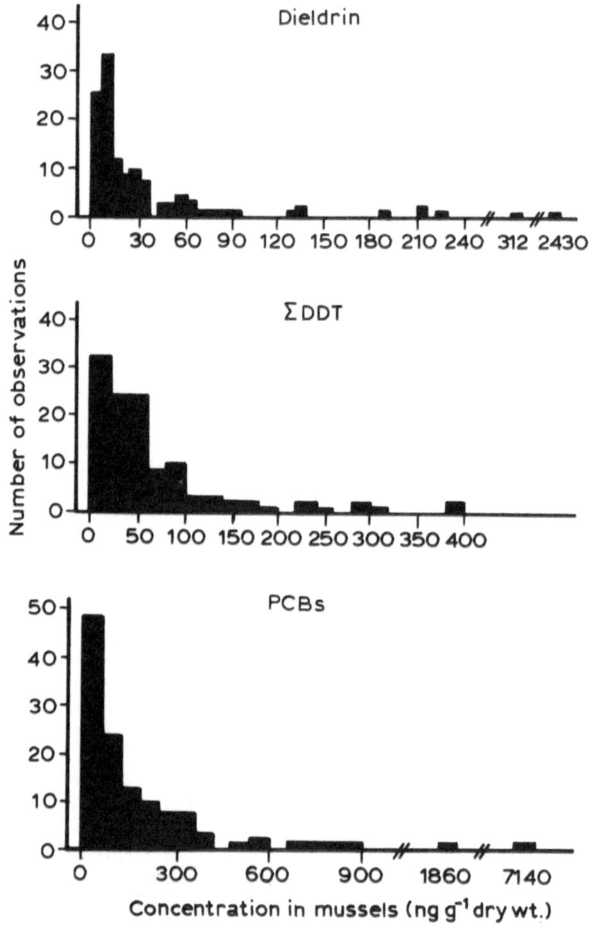

Fig. 61. Concentration-frequency distributions for dieldrin, ΣDDT and PCBs in mussels (*Mytilus edulis*) from the coast of Scotland. Note the differences and discontinuities in the horizontal scales. After Cowan (1981).

as a mothproofing agent in the wool and carpet industries. PCB hot-spots were found at Shandon in the Firth of Clyde and at Invergordon in the Cromarty Firth.

Phillips (1985b) employed the green-lipped mussel *Perna viridis* as a biomonitor of both trace metals and organochlorines in the subtropical waters of Hong Kong. The heavily-urbanised area of Victoria Harbour was shown to be generally contaminated by DDT and its metabolites, probably due to the industrial use of DDT in Hong Kong Island and Kowloon. By contrast, PCBs were found at detectable levels in *P. viridis* from only a few sites, although one of these (in Junk Bay, to the east of Victoria Harbour) was very heavily contaminated, and this led to further studies of PCB kinetics in green-lipped mussels (Tanabe *et al.*, 1987b; Kannan *et al.*, 1989). No HCB was found in these mussels, but isomers of HCH were noted in both this and a later survey (Phillips, 1990c), and varied significantly in abundance with season. Interestingly, the organochlorines found in local mussels in Hong Kong are precisely those which are present in quantity in the breast milk of ethnic Chinese females residing in the Territory (Ip and Phillips, 1989), and it is believed that seafoods represent the major source of such compounds to the Hong Kong population.

While biomonitoring data for hydrocarbons in bivalves are less abundant than those for organochlorines, several surveys may be mentioned here (in addition to the US Mussel Watch work, discussed above). Burns and Smith (1980, 1981) studied hydrocarbons in Port Phillip and Western Port Bays in Australia, and provided a convincing case for the use of bivalves as biomonitors of these compounds, rather than relying on their quantification through the analysis of receiving waters. Chronic contamination of the more urbanised and/or industrialised areas of Port Phillip Bay was noted in these studies, and temporal variations in the hydrocarbon levels of mussels were significant. The use of transplanted mussels for certain purposes was recommended (see also Phillips and Segar, 1986), and techniques for this are readily available (Young *et al.*, 1976; Curran *et al.*, 1986; Green *et al.*, 1986). The regular surveillance of water quality for hydrocarbons through the use of bivalves has also been undertaken at particular potential point sources, such as oil production platforms (Tibbetts *et al.*, 1982), and this type of application is reminiscent of the recommendations of Burns and Smith (1981).

More recently, bivalves have been employed to monitor hydrocarbon distributions in both Spain and Brazil. Risebrough *et al.* (1983) reported significant contamination of several bivalves by hydrocarbons, ΣDDT (including *o,p'*-DDD, surprisingly) and PCBs in the Ebro Delta of north-east Spain; HCB, endrin and γ-chlordane were also present. While many biogenic hydrocarbons were present in the bivalves analysed (presumably being derived through the food chain from phytoplankton), there was clear evidence of anthropogenic petroleum-derived hydrocarbons also; the triterpane and sterane compositions of samples confirmed this. It may be noted here that the ratios of various constituents of the hydrocarbon mixtures in bivalves and other biomonitors provide useful data on the sources of contamination, as crude oils vary substantially in their precise chemical composition (Grahl-Nielsen and Lygre, 1990).

Tavares *et al.* (1988) provided similar data for hydrocarbons, ΣDDT and PCBs in the clam *Anomalocardia brasiliana* from the Todos os Santos Bay, Bahia, Brazil. The northern and eastern portions of this Bay were found to exhibit significant contamination by petroleum-derived hydrocarbons, and this geographical distribution correlated to the pattern of industrialisation in the area. By contrast, the levels of both ΣDDT and PCBs in *A. brasiliana* were low, with relatively little contamination evident.

Further work on the use of clams as biomonitors of hydrocarbons in industrially-contaminated areas has been reported recently by Pereira *et al.* (1992). These authors employed the newly-introduced Asiatic clam *Potamocorbula amurensis* to monitor hydrocarbons in northern San Francisco Bay, which is exposed to significant pollution from several refineries. They demonstrated that the hydrocarbon compositions in sediments and in *P. amurensis* were broadly similar and were dominated by combustion-derived compounds. This suggested not only the presence of considerable contamination of the study area, but also that the clam could act as an efficient biomonitor for hydrocarbons in such circumstances. However, it is notable that no studies were undertaken of the temporal variability in hydrocarbon concentrations in the clams employed in this work.

Among the most recent data involving the use of bivalve molluscs as biomonitors of trace contaminants over large geographical areas are those from the reports of Lauenstein *et al.* (1990) and Sericano *et al.* (1990). Both these sets of data are derived from the so-called National Status and Trends Program of the National Oceanic and Atmospheric Administration (NOAA); this has replaced the original Mussel Watch Program undertaken in the USA in the mid-1970s. The methods employed in the NOAA project are generally similar to those used previously (see Goldberg *et al.*, 1978), involving the sampling of several species of bivalves at various coastal locations for the analysis of metals and trace organic contaminants. Examples of the data from the Gulf coast are shown in Fig. 62, from which it will be noted that pronounced location-based differences in ΣDDT and PCB concentrations were present in the oysters (*Crassostrea virginica*) employed as biomonitors in the Gulf of Mexico. Differences in organochlorine abundance in *C. virginica* also occurred between years (Sericano *et al.*, 1990), but in general these were minor relative to the

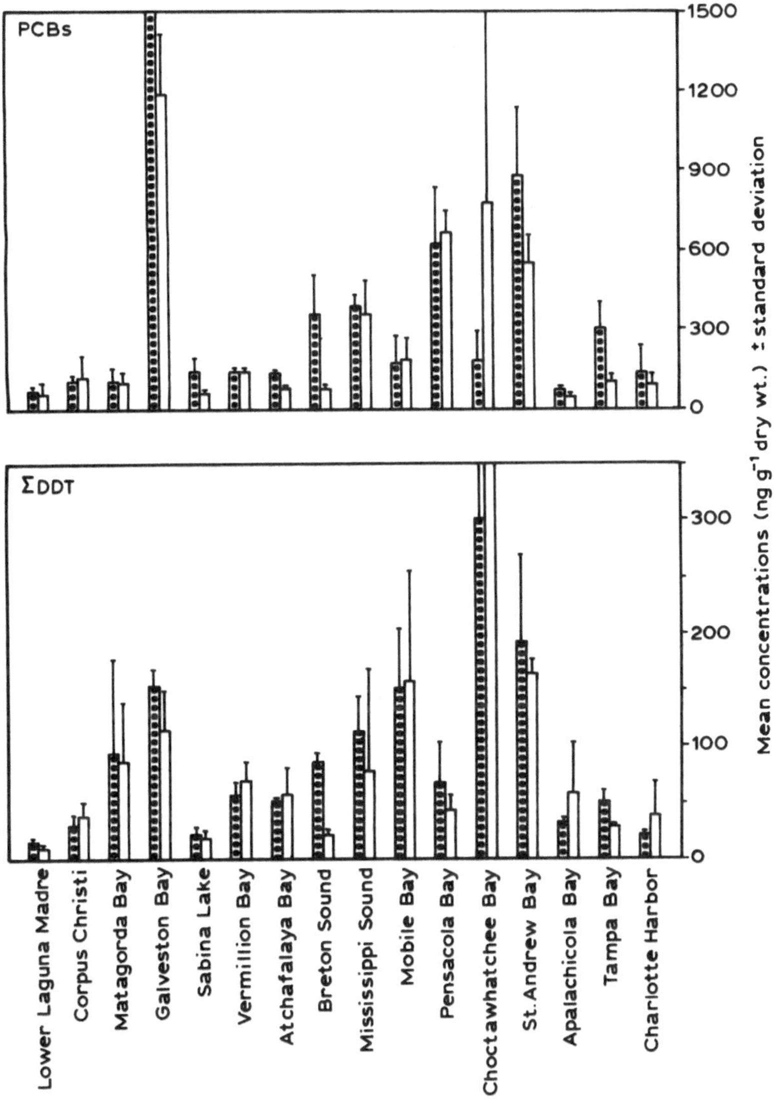

Fig. 62. Mean concentrations of ΣDDT and total PCBs in oysters (*Crassostrea virginica*) sampled at selected sites during 1986 (stippled bars) and 1987 (open bars) in the Gulf of Mexico. After Sericano *et al.* (1990).

site-to-site variations in organochlorine concentrations found in any one year of study.

While the use of finfish as biomonitors of trace organic contaminants in aquatic ecosystems has been widespread, data from such studies should be interpreted with caution in many instances. Phillips (1978, 1980, 1986) has discussed the relative merits of the use of sessile or sedentary and non-sedentary biomonitors, and has emphasised the difficulties involved in interpreting data from the analysis of contaminants in organisms which are known to exhibit significant mobility. Essentially, these involve the inability of researchers using such methods to distinguish between spatial and temporal effects on contaminant bioaccumulation. Thus, there is no guarantee that a particular finfish taken at a given location will accurately reflect the contamination of that site, as its past history with respect to movement (and hence in terms of its spatial and temporal exposure to contaminants) is unknown.

Several authors dealing with the use of finfish as biomonitors of trace metals recognised this in early studies (e.g. Johnels *et al.*, 1967; Dix *et al.*, 1976) and recommended the use of territorial species of finfish to attempt to overcome the problem. In some cases, this recommendation has been followed by researchers using finfish as biomonitors of organochlorines. Thus, for example, Moilanen *et al.* (1982) used pike (*Esox lucius*) as a biomonitor of DDT, PCBs and chlordane in waters of the Turku Archipelago in Finland, and roach (*Rutilus rutilus*) or perch (*Perca fluviatilis*) were chosen as biomonitors of pesticides and PCBs in West German rivers because they were known to be territorial (Brunn and Manz, 1982; Schüler *et al.*, 1985).

In other instances, however, the advantages of using territorial species of finfish have been ignored by researchers, and data resulting from studies of migratory (or at least, mobile) species have been difficult to interpret unequivocally because of this factor. Nevertheless, studies of organic contaminants in finfish species have contributed significantly to our present understanding of the factors which may affect the bioaccumulation of trace organic contaminants in organisms, and these aspects will be emphasised in the following discussions.

The most important parameter affecting the uptake and persistence of organochlorines in biota is undoubtedly lipid content. As noted by Phillips (1978, 1980, 1986), lipid contents and distributions in organisms not only influence the partitioning of organochlorines between tissues in finfish and other species, but also contribute to variations in organochlorine abundance between different individuals of the same species. Furthermore, variations in lipid content often constitute the primary cause of observations connecting changes in organochlorine levels in a given species with other factors, such as size or sex.

Whether the overriding influence of lipid contents on the bioaccumulation of organochlorines by biota is matched by a similar dependence of the uptake and retention of hydrocarbons appears less certain, however. Some authors have indeed reported a covariation between lipids and the accumulation of hydrocarbons (e.g. Geyer *et al.*, 1985), and such a relationship would be expected on theoretical grounds, at least for the more hydrophobic compounds. However, other reports note no such correlation (e.g. Coates *et al.*, 1986), and it is probable that the metabolism of petrogenic hydrocarbons by biota (see Chapter 9) serves to at least partially mask the effects of lipid contents on their bioaccumulation.

Certain examples merit citation here with respect to the impacts of biological and environmental variables on the accumulation of organochlorines by finfish. The factors causing seasonal fluctuations of organochlorines in biota (i.e. lipid changes correlated to gametogenesis and spawning, and temporal variations in the delivery of organochlorines to aquatic environments) have already been considered, using data on bivalves. An additional example for finfish is shown in Fig. 63, from the data of Olsson *et al.* (1978) on roach (*Rutilus rutilus*) from Lake Roxen in Sweden. In this instance, the seasonal increases in PCBs evident in the spring of most years did not depend on changes in lipid contents of the samples, but were thought to be due to the transport of PCB-contaminated sediments to the lake from inflowing streams at the time of ice-melt in Sweden. These data are reminiscent of the results of Södergren *et al.* (1972) mentioned earlier in the present chapter, and serve to emphasise the effects of climate and runoff on contaminant abundance in aquatic ecosystems.

Many examples of the impacts of fish size on organochlorine concentrations in finfish are available, and some of these have been reviewed by Phillips (1980, 1986). The most frequently reported effect involves an increase in organochlorine concentrations with size of the fish within discrete populations, and this is usually correlated to increases in lipid contents of fish with size or age. However, in a few instances, inverse relationships have been noted between fish size and organochlorine concentrations (e.g. see Hattula *et al.*, 1978; van den Broek, 1979; Klauda *et al.*, 1981; Ravid *et al.*, 1985). These may be caused either by the decrease of lipid contents with size in fish (which is unusual, but has been documented for a few species; see Hattula *et al.*, 1978), or through other

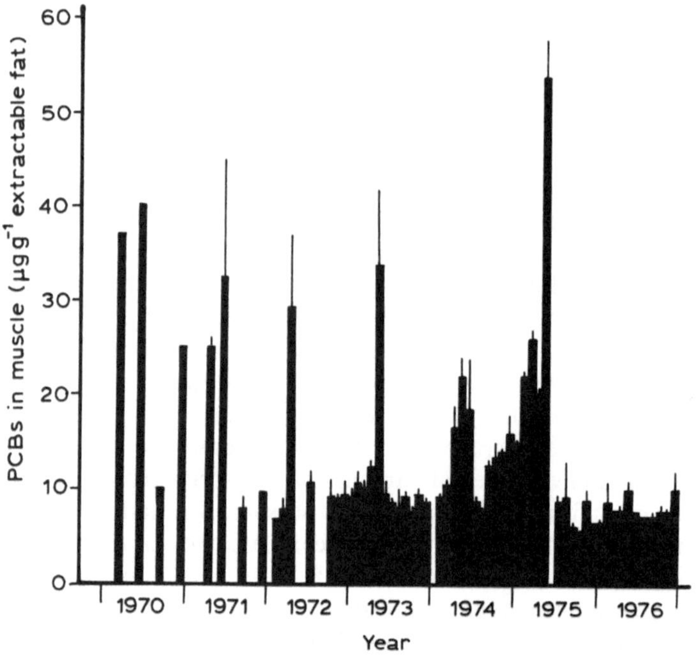

Fig. 63. Concentrations of PCBs (means or means and standard errors, $\mu g\ g^{-1}$ in extractable fat) in the muscle of roach (*Rutilus rutilus*) from Lake Roxen in Sweden between 1970 and 1976. After Olsson *et al.* (1978).

mechanisms, such as a change in diet (and therefore in exposure to organochlorines) with fish size.

In certain instances, the sex of finfish is also known to affect their bioaccumulation of organochlorines. It is generally thought that female fish lose greater quantities of organochlorines to their developing eggs than do males to milt, although the extent to which this affects the concentrations of organochlorines in the other tissues appears to vary between species (see Anderson and Everhart, 1966; Anderson and Fenderson, 1970; Ballschmiter *et al.*, 1981; Zabik *et al.*, 1982). However, DeFoe *et al.* (1978) noted that the sex-based differences in bioaccumulation of PCBs in the fathead minnow (*Pimephales promelas*) were based not on differences in the amounts of PCBs lost to gametes in the two sexes, but on the greater lipid contents of females than males (Fig. 64). This appears to be unusual amongst fish.

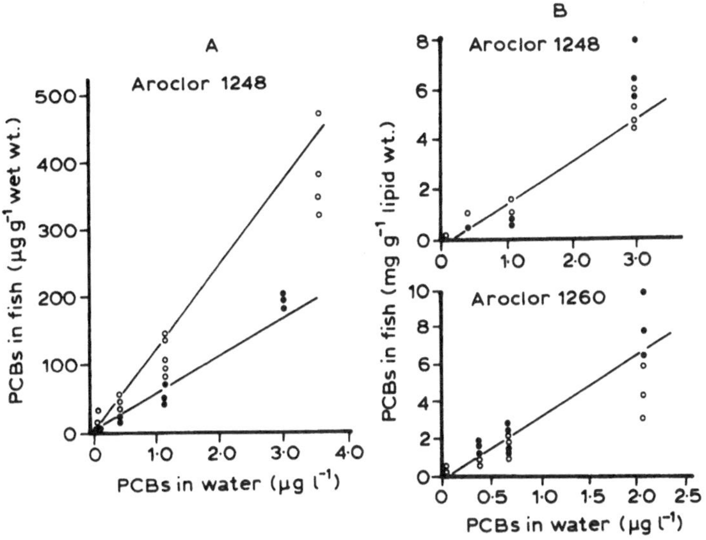

Fig. 64. (A) The net uptake of PCBs from exposure to Aroclor 1248 by male (●) and female (○) fathead minnows, *Pimephales promelas*. Note the wet weight basis for concentrations in the fish. (B) The net uptake of Aroclors 1248 and 1260 by the same fish, based on lipid weights rather than wet weights, demonstrating the effects of lipid. After DeFoe *et al.* (1978).

Environmental factors which influence organochlorine uptake by fish include salinity and water temperature. Murphy (1970) showed that mosquitofish (*Gambusia affinis*) accumulated greater amounts of DDT and its metabolites at lower salinities, although this report does not seem to have been followed up by additional studies. Several authors have reported that DDT and PCBs are accumulated more rapidly by fish at higher water temperatures (Reinert *et al.*, 1974; Edgren *et al.*, 1979; Powell and Fielder, 1983), and the elimination of HCB and mirex from rainbow trout (*Salmo gairdneri*) is also more rapid at higher temperatures (Niimi and Palazzo, 1985).

The early studies of organochlorines in marine mammals cited previously in this Chapter have been significantly extended by more recent work, and a few investigations of hydrocarbons in marine mammals have also been reported. Work on seals concerned not only their potential as biomonitors, but also the changes in organochlorines with sex and age, and the impacts of these compounds on reproduction and survival. Recent "seal plagues" in Scandinavia and Europe have provided additional impetus for such studies, as some authors have speculated that PCBs or other persistent contaminants may have contributed to the fatalities, perhaps through effects on the immune systems of seals. Other factors may also be involved, however (Harwood and Grenfell, 1990; Lavigne and Schmitz, 1990), and no unequivocal link between environmental contamination and the incidence of virus-related deaths of seals has been produced to date.

There can, however, be no doubt that certain seal populations (and perhaps sea lions and dolphins also) suffer significant reproductive impacts from accumulated PCBs. This has been amply demonstrated by Helle *et al.* (1976a,b) for ringed seals (*Pusa hispida*) from the Baltic Sea, and by Reijnders (1986) for common seals (*Phoca vitulina*) in the Wadden Sea (see also data for sea lions in DeLong *et al.*, 1973; Buhler *et al.*, 1975).

The more recent work on changes in organochlorine levels in marine mammals with age and sex has both confirmed and extended the initial conclusions of Gaskin *et al.* (1971) presented earlier in this Chapter (see Addison *et al.*, 1973; Frank *et al.*, 1973; Gaskin *et al.*, 1973; Addison and Smith, 1974; Addison and Brodie, 1977; Reijnders, 1980; Donkin *et al.*, 1981; Tanabe *et al.*, 1981, 1982c; Helle *et al.*, 1983; Ronald *et al.*, 1984; Perttila *et al.*, 1986; Law *et al.*, 1989; Morris *et al.*, 1989; Skaare *et al.*, 1990). At birth, marine mammals exhibit relatively low concentrations of organochlorines, these having been derived transplacentally *in utero* from the mother. Tanabe *et al.* (1982c) have shown, in typically elegant fashion, that the more hydrophobic contaminants are in fact transferred to the foetus rather inefficiently because of the predominance of polar phospholipids in the blood of marine mammals. However, the young suffer considerable further contamination through the accumulation of residues in milk at suckling, and later from solid food (finfish, in most species).

Organochlorine concentrations thus increase significantly with age in young marine mammals. No differences in contaminant levels exist between the sexes at this stage, prior to sexual maturity being attained. Thereafter, however, differences between the sexes begin to emerge, at least in populations (or portions thereof) which are breeding successfully, as mature females lose residues through both parturition and lactation. Sexually mature males thus exhibit higher concentrations of organochlorines than do breeding females of similar age.

This pattern is disturbed where breeding is not successful (Helle *et al.*, 1976a,b; Reijnders, 1986), as the female animals cannot lose residues to the foetus *in utero* or through lactation. As noted in certain finfish species, therefore, the age-related and sex-based differences in organochlorine levels of marine mammals are not of a fundamental biochemical nature (or growth-related, as in mercury in sharks), but are a simple function of the differences between the sexes in the loss of contaminants through reproduction.

Rather less is known of organochlorines in cetaceans (especially the larger species) than in seals, and on biological grounds, certain differences may be expected (i.e. feeding habits and the frequencies of producing young differ between most seals and the larger whales). The smaller database for cetaceans is probably a function of sampling difficulties in many cases; biopsy techniques (see Taruski *et al.*, 1975) have apparently not been frequently used, and relatively little work was completed during the period of widespread commercial whaling. Sex-based differences in organochlorine levels appear to be less marked in larger cetacean species than in seals, and one report concerning sperm whales suggests the existence of higher concentrations of ΣDDT and PCBs in females than males (Aguilar, 1983), although this is probably an artifact of the small sample numbers employed in this study.

While it is clear that marine mammals exhibit significant mobility, and certain species undertake regular and sometimes extended migrations, these animals nevertheless reflect broad differences between regions in the abundance of organochlorines. Thus, for example, the variations in residue levels between seal populations around British coasts, first observed by Holden (1972; see Fig. 55) have been confirmed more recently (Law *et al.*, 1989; Morris *et al.*, 1989). Such differences are considerably more dramatic when viewed on a global scale (Fig. 65), and provide interesting insights into the efficiencies of global atmospheric transport mechanisms for organochlorines. It may be anticipated that the profiles shown in Fig. 65 will alter subtly with time, as the global utilisation of organochlorines continues to shift from the northern hemisphere to the tropics. The use of marine mammals as biomonitors of such temporal changes in the contamination of both coastal and remote environments by organochlorines appears fully justified (Tanabe *et al.*, 1983b; Addison *et al.*, 1984; Loganathan *et al.*, 1990).

Much less is known concerning the concentrations of petroleum-derived hydrocarbons in marine mammals, although this would seem a profitable topic for study. Marine mammals certainly exhibit P-450 linked microsomal enzyme activity which may be induced by hydrocarbons (Engelhardt, 1983; Watanabe *et al.*, 1989), but they nevertheless accumulate significant quantities of compounds such as PAHs (Engelhardt *et al.*, 1977; Shaw *et al.*, 1978; Hellou *et al.*, 1990). At least one study has suggested that

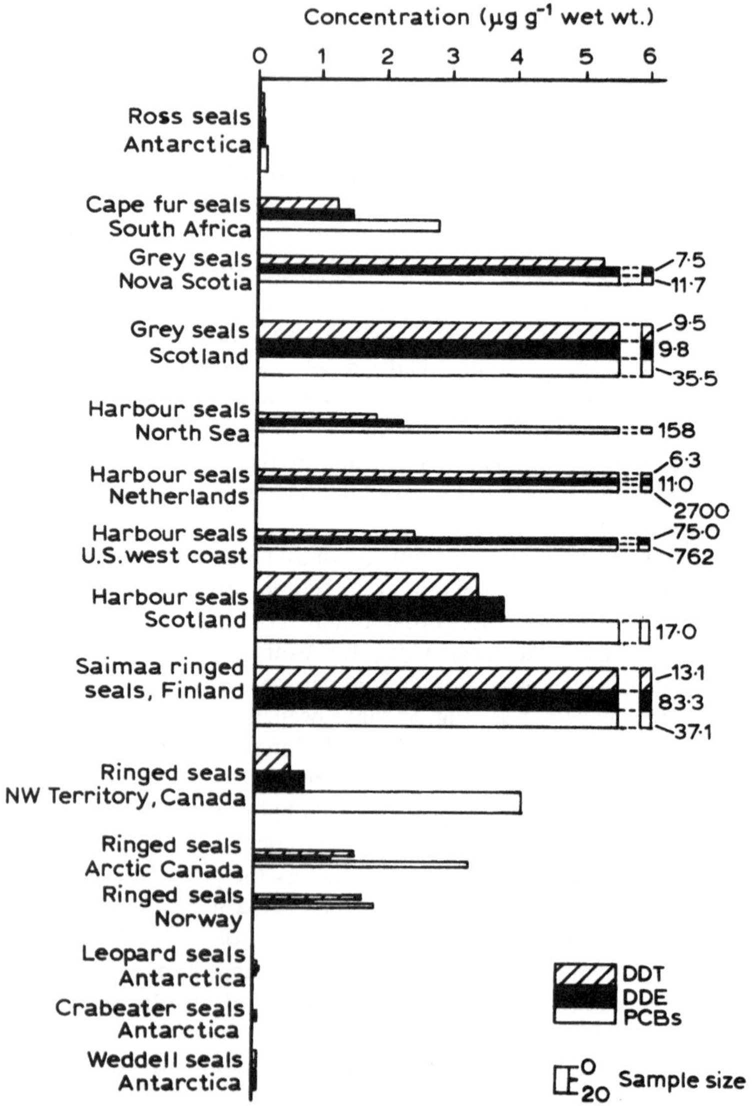

Fig. 65. Comparisons of reported concentrations of DDT, DDE and PCBs ($\mu g\ g^{-1}$ wet weights) in the blubber of seals of various species from different geographical locations. Width of the bar denotes sample size in each case. After McClurg (1984).

this may be of toxicological significance; Martineau *et al.* (1988) have correlated the presence of tumours in Beluga whales from the Gulf of St. Lawrence to levels of benzo(*a*)pyrene in their tissues, although a causative link *per se* remains to be proven.

C. SUMMARY

While the sources of organochlorines and hydrocarbons in the ecosphere are almost completely distinct, their global cycling is similar in many respects. Both groups of contaminants tend to be found at enriched levels in coastal locations subjected to urban and industrial development, but are also present (by virtue of their atmospheric transport) in remote sites. The considerable toxicity of each of these classes of contaminants to aquatic biota has provided no little impetus to studies of their regional and global distributions. Many of these studies have preferred the use of biota to the analysis of either water or sediments, although the available database for biomonitors remains considerably more robust for organochlorines than for petrogenic hydrocarbons. This is partly due to analytical difficulties in separating the complex mixtures of hydrocarbons found in aquatic environments, and it is to be hoped that advances in the analytical arena will give rise to improvements in the biomonitoring database relating to hydrocarbons in the near future.

Many of the early attempts to use biomonitors to quantify organochlorines in aquatic ecosystems relied on the use of bivalve molluscs, and this trend has continued and been extended in recent years to the study of hydrocarbons also. While the lipid contents of bivalves are not great, and hydrophobic contaminants are thus not particularly highly concentrated by these species, their other attributes as biomonitors more than compensate for this, and there can be little doubt of their usefulness to monitor these contaminants in coastal environments in particular.

Both macroalgae and crustaceans have also been employed as biomonitors of trace organic contaminants in inland and coastal waters, but their use has not been as frequent as that of bivalve molluscs. Macroalgae exhibit unusually low lipid contents, but may nevertheless be worthy of additional study, as their capacity to metabolise organochlorines appears to be low. Their use in investigations of hydrocarbons is less promising, however, due to the high levels of biogenic hydrocarbons found in primary producers and the interference of these in the analysis of anthropogenically-derived compounds. Crustaceans also merit further study, and it is notable that almost nothing is known of organic contaminants in barnacles, which are excellent biomonitors of trace metals and radionuclides (see Chapter 5).

Finfish have enjoyed much emphasis as potential biomonitors of organochlorines and hydrocarbons, but this is considered ill-placed in some instances. While territorial species in both inland and coastal waters may reflect real spatial trends in contamination, the use of more mobile or migratory finfish gives rise to problems in data interpretation, and these have not been adequately addressed to date. Nevertheless, studies of finfish have provided much useful information on the effects of both biological and environmental factors on the bioaccumulation of trace organic contaminants, and these have assisted in improving biomonitoring techniques as a whole.

Marine mammals have also been extensively studied, and a robust database is available for organochlorines in these species, although much less is known of hydrocarbons in either pinnipeds or cetaceans. The changes in organochlorine concentrations in seals and dolphins with age and sex have been particularly well documented, and these are of toxicological importance with respect to PCB impacts in some seal populations at least. While none of these species can be employed as can bivalve molluscs to delineate fine-scale geographical trends in the contamination of aquatic environments, they are nevertheless of considerable use to identify gross regional differences in the abundance of organochlorines; further studies are needed on hydrocarbons.

Finally, it should be noted that while the overriding importance of tissue lipids in defining the bioaccumulation of organochlorines by organisms has been noted by many authors, its true biochemical basis has not been adequately studied. Certain authors have attempted investigations at a gross level of the relationships between the bioaccumulation of trace organic contaminants and lipid composition (e.g. Boon and Duinker, 1985; Kawai *et al.*, 1988), but much remains unknown and this would be a most profitable area for further study.

Chapter 7

Monitoring Trace Contaminants in Freshwater Ecosystems

A. INTRODUCTION

Data concerning the monitoring of trace contaminants in freshwater ecosystems have been included elsewhere in this text, where these are relevant to particular topics under discussion. However, the present chapter provides a consolidated synthesis of studies to date on trace metals, organochlorines and hydrocarbons in freshwater environments, and contrasts the approaches taken in such environments with those used in marine or estuarine ecosystems.

There can be no doubt that studies of trace contaminants in freshwater organisms have contributed markedly to the development of the biomonitoring concept. Thus, for example, the classic paper by Johnels *et al.* (1967) on the use of pike (*Esox lucius*) as a biomonitor of mercury laid a foundation for much of the later work of a similar nature in marine waters (e.g. see Dix *et al.*, 1976; Krom *et al.*, 1990). Similarly, the studies of Hunt and Bischoff (1960) on DDD in the biota of Clear Lake in California provided much useful early data on species of potential use as biomonitors of organochlorines in freshwater environments. In addition to such field investigations, laboratory-based studies employing freshwater species to elucidate aspects of the kinetics of trace contaminants have been numerous, and at least some of these have been of general relevance to marine studies also.

A brief review of some of the more important literature published to date on trace contaminants in freshwater environments is presented here; laboratory-based studies being considered first, followed by field data. Thereafter, conclusions are provided on the present state of the art in monitoring trace contaminants in freshwater species.

B. LABORATORY STUDIES

(i) Trace Metals

The uptake of trace metals by freshwater organisms has been subjected to study by many authors. More data are available for fish species than for invertebrates, and surprisingly, work on freshwater mussels or other molluscs has been sparse.

As noted for marine environments, the net uptake of trace elements by primary producers in freshwater ecosystems has been shown to be significant. Thus, for example, Conway and Williams (1979) found that cadmium was rapidly accumulated by the freshwater diatoms *Asterionella formosa* and *Fragilaria crotonensis*. Interestingly, data from studies in the light and the dark suggested that the uptake process might be of a partially active nature in *A. formosa* (perhaps indicating that cadmium was taken up through a major ion pump; see Chapter 5 and Simkiss and Taylor, 1989a). *A. formosa* was also more sensitive to the toxic effects of the metal than was *F. crotonensis*. Wang and Wood (1984) noted that nickel was also accumulated efficiently by freshwater microalgae, this process being affected by pH and also by the presence of cobalt or humic acids. While such data are relevant to the entry of trace elements into the food web (see also Vighi, 1981; Tarifeño-Silva *et al.*, 1982), no concerted attempts have been made to date to employ microalgae as biomonitors of trace metals in freshwater environments. The differential uptake of metals by different species would be likely in any event to render any such efforts meaningless, were they to be based upon field collections of natural algal assemblages at different sites.

Studies on the uptake of trace metals by larger freshwater plant species have been driven either through a desire to use these species as biomonitors, or because of their potential for the treatment of metal-contaminated wastewaters. The kinetics of metal uptake have been studied in the water hyacinth *Eichhornia crassipes* by several authors (e.g. see Tokunaga *et al.*,

1976; Cooley and Martin, 1979; Chigno *et al.*, 1982; O'Keeffe *et al.*, 1984; Hardy and O'Keeffe, 1985; Hardy and Raber, 1985). Similarly, Nasu *et al.* (1983) have investigated the kinetics of cadmium and copper uptake in the duckweed *Lemna paucicostata*, and the uptake of ^{54}Mn by the aquatic moss *Rhynchostegium riparioides* has been studied (Hébrard and Foulquier, 1975; species named *Platyhypnidium riparioides* by these authors). This work has clearly shown that such freshwater macrophytes are capable of accumulating high concentrations of trace metals. However, in several instances, it has been demonstrated that significant interactions between metals occur at uptake, and this is likely to severely limit the usefulness of such species as biomonitors (see Phillips, 1980, for a detailed discussion of the impacts of metal interactions on biomonitoring studies).

As noted previously, laboratory investigations of the kinetics of trace metals in freshwater species of bivalves or other molluscs have been rare; this is remarkable, given the very considerable attention paid to their counterparts in estuarine and marine ecosystems. Millington and Walker (1983) undertook both laboratory and field studies of the uptake of zinc by the Australian freshwater mussel *Velesunio ambiguus*. They noted that this species exhibits avoidance of very high concentrations of zinc in solution (>20 mg litre^{-1}), by curtailing siphoning, reducing movement and by valve closure. This is similar to findings for the marine mussel *Mytilus edulis* exposed to copper, but the response of *M. edulis* occurs at much lower (and more environmentally meaningful) concentrations of copper (Davenport, 1977; Davenport and Manley, 1978; Davenport and Redpath, 1984). In addition, Millington and Walker (1983) found that the concentrations of zinc accumulated by *V. ambiguus* in laboratory dosing studies were extremely variable, and there was evidence that this species might regulate its tissue zinc levels to some degree (see Maher and Norris, 1990, and the discussion of field data for *V. ambiguus* below).

Graney *et al.* (1983, 1984) investigated the potential of the Asiatic clam *Corbicula fluminea* to act as a biomonitor of trace elements, undertaking careful studies of the effects of several factors on the accumulation of cadmium in particular. The factors affecting cadmium uptake included pH, substrate type and temperature; despite their impacts, however, these authors concluded that *C. fluminea* exhibited promise as a biomonitor in freshwater ecosystems. Evidence from both these studies and the work of Tatum (1986) suggests that *C. fluminea* accumulates significant quantities of metals from contaminated sediments.

Cadmium kinetics have also been studied in the freshwater mussel *Dreissena polymorpha* in Germany (Bias and Karbe, 1985). These investigations revealed very considerable uptake of radioactive cadmium isotopes by the shell of *D. polymorpha*, with less marked uptake into soft parts (Fig. 66(A)). However, the cadmium adsorbed to the shell was lost relatively rapidly upon a return to clean water, whereas that in the soft parts exhibited a long half-life (Fig. 66(B)). This was consistent with the finding that most of the cadmium in the shell of the mussel was bound to the periostracum, which exhibited concentration factors for the element of greater than 70,000 under some conditions. Similar findings have been reported for estuarine and marine species of molluscs by several authors (see Phillips, 1980).

Among gastropods, the uptake of lead from solution has been studied in the freshwater gastropod *Viviparus viviparus* by Fantin *et al.* (1982), who noted that the mantle accumulated the highest concentrations of lead in this species. Histological effects of lead contamination were observed in the digestive gland upon exposure of *V. viviparus* to high external levels of lead in solution.

Much more information is available for the kinetics of metal uptake and excretion in freshwater fish species. Salmonids in particular have enjoyed much attention, and the rainbow trout *Salmo gairdneri* has been the most popular species for study. Authors have investigated the kinetics of several elements in *S. gairdneri*, including cadmium (Pärt and Wikmark, 1984; Pärt *et al.*, 1985), chromium (van der Putte *et al.*, 1981; van der Putte and Pärt, 1982), copper (Dixon and Sprague, 1981), lead (Wong *et al.*, 1981), selenium (Kaiser *et al.*, 1979), and zinc (Lovegrove and Eddy, 1982). However, the most extensive database by far concerns the kinetics of inorganic and organic forms of mercury in *S. gairdneri* (e.g. see MacLeod and Pessah, 1973; Phillips and Buhler, 1978; Boudou *et al.*, 1980; Rodgers and Beamish, 1983; Boudou and Ribeyre, 1984; Ribeyre and Boudou, 1984).

Other species of trout have also been tested for trace metal bioaccumulation (e.g. see Huckabee *et al.*, 1978; MacCrimmon *et al.*, 1983; O'Grady and Abdullah, 1985), and further studies of this nature have been undertaken in species such as the sunfish *Lepomis gibbosus* and *L. macrochirus* (Merlini and Pozzi, 1977; Cember *et al.*, 1978; Anderson and Spear, 1980; Lemly, 1982); the largemouth bass *Micropterus salmoides* (Lemly, 1982); and the pike *Esox lucius* (Lockhart *et al.*, 1972).

It appears from these studies that the tissue distribution of accumulated trace metals in freshwater fish is generally similar to that in marine fish species. Most metals are accumulated preferentially in the liver, kidney, spleen and gills of exposed fish, and concentrations generally remain low

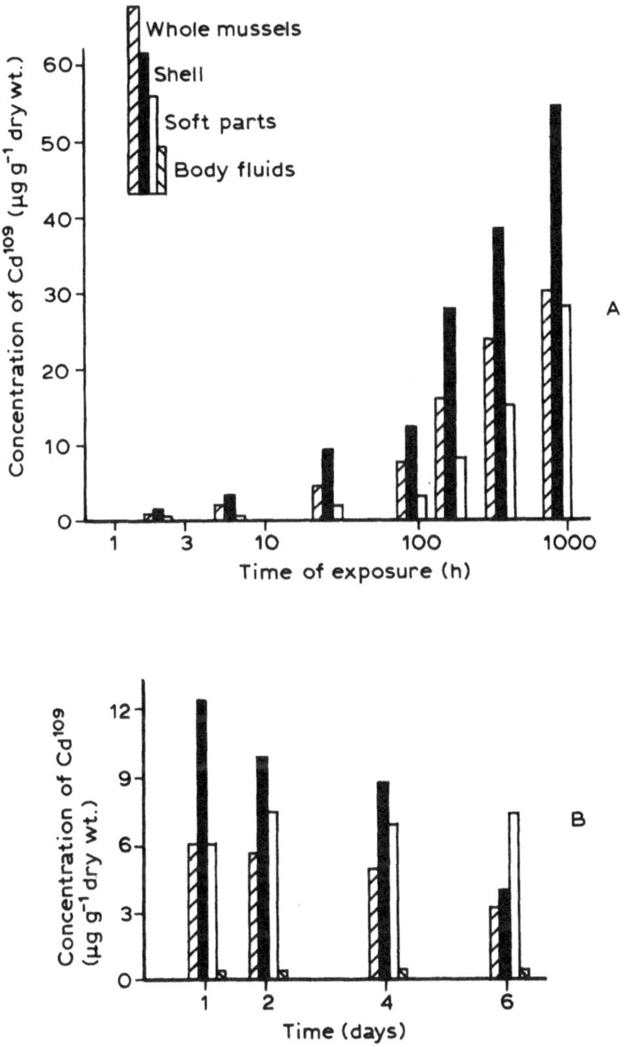

Fig. 66. Accumulation (A) and elimination (B) of ^{109}Cd by whole animals, shells and soft tissues of the freshwater mussel *Dreissena polymorpha*. Data on accumulation are for mussels exposed to 50 μg^{-1} of the nuclide in solution. Data on elimination are for a separate group of mussels exposed to 4.85 μg^{-1} for 43 days. All concentration data are shown as $\mu g\,g^{-1}$ dry weights. After Bias and Karbe (1985).

(and some metals may be subject to regulation) in the axial muscle tissues. However, certain forms of trace metals are significantly accumulated in the muscle tissues of freshwater fish species, and these include lipophilic forms such as tetramethyllead (Wong *et al.*, 1981) and methylmercury (e.g. Lockhart *et al.*, 1972; Ribeyre and Boudou, 1984). Several factors have been shown to influence the bioaccumulation of metals by freshwater fish, including age or size of the fish (Kaiser *et al.*, 1979; Anderson and Spear, 1980; Boudou and Ribeyre, 1984); diet (Phillips and Buhler, 1978; MacCrimmon *et al.*, 1983); and water quality and temperature (MacLeod and Pessah, 1973; Merlini and Pozzi, 1977; Cember *et al.*, 1978; Boudou *et al.*, 1980; Rodgers and Beamish, 1983). Such factors are important in terms of attempts to employ freshwater fish species as biomonitors of trace metals, as will be discussed below.

(ii) Trace Organic Contaminants

Investigations of the uptake of trace organic contaminants in freshwater organisms have been restricted largely to fish, and have almost exclusively involved organochlorines; very little work has been published on the uptake kinetics of hydrocarbons in freshwater biota.

Several studies on the bioaccumulation and metabolism of organochlorines by invertebrates nevertheless deserve mention here. Early work by Södergren (1968) on the uptake of DDT by the unicellular green alga *Chlorella* sp. was extended by Wang *et al.* (1982) to investigations of the uptake of three PCB homologues by *Chlorella fusca*. Organochlorines are taken up rapidly by unicellular freshwater algae, but food chain transfer probably does not predominate in the contamination of higher trophic levels, as lipid-water partitioning is thought to be the most important uptake mechanism in most cases (see Chapter 6 and Hamelink *et al.*, 1971; Reinert, 1972; Hamelink and Waybrant, 1976). The uptake of PCBs by freshwater algae follows the expected pattern (Wang *et al.*, 1982), homologues of higher degrees of chlorination being accumulated to greater extents in general (reflecting their lower water solubilities and higher octanol:water partition coefficients; see Chapter 6).

The bioaccumulation of organochlorines is also rapid in invertebrate animals (e.g. Crosby and Tucker, 1971; Johnson *et al.*, 1971), concentration factors exceeding 10^4 being attained subsequent to short-term exposures to DDT or other organochlorines in solution. Invertebrates exhibit the capability to metabolise certain of the pesticides, such as DDT and aldrin; however, the products of this metabolism (DDE and dieldrin in the cases cited) may nevertheless be of significant persistence in the organism (e.g. see Johnson *et al.*, 1971; Khan *et al.*, 1972).

In at least some instances, pesticides may be taken up by invertebrates from contaminated sediments in addition to their uptake from solution. Thus, for example, the eastern oyster *Crassostrea virginica* has been shown to accumulate significant quantities of the pesticide Kepone from contaminated sediments of the James River, under conditions of low salinity (Morales-Alamo and Haven, 1983). Once again, this attribute has also been noted in marine environments, and sediments are viewed as both a sink and a source for organochlorines in aquatic ecosystems.

Investigations of the uptake of organochlorines in fish commenced in the early 1960s, with Holden (1962) reporting on the accumulation of DDT from solution by the brown trout *Salmo trutta*. He noted that DDT was concentrated preferentially in tissues with a high lipid content, and this general rule has since been confirmed in many species from both freshwater and marine environments (see reviews by Phillips, 1978, 1980, 1986). As is the case for trace elements, work on organochlorines in freshwater fish has been dominated by studies involving salmonids (e.g. see Macek and Korn, 1970; Macek *et al.*, 1970; Branson *et al.*, 1975; Tooby and Durbin, 1975; Niimi and Cho, 1981; Skea *et al.*, 1981; Niimi and Oliver, 1983; Niimi and Palazzo, 1985), and these have incorporated work on a variety of compounds, including DDT, HCB, lindane (γ-HCH), mirex, and PCBs.

Laboratory investigations of PCB kinetics in fish, including both salmonids and other species, have provided particularly useful data on the persistence of individual PCB homologues in biota. Several early studies indicated that freshwater fish accumulate PCBs efficiently from solution or from food, and that the toxic impacts of PCBs depend on the life-stage of the fish and other factors (Nebeker *et al.*, 1974; Frederick, 1975; Hansen *et al.*, 1975; Sanborn *et al.*, 1975). While PCB homologues of higher chlorination are generally accumulated more efficiently than those of lower chlorination (e.g. see Sanborn *et al.*, 1975; Bengtsson, 1980; Bruggeman *et al.*, 1981; Niimi and Oliver, 1983), the position of the chlorine atoms on the biphenyl molecule is of importance in defining the precise levels of bioaccumulation (Bruggeman *et al.*, 1981; Niimi and Oliver, 1983) and biological selectivity may exist in certain instances (see Hansen *et al.*, 1976, and discussion in Phillips, 1980).

In addition, the predominance of factors such as tissue lipid levels in determining the degree of bioaccumulation (see Chapter 6) should be re-emphasised here. An interesting example of lipid impacts was reported by DeFoe *et al.* (1978) for the fathead minnow (*Pimephales promelas*), where female fish accumulated greater quantities of PCBs than did males, due to their higher lipid contents.

Several factors other than lipid contents have also been shown to influence the bioaccumulation of organochlorines by freshwater fish. These include: animal size; water temperature; the presence or absence of other contaminants; metabolism of the compound taken up; and in certain cases, the presence of resistance created by previous exposure to a pesticide (Ferguson *et al.*, 1966; Macek *et al.*, 1970; Murphy, 1971; Watkins and Yarbrough, 1975; Niimi and Cho, 1981; Rao and Murty, 1982; Niimi and Palazzo, 1985).

The interactive effects of one contaminant on the uptake of a second have been suggested to be of particular importance to biomonitoring studies (see Chapter 6 and Phillips, 1978, 1980, 1990b), as this factor cannot be accounted for in any simple fashion in the field sampling of biomonitors. An example of such an effect, involving the uptake of DDT and dieldrin by the pyloric caeaca of rainbow trout (*Salmo gairdneri*), is shown in Fig. 67 (see Macek *et al.*, 1970). Fortunately, such examples are apparently rare, at least on the basis of our current knowledge of organochlorine bioaccumulation by aquatic organisms.

The alteration of pesticide uptake due to the acquired or inherited resistance of organisms is also of considerable relevance to biomonitoring, and merits further discussion. Ferguson *et al.* (1966) investigated the bioaccumulation of endrin by mosquitofish (*Gambusia affinis*) from two rivers in Mississippi differing in their contamination by this pesticide. They found that the rates of net uptake of endrin in the two fish populations were indistinguishable at comparable external endrin concentrations, but that the resistant mosquitofish simply tolerated accumulated residues better, probably through the exclusion of residues from the brain. By contrast, Watkins and Yarbrough (1975) found in similar studies that sensitive and resistant populations of *G. affinis* differed substantially in their net uptake of aldrin and dieldrin residues in certain tissues at high (but not at low) exposure concentrations (Fig. 68).

It is clear from these and other data that several mechanisms may be operative in producing resistance in fish populations which are chronically exposed to pesticides. These include the exclusion of residues from primary target organs such as the brain; the more rapid excretion of residues; and a decreased uptake of residues from solution. The last two of these mechanisms have important implications for the use of such species as biomonitors of pesticides in freshwaters, and this factor should

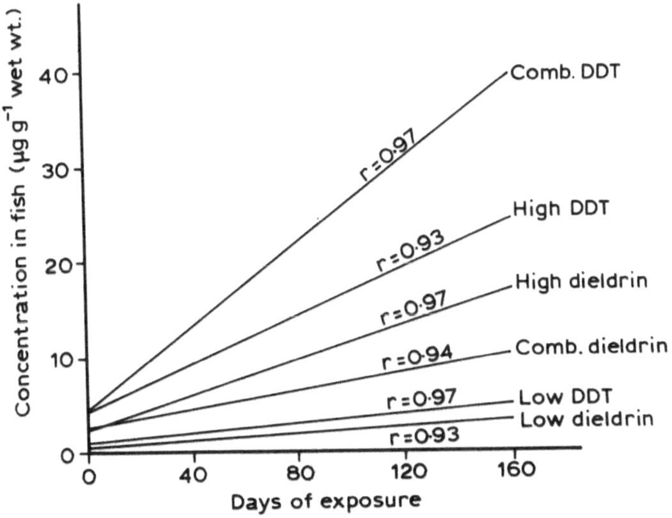

Fig. 67. Regressions and their coefficients of correlation (*r*) describing the accumulation of DDT and dieldrin in the pyloric caeca of rainbow trout, *Salmo gairdneri*, exposed to low or high doses of the pesticides (0.2 or 1.0 mg kg^{-1} $week^{-1}$) in food, or to a combination of the two pesticides at the high concentration. After Macek *et al.* (1970).

be considered in any decisions made on the selection of species for biomonitoring studies in chronically contaminated freshwater ecosystems.

C. FIELD INVESTIGATIONS

(i) Trace Metals

Concentrations of trace metals vary markedly with time in many freshwater environments. Such fluctuations may be even more extreme in fresh waters than in estuarine or marine ecosystems (e.g. see Duinker *et al.*, 1980; Habib and Minski, 1982; Jones, 1986), as temporal variations in river flows interact with changes in the magnitudes of sources of trace elements. In such a temporally fluctuating environment, an excellent case may be made for the use of biomonitors to elucidate average conditions of contamination. This section reviews the more important literature produced to date on field studies of biomonitors of trace metals in fresh waters, and contrasts the approaches taken in such environments to those employed in estuarine or marine ecosystems.

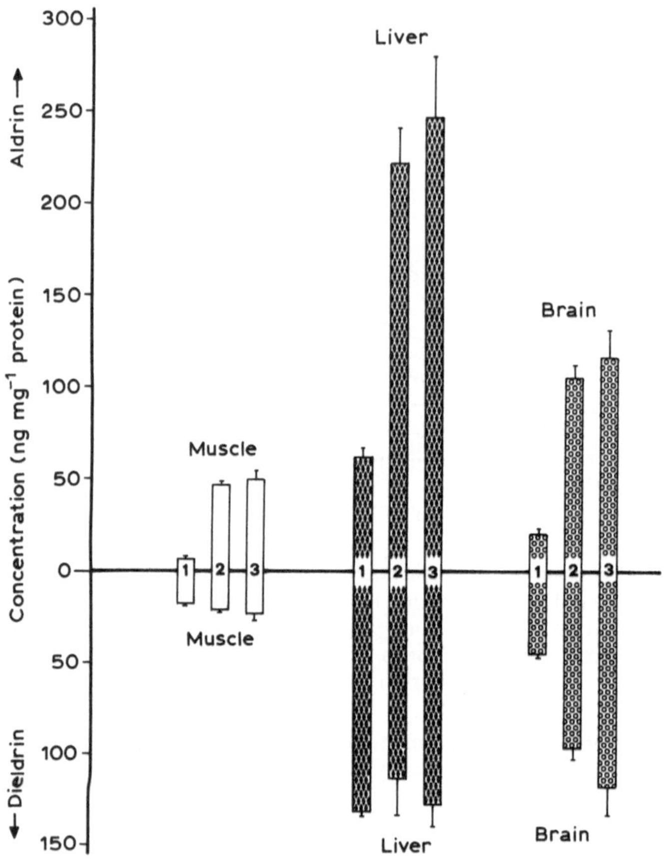

Fig. 68. Concentrations (means of five determinations, quoted as ng mg^{-1} protein ± standard errors) of ^{14}C-aldrin and ^{14}C-dieldrin in mosquitofish (*Gambusia affinis*) from populations differing in sensitivity to these pesticides. Population (1) is of resistant asymptomatic fish; (2) is of susceptible asymptomatic fish; and (3) is of susceptible fish showing symptoms of toxicity. After tabulated data in Watkins and Yarbrough (1975).

The use of aquatic plants of various types as biomonitors of trace metals in fresh waters is essentially comparable to that of macroalgae in estuarine or marine environments, and several authors have published data on such studies. The types of plants employed have included filamentous algae, red algae such as *Lemanea fluviatilis*, green algae of various species, vascular plants and aquatic mosses. Whitton *et al.* (1981) discussed the advantages of the use of such macrophytes as biomonitors of metals in freshwater

Table 23 Species and genera of aquatic plants recommended for use as biomonitors of trace elements in freshwater ecosystems (after Whitton *et al.*, 1981)

Species/genus	*Advantages and disadvantages*
Cladophora glomerata	Geographically widespread green alga; occurs in lakes and rivers. Growth often seasonal, however. Metal uptake kinetics reasonably well-studied.
Enteromorpha species	Green alga, abundant in Europe, but taxonomy uncertain. Seasonal growth in some parts of its range.
Lemanea fluviatilis	Red alga, widespread in Europe and also found in North America. Tolerant of high concentrations of metals. Adult plants are seasonal in many areas. Some taxonomic uncertainty exists.
Nitella flexilis	Widespread in the northern hemisphere in both rivers and lakes. Some data available on trace metal accumulation.
Amblystegium riparium	Cosmopolitan aquatic moss, found in both still and flowing waters. Tolerant of high nutrient levels.
Fontinalis antipyretica	Very common moss in the northern hemisphere; generally available year-round. Trace metal accumulation studied in several locations.
Fontinalis squamosa	Exhibits a restricted distribution compared to *F. antipyretica*; generally found in fast-flowing soft waters.
Rhynchostegium riparioides	Common northern hemisphere moss, found in both flowing and still waters; prefers nutrient-rich environments. Challenging to identify taxonomically.
Scapania undulata	Liverwort, of widespread distribution. Tolerates high concentrations of metals and generally available year-round.
Potamogeton pectinatus	Very cosmopolitan submerged flowering plant, found in both Europe and North America. Prefers nutrient-rich waters and tolerates brackish salinities; occurrence is seasonal, however.

ecosystems, essentially paraphrasing those previously identified in estuarine and marine studies of biomonitors (see Chapter 4 and Butler *et al.*, 1971; Haug *et al.*, 1974; Phillips, 1977, 1978, 1980). They recommended ten genera or species of potential use in the United Kingdom and elsewhere, each with its own inherent advantages and disadvantages (Table 23). The use of several of these species is discussed in greater detail below.

Stokes *et al.* (1983) investigated the uptake of mercury by filamentous algae collected on artificial substrates placed in five Canadian lakes, and also studied mercury levels of algae from natural substrates. There was evidence that the degree of contamination of the algae taken from artificial substrates correlated to the concentrations of mercury found in yearling perch from the same sites, suggesting that the collected algae were reflecting the same bioavailable source of mercury as the fish. However, although all samples (from both artificial and natural substrates) were dominated by the same two genera of filamentous green algae (*Mougeotia* spp. and *Spirogyra* spp.), no account was taken of the impacts of species differences on mercury accumulation at the various sites investigated. Studies in marine environments would suggest that this approach may be invalid, as species are known to vary considerably in their capacities to accumulate metals. A similar problem exists with attempts to monitor trace elements by the use of aufwuchs (e.g. see Newman *et al.*, 1985).

Studies on the red alga *Lemanea fluviatilis* also suffer from taxonomic uncertainty surrounding this species (Harding and Whitton, 1981). However, *L. fluviatilis* appears to be extremely tolerant of high concentrations of metals, and may therefore be of particular use in streams draining metalliferous areas. Harding and Whitton (1981) undertook careful studies of the correlation between the concentrations of cadmium, lead and zinc in this alga and those in its external aquatic environment. This work provided good evidence of the capacity of *L. fluviatilis* to act as an efficient biomonitor of the three metals (Fig. 69), although the uptake of zinc by the alga appeared to be influenced somewhat by the ambient concentrations of calcium and/or magnesium. Amounts of metals in the growing tips of *L. fluviatilis* were considered to act as indicators of recent changes in water quality, i.e. these portions of the alga exhibited a short time-integration capacity compared to the older portions of the plant (see also Harding *et al.*, 1981; Wehr *et al.*, 1983). This finding is similar to the conclusions reached from investigations of macroalgae in marine waters (e.g. see Bryan and Hummerstone, 1973a).

Welsh and Denny (1980) considered that algae of the genus *Nitella* were useful monitors of copper and lead in fresh waters, and both *N. flexilis* and the rooted grass *Glyceria fluitans* were studied in the Derwent Reservoir in northern England by Harding and Whitton (1978) and Harding *et al.* (1981). They found that *N. flexilis* responded to the concentrations of metals in solution, whereas *G. fluitans* appeared to take up metals from sediments also. Some of the latter tendency may have been due to the adsorption of inorganic particulates containing metals to parts of the plant. It has been noted in studies on marine macroalgae that samples may be contaminated through the surface adsorption of small particulates, and this may also be a problem in the use of freshwater macrophytes as biomonitors of trace elements. In addition, Patrick and Loutit (1977) have shown that bacterial epiphytes present on the surface of freshwater plants may contribute significantly to their total concentrations of metals. The methodological literature on the use of freshwater macrophytes as biomonitors of trace metals does not adequately address this problem (e.g. see Wehr *et al.*, 1983), and further effort is needed to resolve this aspect.

Whitton *et al.* (1981) noted that, while several studies had been reported on trace elements in green algae such as the genus *Cladophora*, the data remained fragmentary and no standard sampling and analytical methodology existed for such investigations. They recommended that sampling should involve only the terminal growing tips of the plant. This implies that the time-integration of metal levels would be limited, perhaps only involving days rather than longer periods; however, the precise time involved would depend on the growth rates of the algae at the various study sites (Phillips, 1980).

The liverwort *Scapania undulata* has been employed by several authors to study trace metal levels in freshwater environments (e.g. Burton and Peterson, 1979; Whitton *et al.*, 1982; Caines *et al.*, 1985), and offers the advantage of a high tolerance to metals in solution. Correlations between concentrations of cadmium, lead and zinc in water and those in the liverwort appear robust over a wide range of exposure levels (Whitton *et al.*, 1982), but both pH and calcium concentrations may affect the uptake of certain metals by the plant. Other species of aquatic macrophyte which have enjoyed consideration as biomonitors of trace metals in fresh waters include: pondweeds (*Potamogeton* spp.) in Finland; the floating leaves of *Trapa natans* and *Polygonum amphibium* in Lake Maggiori in Italy; and the macrophyte *Elodea densa* in Canada (Gommes and Muntau, 1981; Aulio and Salin, 1982; Mortimer, 1985).

As noted in Table 23, aquatic mosses or bryophytes have also been the subject of no little study as potential biomonitors of metals in freshwater ecosystems. This is a logical extension of the use of lichens in terrestrial environments to monitor trace metals in the atmosphere (see reviews by Nieboer *et al.*, 1978; Tuthill *et al.*, 1982), but studies in aquatic ecosystems have generally lagged behind those in terrestrial environments. The first report of the use of aquatic bryophytes as biomonitors was from New Zealand, where Whitehead and Brooks (1969) employed mosses to study the distribution of uranium in streams. They demonstrated a relationship between the uranium levels in the mosses used and those in the stream

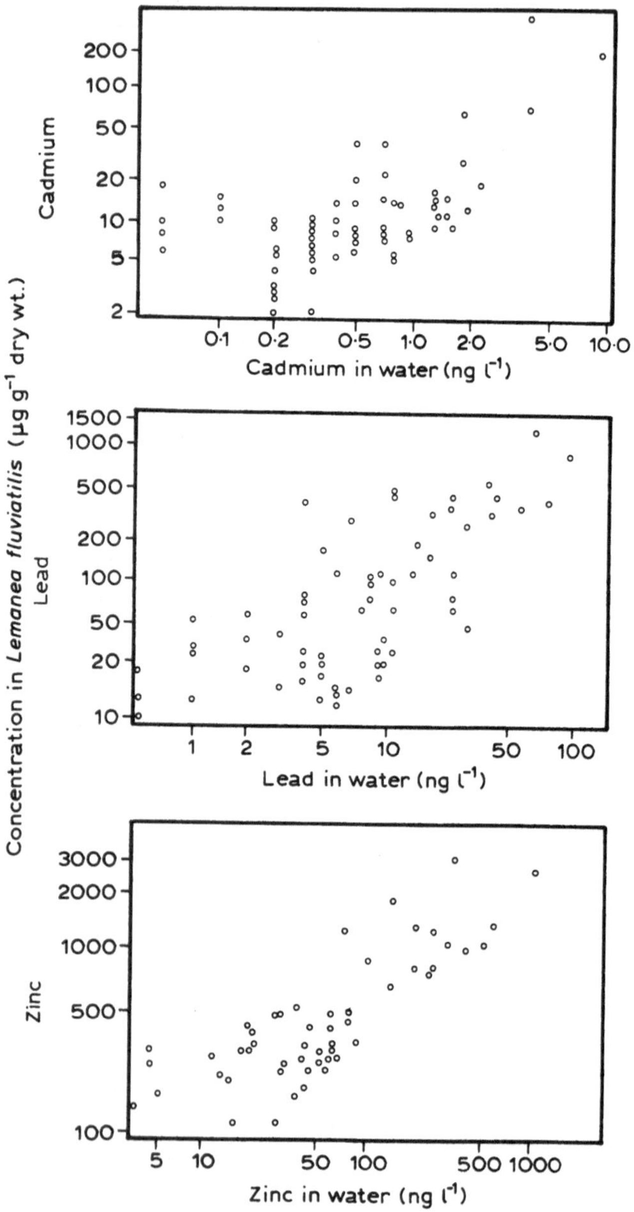
Concentration in Lemanea fluviatilis (µg g⁻¹ dry wt.)
Cadmium
200
100
50
20
10
5
2
0·1
0·2
0·5
1·0
2·0
5·0
10·0
Cadmium in water (ng l⁻¹)
Lead
1500
1000
500
200
100
50
20
10
1
2
5
10
50
100
Lead in water (ng l⁻¹)
Zinc
3000
2000
1000
500
100
5
10
50
100
500
1000
Zinc in water (ng l⁻¹)

waters, and noted that bryophytes had considerable promise for monitoring in such circumstances. A few isolated studies were then reported in the 1970s on trace elements in aquatic mosses in mining areas of Wales (Maclean and Jones, 1975; Burton and Peterson, 1979), and Empain (1976a,b) undertook similar investigations in Belgian streams. The species of most common use were *Fontinalis antipyretica* and *Rhynchostegium riparioides* (Table 23 and Whitton *et al.*, 1981), and later studies on these species and others have also been reported.

Say *et al.* (1981) used four species of aquatic mosses in the River Etherow near Manchester in the UK, and found a significant correlation between the concentrations of zinc in waters of this river and those in the terminal shoot tips of the mosses employed (Fig. 70). They noted that this correlation was dependent mostly on the data for *R. riparioides* and *Fontinalis squamosa*, as the other two species used (*F. antipyretica and Amblystegium riparium*) could be found at only a few sites. Similar data for zinc uptake by *R. riparioides* were reported by Wehr *et al.* (1981) for the River Team in north-east England.

However, the most critical series of publications on the use of these species as biomonitors of trace metals in freshwater ecosystems are those of Say and Whitton (1983) and Wehr and Whitton (1983a). The former authors reviewed the methods employed to monitor metals in such environments using *F. antipyretica* (see also Wehr *et al.*, 1983), and noted that positive correlations did not always exist between metal levels in water and those in the moss. Thus, for example, cadmium and lead concentrations in the terminal shoot tips of *F. antipyretica* tended to correlate better with levels of the same metals in sediments rather than with those in the water column. In addition, both manganese and zinc appeared to affect the uptake of lead by this species, tending to reduce its value as a biomonitor of lead in streams.

Wehr and Whitton (1983a) produced a similar review for *R. riparioides*, also concluding that zinc influenced the bioaccumulation of lead by this species. Notwithstanding such problems with the monitoring of lead, however, it appears that aquatic mosses have considerable potential as biomonitors of metals in freshwater ecosystems. Temporal variability in metal levels has been documented in several species (Wehr and Whitton,

Fig. 69. Relationships between metal concentrations ($\mu g\ g^{-1}$ dry weight) in the terminal 2 cm of filament tips of the red alga *Lemanea fluviatilis* and concentrations of the same metals in filtered (lead; zinc) or total (cadmium) water samples from the same area of origin. After Harding and Whitton (1981).

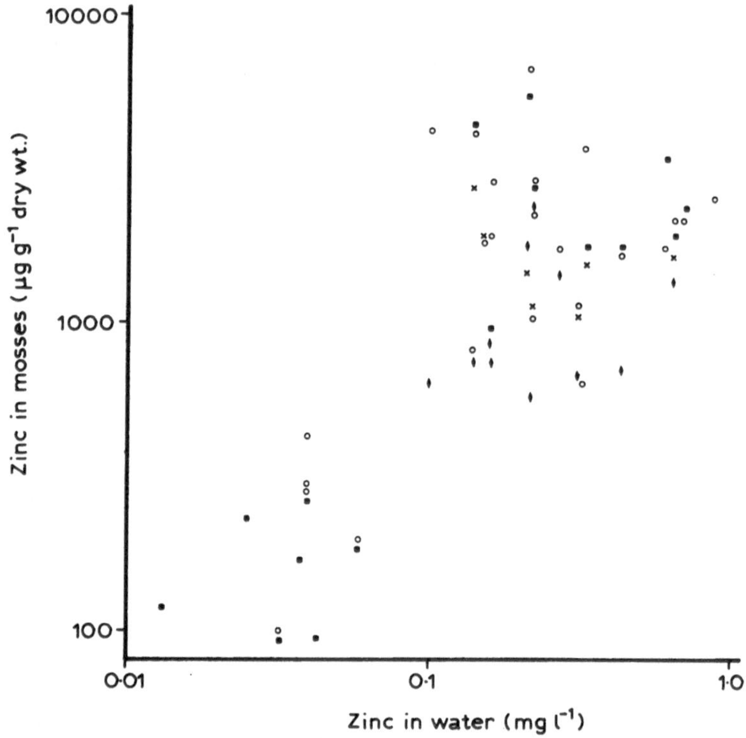

Fig. 70. Relationship between the concentrations of zinc in various aquatic mosses ($\mu g\ g^{-1}$ dry weight) and those in water from the same site: x, *Amblystegium riparium*; ◆, *Fontinalis antipyretica*; ■, *Fontinalis squamosa*; o, *Rhynchostegium riparioides*. Correlation coefficient $r = 0.6678$; $p < 0.001$. After Say *et al.* (1981).

1983b), but this is expected, given the fluctuations in water quality in streams. In addition, it is known that pH influences metal accumulation (see Caines *et al.*, 1985, and Chapter 5). However, the cosmopolitan distribution of aquatic bryophytes of several species (Table 23) and the possibility of their transplantation from one site to another if required (e.g. Mouvet, 1984) argue for their continued development for biomonitoring purposes.

Compared to their extremely widespread use for monitoring trace elements in estuarine and marine environments, bivalve molluscs have been relatively little used as biomonitors of metals in freshwater ecosystems. This appears to be partly due to the absence of a particularly cosmopolitan genus or species (cf. the genera *Mytilus* and *Crassostrea* in marine waters), and the species employed in studies of fresh waters have varied widely with location. Manly and George (1977) used the freshwater mussel *Anodonta anatina* for work in the River Thames in England, correlating metal levels in the mussel tissues to the presence of contaminated discharges from wastewater treatment plants in urban areas. They noted that the concentrations of many metals varied with tissue weights of these mussels, but nevertheless concluded that *A. anatina* exhibited promise as a biomonitor. A related species, *A. cygnea*, has been used by Salanki *et al.* (1982) to monitor trace metals in Lake Balaton in Hungary.

A point source discharge of copper from an electroplating works in the Muskingum River, USA, was studied by Foster and Bates (1978), using transplanted samples of the mussel *Quadrula quadrula*. As the electroplating effluent was by far the greatest source of copper in the river, these studies provided a robust test of the biomonitoring potential of *Q. quadrula* for copper, and the results (Fig. 71) suggested that the mussel was of considerable value in monitoring this element.

In Australia, *Velesunio ambiguus* has been the mussel species most commonly used for investigations of metals in fresh waters, although *V. angasi* and *Alathyria jacksoni* have also been subject to some study (Jones and Walker, 1979; Millington and Walker, 1983; Maher and Norris, 1990; R.D. Simpson, unpublished data). As noted above, Millington and Walker (1983) considered that the net uptake of zinc in *V. ambiguus* did not reflect exposure concentrations, and this was the case in both laboratory and field studies. Intra-population variability in the concentrations of iron, manganese and zinc in this species could be only partly explained by the influences of tissue weights and age, and the mussel did not appear to be a promising candidate for biomonitoring studies of metals, perhaps because at least some elements were partially regulated (see Chapter 5).

Yet more species of freshwater bivalves have been employed elsewhere to study trace metals. In Holland, *Dreissena polymorpha* has been preferred for study (del Castilho *et al.*, 1984), while on the American continent, the clam species *Elliptio complanata* and *Corbicula fluminea* have been used (Cory and Dresler, 1981; Elder and Mattraw, 1984; Dermott and Lum, 1986; Luoma *et al.*, 1990). The last of these species in particular shows considerable promise for such investigations, and has also been employed in the Toxic Substances Monitoring Program in California (CSWRCB, 1986) and in attempts to measure sublethal stress on freshwater biota (Cantelmo-Cristini *et al.*, 1985; Foe and Knight,

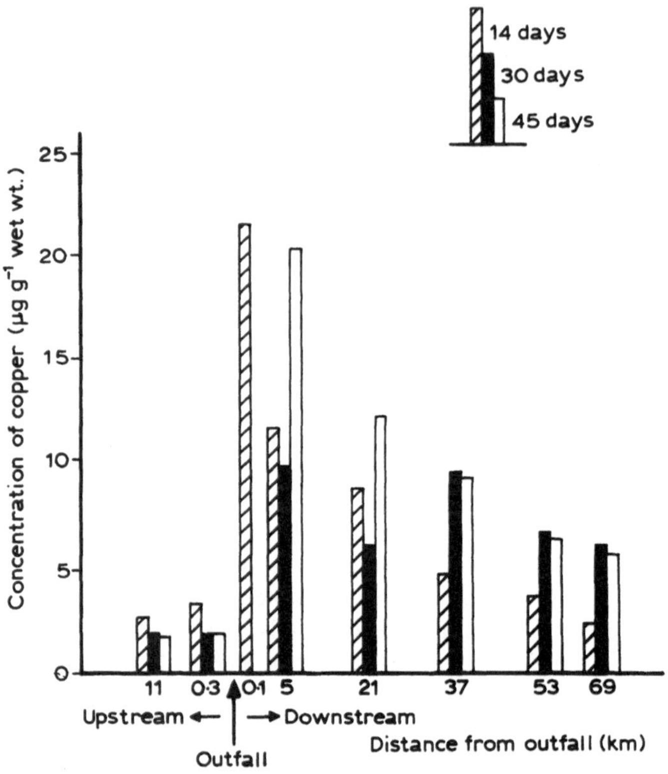

Fig. 71. Concentrations of copper ($\mu g\ g^{-1}$ wet weight) attained by soft tissues of caged mussels (*Quadrula quadrula*) exposed *in situ* to an electroplating effluent discharged to the Muskingum River, USA. After Foster and Bates (1978).

1987). Further critical investigations are required of the biomonitoring capability of *C. fluminea* before it can be accepted for routine use, however, and these should include study of the taxonomy of the genus (see Hillis and Patton, 1982; McLeod, 1986).

A few studies have also been reported on the use of freshwater gastropods to monitor trace metals in freshwater ecosystems. Moriarty and French (1977) encountered difficulties in monitoring mercury contamination of the rivers draining to the Wash in eastern England, principally because of the existence of intermittent sources of the element. They proposed that aquatic gastropods could be employed to smoothe out the temporal fluctuations in mercury concentrations in the rivers, providing a time-integrated measure of conditions. Newman and McIntosh (1982) considered that gastropods of several species were useful biomonitors of lead in freshwater lakes of New Jersey, USA, although it should be noted that only two sites were studied and that significant seasonal fluctuations in lead levels were noted in the species used. Everard and Denny (1984), working in Ullswater in Cumbria, UK, provided evidence for the use of the freshwater gastropod *Lymnaea peregra* for monitoring lead levels. This species apparently synthesised granules containing lead and laid these down in the soft tissues under conditions of high exposure, excreting them when the exposure to lead was reduced.

While various other types of freshwater organisms have also been employed to attempt to monitor trace metals (ranging from crustaceans to insects; see Salanki *et al.*, 1982; Zauke, 1982; Lynch *et al.*, 1988; Hare *et al.*, 1989), the most frequently used species have undoubtedly been fish. Unfortunately, however, the care and attention lavished on the classic early studies such as that of Johnels *et al.* (1967) have not always been apparent in the more recent work, and the utility of fish species to monitor trace metals other than mercury in freshwater environments remains open to doubt. In addition, the effects of fish mobility on monitoring data are once again relevant in discussions of this topic.

Considerable work on pike (*Esox lucius*) has been undertaken subsequent to the early studies of Johnels *et al.* (1967), and this species appears to be a useful biomonitor of mercury and of organic contaminants (see below) in freshwater ecosystems (Scott and Armstrong, 1972; Olsson and Jensen, 1975; Hattula *et al.*, 1977, 1978; Armstrong and Scott, 1979; Bull *et al.*, 1981; Göthberg, 1983; Björklund *et al.*, 1984). The territorial nature of *E. lucius* is critical in its use to monitor contamination, and several studies have shown that consistent differences exist between the concentrations of mercury in pike taken from different parts of the same water body (e.g. Hattula *et al.*, 1977). Björklund *et al.* (1984) correlated the mercury levels in *E. lucius* from 220 lakes in Sweden to the presence of local sources of the element; to regional deposition of mercury from the atmosphere; and to the increasing acidification of lakes, created by acid rain.

Several fish species in addition to pike have been employed as biomonitors of mercury in freshwater environments; these include salmonids, bass, sunfish and others (e.g. Kelso and Frank, 1974; Hattula *et al.*, 1977; Armstrong and Scott, 1979; Moore and Sutherland, 1980; Akielaszek and Haines, 1981; Wren and MacCrimmon, 1983). While the bioaccumulation of mercury in such species depends on age or weight of the fish as well as on the pH of the water, these factors can be accounted for at sampling or data interpretation, and the method appears robust (especially for species exhibiting limited mobility) and of general application.

By contrast, the use of freshwater fish to monitor trace elements other than mercury in either freshwater or marine ecosystems is not recommended (see Chapter 5). There is ample evidence of the regulation of certain metals in fish muscle tissues at least (Bryan, 1976; Phillips, 1977, 1980; Wiener and Giesy, 1979). Concentrations of metals in other tissues may be partially regulated, especially in the case of essential elements such as copper and zinc. This does not, however, imply that external factors (other than the ambient abundance of metals) have no influence on the bioaccumulation of elements by freshwater fish. For example, several authors have documented changes in the concentrations of metals in such species with age, size or tissue weight of the fish (e.g. Tong *et al.*, 1974; Wilson *et al.*, 1980; Badsha and Goldspink, 1982; Moriarty *et al.*, 1984). The existence of partial regulation of metals (other than mercury) in freshwater fish implies that these species should not be relied upon as biomonitors of trace elements in most circumstances.

(ii) Trace Organic Contaminants

Rivers and lakes are commonly the primary recipients of land-derived organochlorines or hydrocarbons, released either accidentally or through the systematic use of the compound (e.g. as pesticide formulations for agriculture). As noted in Chapter 6, the concentrations of trace organic contaminants in freshwaters are therefore highly variable with time (e.g. see Ochiai and Hanya, 1976; Duinker *et al.*, 1980; Yamato *et al.*, 1980; Boryslawskyj *et al.*, 1985), and the use of biomonitors is recommended to provide a time-integrated picture of contamination under such conditions. It is notable that, while biomonitors have been quite widely employed to measure the abundances of pesticides and PCBs in fresh waters, their use to monitor petroleum hydrocarbon levels in such environments has been extremely rare. This section therefore concentrates almost exclusively on the database concerning organochlorines in biomonitors of freshwater ecosystems.

In addition, much research has been undertaken in the American continent in particular on the dynamics of trace organic contaminants in particular areas of contamination, involving such compounds as pesticides and PCBs in the Great Lakes (e.g. see Thomann, 1981; Bierman and Swain, 1982; Jensen *et al.*, 1982; Zabik *et al.*, 1982; Fisher *et al.*, 1983;

Thomann and Connolly, 1984), PCBs in the Hudson River (Bopp *et al.*, 1981), and Kepone in the James River (Huggett and Bender, 1980). Certain of these studies have also included work on the contamination of biota, and have contributed to our present understanding of the accumulation of trace organic compounds by organisms.

Data on the bioaccumulation of organochlorines by freshwater species of algae are sparse, matching the situation for estuarine and marine waters. There is no question that both microalgae and larger species accumulate significant quantities of organochlorines (e.g. see Godsil and Johnson, 1968; Hannon *et al.*, 1970; Särkkä *et al.*, 1978; Lederman and Rhee, 1982; Mowrer *et al.*, 1982). However, planktonic algae are unsuitable as biomonitors because of variations in the accumulation of contaminants by different species and the fact that natural assemblages of plankton invariably contain a wide range of species, differing between locations. Macroalgae have apparently not been preferred for study because of their low lipid contents, although (as noted in Chapter 6) they may provide useful clues to the cycling of certain compounds such as the lower-chlorinated PCBs. No attempts have been made to employ aquatic mosses as biomonitors of trace organic contaminants, although by analogy to the trace metal studies cited previously, this would appear to be a profitable topic for future investigations.

Most surprisingly, invertebrate animals have also been little-used to date in investigations of organochlorines in freshwater ecosystems. The absence of cosmopolitan species of bivalves or other invertebrates from fresh waters (e.g. counterparts to the estuarine and marine bivalve species of the genera *Mytilus* and *Crassostrea*) may constitute one reason for this. However, it is likely that fish have been preferred as potential biomonitors for these compounds because of the additional concerns raised by the consumption of fish by humans, and the need to protect public health. As noted by Phillips and Segar (1986), these objectives should be separated, as programmes intended for biomonitoring purposes and for the protection of public health should differ radically in their design.

A few examples of the use of invertebrates to monitor organochlorines in fresh waters may nevertheless be cited here. Södergren *et al.* (1972) employed amphipods (*Gammarus pulex*) to monitor the levels of DDT and its metabolites and PCBs in Swedish streams, and their fluctuations with season. Other authors have preferred the larvae or adults of aquatic insects for such studies, however (e.g. Kellog and Bulkley, 1976; Bush *et al.*, 1985), and the intent here perhaps relates more closely to defining the bioavailable fraction of contaminants in the ecosystem rather than to the use of biomonitors *per se*.

More closely related to the latter concept are the few studies of organochlorines in molluscs from freshwater environments. Godsil and Johnson (1968) used both native and transplanted clams of the genus *Gonidea* to study pesticides in the waters of the Tule Lake National Wildlife Refuge in California. The abundance of organochlorines in these clams was shown to fluctuate significantly with season, in concert with application periods for the various compounds for agriculture in the catchment. Bedford *et al.* (1968) undertook similar studies with three species of mussel (*Lampsilis siliquoidea*, *L. ventricosa* and *Anodonta grandis*) in the Red Cedar River in Michigan, and concluded that freshwater mussels exhibited considerable promise as biomonitors of pesticides. Nadeau and Davis (1976) preferred gastropods as biomonitors of PCBs in the Hudson River, however, using species of the genera *Helisoma*, *Physa* and *Limnacea*. Most recently, the Toxic Substances Monitoring Program in California has introduced analyses of the freshwater clam *Corbicula fluminea* to supplement those of fish for organochlorines. As noted previously, it is thought that this species may have particular promise as a cosmopolitan biomonitor of trace contaminants in fresh waters (CSWRCB, 1986). There can be no doubt that the use of invertebrates (and bivalve molluscs in particular) as biomonitors in freshwater ecosystems has been sadly ignored, and requires considerable further effort.

By contrast to the situation with invertebrates, fish have been most frequently employed to study organochlorines in fresh waters. The early studies of Hunt and Bischoff (1960) on DDD in Clear Lake in California employed a range of fish species, and the later work undertaken on inland waters of the USA for national or State pesticide monitoring programmes followed this trend (e.g. see Lyman *et al.*, 1968; Henderson *et al.*, 1969, 1971; Morris and Johnson, 1971; Frank *et al.*, 1974; CSWRCB, 1986). The problem with this approach involves the frequent need to compare data for different species at multiple locations, which gives rise to difficulties in interpretation. The use of a single cosmopolitan species is therefore preferred for true biomonitoring studies, and species of the genus *Esox* and of salmonids have been most frequently employed as biomonitors of organochlorines in fresh waters.

The use of the northern pike *Esox lucius* as a biomonitor has been discussed above with respect to mercury, and this or related species have also been employed in Sweden and America for investigations of organochlorines in freshwater ecosystems (e.g. see Kelso *et al.*, 1970; Zitko, 1971;

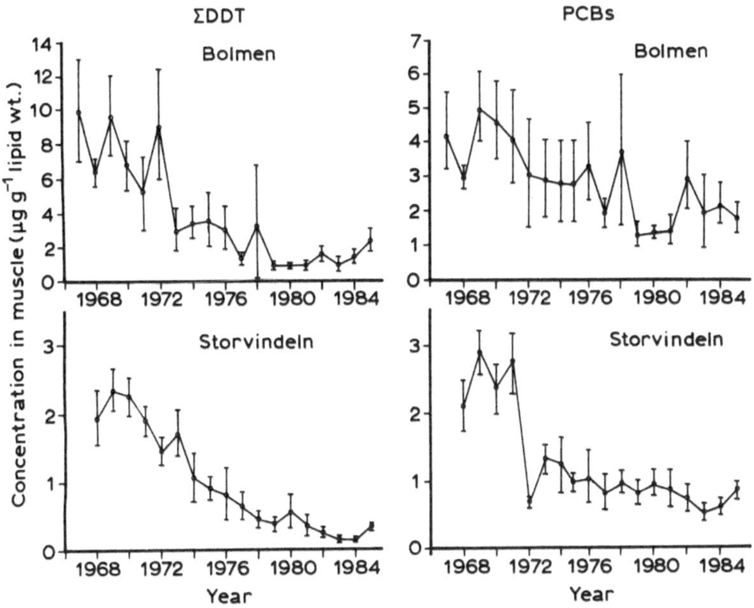

Fig. 72. Concentrations of ΣDDT and PCBs ($\mu g\, g^{-1}$ lipid weight; means and 95% confidence intervals) in the axial muscle of pike (*Esox lucius*) from Lake Bolmen in southern Sweden and Lake Storvindeln in northern Sweden, between 1967 and 1985. After Olsson and Reutergårdh (1986).

Olsson and Jensen, 1975; Hattula *et al.*, 1978; Tsui and McCart, 1981; Olsson and Reutergårdh, 1986). While organochlorine concentrations may vary with fish size and possibly sex of the fish in certain instances (e.g. see Olsson and Jensen, 1975), these variables can be accounted for in the sampling design of biomonitoring programmes. At least partly due to their territorial habits, pike appear to be particularly useful biomonitors of organochlorines, reflecting both spatial and temporal variations in the abundances of pesticides and PCBs. In Sweden, their use to monitor long-term trends in the contamination of freshwater environments is particularly notable, and data from Olsson and Reutergårdh (1986) on the decrease in ΣDDT and PCB levels in *E. lucius* since the Swedish restrictions on the use of these compounds are shown in Fig. 72.

Most of the work on salmonid species has been undertaken on the American continent, particularly in the Great Lakes. These studies were driven at least partly by concerns over the impacts of organochlorines on the lucrative sport fishery for these species, and more recently, by the need

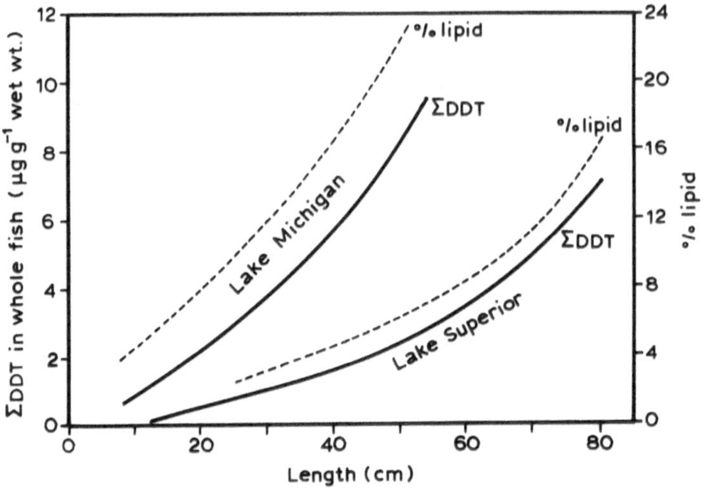

Fig. 73. Variations in the concentrations of ΣDDT (solid lines; left axis) and percentage lipid (dashed lines; right axis) with length in lake trout (*Salvelinus namaycush*) from Lake Michigan and Lake Superior. After Reinert (1970).

to protect public health from PCBs in fish (e.g. see Burdick *et al.*, 1964; Anderson and Everhart, 1966; Sprague *et al.*, 1971; Armstrong and Lutz, 1977a,b; Rohrer *et al.*, 1981; Jensen, 1984). The causes of variability in concentrations of organochlorines among individuals of a species have been documented by several authors (e.g. see Anderson and Fenderson, 1970; Reinert, 1970; Reinert and Bergman, 1974). The most important factors are fish size, which generally correlates to lipid content in salmonids (Fig. 73), and seasonal fluctuations, which are also related to lipid weight changes (Table 24; see also Phillips, 1978, 1980, 1986). Both of these may be accounted for in the design of sampling programmes and/or by the calculation of data based upon lipid weights of samples for interpretive purposes (see Chapter 6 and Phillips, 1980).

A further database which merits mention concerning the impacts of fish size on organochlorine levels in salmonids derives from studies at Cayuga Lake, New York State. This lake is regularly stocked with tagged lake trout (*Salvelinus namaycush*), and the concentrations of ΣDDT and PCBs in the tagged fish have been studied by several authors over a 10-year period (Bache *et al.*, 1972; Youngs *et al.*, 1972; Wszolek *et al.*, 1979). The results of these investigations are shown in Fig. 74, and clearly demonstrate both the increase in concentrations of organochlorines (calculated

Table 24 Mean concentrations of ΣDDT ($\mu g\,g^{-1}$ by wet and lipid weights) and percentage lipids in tissues of coho salmon (*Oncorhynchus kisutch*) sampled on three occasions in Lake Michigan. Data for residue levels by lipid weights calculated by Phillips (1978). (After Reinert and Bergman, 1974)

Sample	*Collection date*	*ΣDDT (wet weight)*	*Lipid (per cent)*	*ΣDDT (lipid weight)*
Whole fish[a]	21 August 1968	12.3	13.2	93.2
	18 October 1968	12.6	7.1	177.5
	31 January 1969	12.3	2.8	439.3
Whole muscle	21 August 1968	14.9	16.2	91.9
	18 October 1968	16.3	7.1	230.0
	31 January 1969	18.4	2.3	800.0
Loin region	21 August 1968	5.6	5.4	103.7
	18 October 1968	4.7	2.3	204.3
	31 January 1969	10.1	1.7	594.1
Dorsal region	21 August 1968	61.1	62.1	98.4
	18 October 1968	106.1	39.2	270.7
	31 January 1969	59.6	8.2	726.8
Medial region	21 August 1968	41.1	44.5	92.4
	18 October 1968	59.8	22.4	266.9
	31 January 1969	41.3	7.7	536.4
Ventral region	21 August 1968	66.0	67.0	98.5
	18 October 1968	106.4	41.5	256.4
	31 January 1969	68.5	8.5	805.9
Brain	21 August 1968	1.2	8.7	13.8
	18 October 1968	2.4	5.8	41.4
	31 January 1969	5.0	7.7	64.9
Eggs	21 August 1968	7.4	10.9	67.9
	18 October 1968	10.2	8.6	118.6
	31 January 1969	7.9	9.4	84.0
Adipose fat[b]	21 August 1968	88.9	85.6	103.9

[a] Average weights of whole fish were 3514, 3486 and 3770 g, respectively for each sampling date. Two 25 mm-thick steaks and the brain had been removed from whole fish samples.
[b] Fish collected in October 1968 and January 1969 lacked adipose fat.

on a wet weight basis) with fish age, and the general temporal decrease in residue levels through the 1970s, consequent to the restrictions placed on the use of DDT and PCBs in the USA. As noted by Phillips (1986), it is interesting that the ΣDDT levels decreased much more markedly than the concentrations of PCBs over this period; this reflects the extreme persistence of the higher-chlorinated PCBs in particular, in freshwater ecosystems.

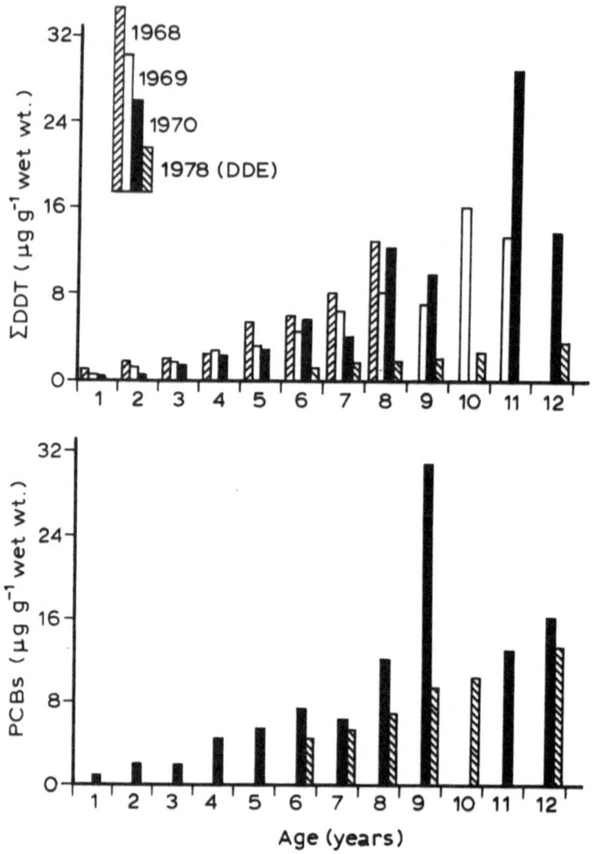

Fig. 74. The effects of fish age on concentrations of PCBs and ΣDDT (1968, 1969 and 1970) or DDE (1978) in lake trout, *Salvelinus namaycush*, from Cayuga Lake, New York State, USA. Data for 1968–1970 refer to whole eviscerated fish, after Bache *et al.* (1972) for PCBs and Youngs *et al.* (1972) for ΣDDT. Data for 1978 are for edible flesh only, after Wszolek *et al.* (1979).

Several studies on trace organic contaminants in other freshwater fish species deserve brief mention here. Channel catfish (*Ictalurus punctatus*) are of widespread distribution in the USA, and have been employed in several studies of pesticides in fresh waters, of Iowa in particular (Morris and Johnson, 1971; Bulkey *et al.*, 1976; Kellogg and Bulkley, 1976). Dieldrin concentrations increase with age in the muscle tissue of this species, but this is only partly the result of changes in lipid weights of muscle tissues; alterations in the diet of larger fish are likely to be the cause of some of the age-related variations in pesticide levels (Bulkley *et al.*, 1976). Elsewhere, roach (*Rutilus rutilus*) have been used as biomonitors of organochlorines in fresh waters (e.g. Olsson *et al.*, 1978; Schüler *et al.*, 1985) and have been of particular use in defining the seasonal variation in pesticide concentrations in such environments. It should also be emphasised here that studies in tropical freshwater environments have been very rare, although the use of *Tilapia* to monitor pesticides in Lake Mariut in Egypt is notable (Saad *et al.*, 1982).

PCB contamination of the Hudson River in New York was derived principally from two facilities (at Fort Edward and Hudson Falls, in the upper Hudson) owned by the General Electric Company, and manufacturing capacitors (Brown *et al.*, 1985). Average daily discharges of PCBs (predominantly Aroclors 1242 and 1016) were calculated to amount to 14 kg over a 30-year period to the mid-1970s, but these were decreased in a stepped fashion to less than 1 g daily by 1977. The massive contamination which resulted from the earlier discharges has elicited several studies, these using striped bass (*Morone saxatilis*) and other species (Cahn *et al.*, 1978; Brown *et al.*, 1985; Connell, 1987). Anadromous species such as *M. saxatilis* must be employed with care in such studies, as their migrations between the freshwater and marine environments clearly affect their bioaccumulation of contaminants.

Brown *et al.* (1985) provided most interesting details of the temporal decline in PCB levels in the Hudson River pursuant to the imposition of controls on the discharge of PCBs by the General Electric Company facilities. The flux of PCBs downriver from the headwaters decreased markedly between 1977 and 1983, and this was reflected particularly well by residue levels in pumpkinseed, *Lepomis gibbosus* (Fig. 75). Brown *et al.* (1985) showed that PCB concentrations declined substantially over this period in other fish species also, including striped bass (*Morone saxatilis*), brown bullhead (*Ictalurus nebulosus*), goldfish (*Carassius auratus*), and largemouth bass (*Micropterus salmoides*).

It is unfortunate that many of the studies of temporal declines in PCB levels in freshwater species of biomonitors do not include data on individual PCB homologues. Such data are important not only in improving our understanding of PCB kinetics in aquatic ecosystems, but also in defining the toxicological risk posed by PCB residues in the environment (as the non-*ortho* coplanar PCBs are those of greatest toxicity and persistence; see Tanabe, 1988). A notable exception to this lack of attention to detail is provided by the excellent publication of Loganathan *et al.* (1989) on organochlorine residues in the lizard goby *Rhinogobius flumineus* from

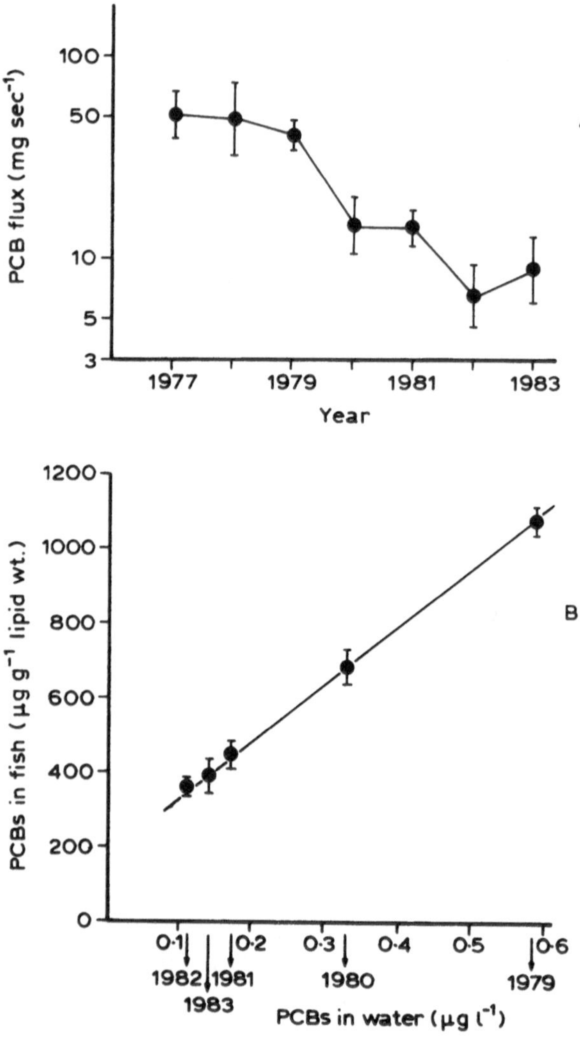

Fig. 75. (A) Flux rates of PCBs (log of annual means, with 95% confidence intervals) in the upper Hudson River at Stillwater. (B) Relationship between mean PCB concentrations (±95% confidence intervals) in summer samples of yearling pumpkinseed (*Lepomis gibbosus*) and those in water samples from Stillwater in the upper Hudson River. After Brown *et al.* (1985).

the River Nagaragawa in Japan. These authors demonstrated that the concentrations of ΣDDT, HCH and PCBs in *R. flumineus* declined markedly from the early 1970s onwards, concurrent to the reductions in the production and use of these compounds in Japan. By contrast, the levels of chlordane compounds in *R. flumineus* from the same river increased sharply in the late 1970s and remain high to date (Fig. 76). The changes in concentrations of the various types of PCBs in the fish are presented in Fig. 77(A). These clearly demonstrate the shift towards a predominance of higher-chlorinated PCBs with time since the imposition of the restrictions on the production and use of PCBs, correlating to both the longer half-lives of the more highly chlorinated components in biota (Fig. 77(B)) and their greater general environmental persistence. It appears probable that the particularly toxic coplanar PCBs are sufficiently refractive to metabolic breakdown that they alter very little in concentration after restrictions are placed on the environmental discharge of PCBs. Thus, although the concentrations of total PCBs may undergo significant reductions in environmental media during such periods, the levels of the most toxic components may remain almost unchanged. This implies that the toxicological risk from previously discharged PCBs may remain with us for some considerable period in even western nations, which introduced restrictions on the use of these substances almost two decades ago.

D. SUMMARY

The fluctuation in concentrations of trace contaminants over time in freshwater ecosystems is commonly even more extreme than that in estuarine or marine environments, as it is driven by temporal changes not only in the magnitude of contaminant sources but also in the volume of the receiving waters. In such circumstances, the use of biomonitors to provide a time-integrated picture of contaminant bioavailability may be recommended, as the identification of average contamination conditions through the analysis of water samples (or sediments) is most challenging.

Considerable effort has been expended to develop the biomonitoring concept in fresh waters; both laboratory-based studies and field investigations have contributed significantly to this effort. The existing database extends to trace metals and organochlorines in many species, but very few studies have been reported on petroleum-derived hydrocarbons in freshwater biota.

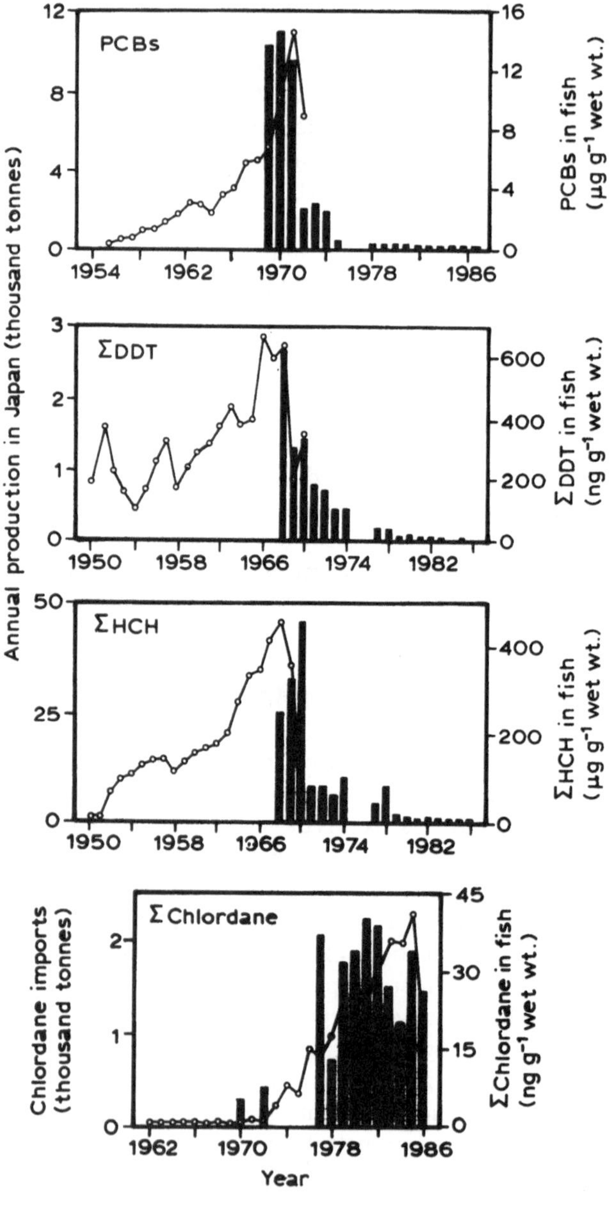
PCBs
ΣDDT
ΣHCH
ΣChlordane
Annual production in Japan (thousand tonnes)
PCBs in fish (μg g⁻¹ wet wt.)
ΣDDT in fish (ng g⁻¹ wet wt.)
ΣHCH in fish (μg g⁻¹ wet wt.)
Chlordane imports (thousand tonnes)
ΣChlordane in fish (ng g⁻¹ wet wt.)
1954 1962 1970 1978 1986
1950 1958 1966 1974 1982
1962 1970 1978 1986
Year

The use of unicellular algae to monitor trace contaminants in fresh waters is not recommended. Although such species accumulate both trace elements and organochlorines rapidly (and may in some instances be important in the transfer of these contaminants through food webs), they are present in freshwater environments as assemblages of mixtures of species which vary from location to location. The different bioaccumulation capacities of each species thus interfere with any attempt to use microalgae as biomonitors of the environmental abundance of trace contaminants at different sites, or at a single location over time.

Freshwater macrophytes of various types have been quite extensively employed as biomonitors of trace metals, although they have been almost completely ignored in studies of organochlorines, perhaps because of their low lipid contents. The types of plants used in studies of trace elements have ranged from algae of several genera to vascular plants and aquatic mosses. Each of these types of macrophyte exhibits inherent advantages and disadvantages for biomonitoring, and no one species may be recommended for use in all situations. However, aquatic mosses (particularly *Fontinalis antipyretica* and *Rhynchostegium riparioides*) exhibit particular promise as biomonitors of trace metals, and may also be worthy of study with respect to the quantification of organochlorines in fresh waters.

Surprisingly little work has been undertaken on the bioaccumulation of trace contaminants by molluscs in freshwater environments. This is partly due to the absence of particularly cosmopolitan species in fresh waters, which could be employed as counterparts to bivalves of the genera *Mytilus* or *Crassostrea* in estuarine and marine waters. Species of the genera *Anodonta, Corbicula, Gonidea, Lampsilis* and *Velesunio* have been used to study either metals or trace organic contaminants. Generally, such investigations have been successful in demonstrating the efficacy of the species used as biomonitors, although the Australian species *Velesunio ambiguus* appears to partially regulate trace metals and its use cannot be recommended. The freshwater clam *Corbicula fluminea* exhibits particular promise for biomonitoring work (and possibly also for investigations of the effects of contaminants), and is worthy of additional investigation, particularly in view of its widespread geographical distribution. Studies on gastropods have been rare, but examples have been reported of their use

---

Fig. 76. Annual production or importation rates of organochlorines in Japan (circles) and their mean concentrations in whole lizard gobies (*Rhinogobius flumineus*; bars) from the River Nagaragawa, Japan. All compounds are shown as sums of their individual isomers or congeners. After Loganathan *et al.* (1989).

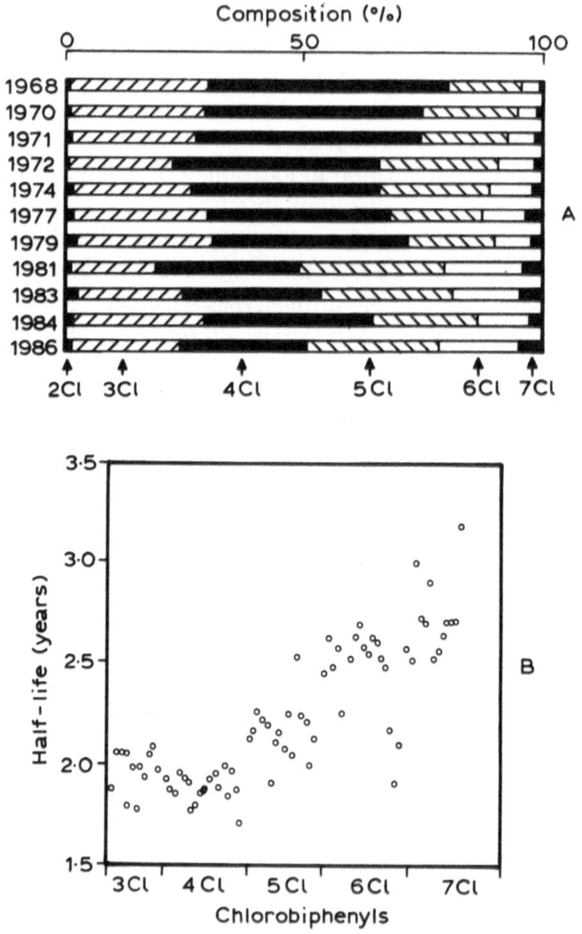

Fig. 77. (A) Temporal changes between 1968 and 1986 in the gross composition of PCBs in the lizard goby *Rhinogobius flumineus* from the River Nagaragawa, Japan. (B) Half-lives of individual PCB isomers and congeners in the same fish species. After Loganathan *et al.* (1989).

to quantify both trace elements and organochlorines in freshwaters, and further developmental work is needed.

More work has been reported on the use of freshwater fish as biomonitors of trace contaminants than for any other taxon. These investigations have been driven mainly by concerns over the preservation of commercial freshwater fisheries (particularly those for salmonids) and the protection of public health. While perfectly legitimate, these objectives are distinguishable from the need to develop biomonitors for the quantification of trace contaminants in fresh waters; as a result, the sampling programmes which have been employed in studies using fish were often not ideal for biomonitoring purposes. Migratory species in particular should not be considered for biomonitoring investigations, as the impacts of space and time on the accumulation of contaminants by such species cannot be separated. In addition, trace elements other than mercury tend to be regulated in at least the axial muscle tissues of freshwater (and marine) fish, and these species should not therefore be employed as biomonitors of metals other than mercury.

Despite these caveats, territorial species of fish (or at least, those species exhibiting limited mobility or migration) are of potential use as biomonitors of certain trace contaminants. Investigations on pike (*Esox lucius*) are of particular note, and certain other salmonid and rough fish species (e.g. roach, sunfish and catfish) have also been employed to good effect for monitoring mercury and organochlorines in fresh waters. The impacts of lipid levels, size or age, season and other factors on the bioaccumulation of such contaminants by fish have been adequately studied in general, and sampling programmes can therefore be designed to account for these variables and provide an accurate indication of the environmental abundance and bioavailability of mercury and organochlorines in fresh waters.

Chapter 8

The Development of *De Novo* Biomonitoring Programmes for Trace Contaminants

A. INTRODUCTION

Although the use of biomonitoring techniques for studying trace contaminants in aquatic ecosystems has become widespread in certain areas of the world, many regions are yet to be studied through such methods. Significant proportions of the fresh and marine waters of the western nations have been insufficiently characterised in terms of their exposure to trace metals, organochlorines or petroleum-derived hydrocarbons. However, the greatest areas of unknown in terms of aquatic contamination involve the inland and coastal waters of subtropical and tropical areas (Phillips, 1991). There is an urgent need to produce reliable data on these ecosystems, as the nations of the tropical zone (many of which are presently in the developing category) are experiencing rapid increases in both populations and industrial expansion (see Chapter 2), and rely heavily on seafoods as a source of protein. It may be accepted that the problems of aquatic contamination suffered in the western nations of the northern hemisphere will also emerge in the developing nations, particularly as pollution controls in the latter are in their infancy (Beanlands and Si, 1988; Phillips and Tanabe, 1989).

There is therefore a need to design and implement new biomonitoring programmes, some of these in areas remote from previous study. This chapter discusses the steps required to develop new monitoring programmes for trace contaminants in aquatic environments, providing information in the sequence in which decisions will generally be made.

B. MONITORING OBJECTIVES

The establishment of clear monitoring objectives is perhaps the most important facet of any attempt to devise a programme to study trace contaminants in aquatic ecosystems. Phillips and Segar (1986) have argued that the definition of unambiguous objectives is fundamental to biomonitoring programmes, as their detailed design cannot be undertaken efficiently if the precise objectives of the monitoring effort are unclear.

The most likely reasons for establishing a monitoring programme for trace contaminants in aquatic environments are as follows:

- A desire to delineate spatial trends in contamination.
- A need to define changes in contamination over time, perhaps allied to regulatory initiatives.
- A requirement to protect commercial fisheries, other biological aspects of ecosystems, or public health.
- A wish to identify new contaminants of concern in aquatic environments.

It is important to acknowledge that, while biomonitoring programmes can be designed which satisfy one or more of these objectives, the optimal programme design differs for each of the objectives listed (NAS, 1980; Phillips and Segar, 1986). Thus, for example, a programme to study spatial trends in contamination should differ significantly in design from one intended to monitor temporal trends in the contamination of any particular water bodies. Similarly, programmes intended to monitor public health should be fundamentally distinct in design from those employed to delineate spatial or temporal trends in aquatic contamination.

Unfortunately, however, very few of the national or international biomonitoring programmes completed to date have been based from the outset on clear objectives. Even when publications on such programmes discuss this topic (which is rare), the objectives cited are often contradictory or vague. Certain examples may be considered here to emphasise this point. The examples employed have been taken from studies in the USA, which are perhaps the best-known of the national biomonitoring programmes undertaken to date.

It is clear from the reports of Butler (1966, 1969a,b, 1971, 1973) that the objectives of the pesticide biomonitoring programme employing bivalve molluscs in estuaries of the USA in the late 1960s and early 1970s developed from an initial concern over pesticide toxicity and potential impacts on aquatic wildlife. Butler (1966) stated:

> Despite two decades of research, the extent and importance of pesticide pollution in estuaries are poorly understood. Laboratory studies of their acute and chronic toxicity indicate that pesticides may be the cause of ill-defined but significant mortality, loss of production, and, perhaps, changes in the direction of natural selection in estuarine fauna. Preliminary investigations show the need for a continuing surveillance program to identify the seasonal and geographical distribution of pesticide pollution in estuaries.

This emphasis on both spatial and temporal trend monitoring in the early work on pesticides was satisfied by the study of contaminants in bivalves from some 170 sites, samples being taken at monthly intervals (Butler, 1969a, 1971). The data were of little direct relevance to public health, and no emphasis was placed upon a need to define new contaminants of concern; the need to "... determine the extent of [the] pollution threat to commercial fisheries..." (Butler, 1971) dominated.

The later national Mussel Watch Program in the USA designed by Goldberg and co-workers also incorporated spatial and temporal trends in aquatic contamination as major objectives. Goldberg *et al.* (1978) commented:

> ... the immediate goals of the Mussel Watch are to assess pollutant levels and their changes with time in a given coastal zone, and to compare these levels and changes among different zones.

However, other objectives also existed in this programme. Farrington *et al.* (1983), commenting on the same studies, stated:

> ... we seek to maximize the detection of Kepone-type incidents in the early stages and we think that the Mussel Watch concept offers promise in this regard.

In addition, a sample bank was established, providing material which could be analysed for newly-discovered contaminants of concern at a later time, utilising novel analytical techniques which were unavailable in the mid-1970s (Goldberg *et al.*, 1978).

Phillips and Segar (1986) have argued that, while these objectives are clearly laudible, they were not fully met by the programme design. Thus, only two sites were subjected to monthly monitoring, and bivalves were collected from the remaining locations at widely differing times of year (Goldberg *et al.*, 1978). This design differs radically from the programme of Butler discussed above, yet the stated objectives were essentially identical. The later programme can only be considered to have established true spatial trends in contamination if it is assumed that seasonal changes in contamination do not vary between sites. However, most data from investigations elsewhere show highly significant differences between sites in contaminant seasonality in bivalves (Phillips, 1980), and there is no evidence to suggest that the seasonal changes in contamination found at the two sites studied in the original US Mussel Watch Program (Narragansett Bay and Bodega Bay) were typical of those occurring at the other locations included in the programme.

In addition, the probability of successful detection of "Kepone-type incidents" (Farrington *et al.*, 1983) would be low using the methodology employed in the US Mussel Watch Program, at least in certain circumstances. Thus, the reliance of this programme upon single annual samples from each study site (excepting the two locations subjected to monthly collections for studies of seasonal fluctuations) implies that the detection of contaminants of short half-life in the bivalves studied would be unlikely, at least if the ambient contamination were of an episodic nature. The distinction here is whether monitoring programmes attempt to provide a "snapshot" of contamination conditions at any one point in time (which requires samples to be taken concurrently at all sites) or prefer to elucidate average conditions over time (which in the case of contaminants of short half-life in biomonitors requires the time-bulking of samples; see Phillips and Segar, 1986).

The more recent National Status and Trends Program of the National Oceanic and Atmospheric Administration (NOAA), which replaced the US Mussel Watch Program of the mid-1970s, has included certain changes to the sampling philosophy. Thus, for example, Sericano *et al.* (1990) reported that the bivalve samples taken from sites in the Gulf of Mexico were collected over either a 2- or 3-month period, which is a significant improvement over the extended sampling period reported by Goldberg *et al.* (1978). The precise relevance of this to the data resulting from biomonitoring surveys will be discussed later in this chapter; however, it may be noted here that it is of importance where spawning times differ between locations (which is a common finding for many bivalve mollusc populations; see Seed, 1975; Phillips, 1980).

Interestingly, the national Mussel Watch studies in the USA have been complemented in some States by regional programmes, and certain important differences may be noted between the national and regional methodologies. The best known of the State Mussel Watch Programs is that of California, where significant contaminant sources are present from both contemporary and historical activities (Martin and Castle, 1984; Martin, 1985; Smith *et al.*, 1986; Martin *et al.*, 1988; Phillips and Spies, 1988).

The Californian programme employs transplanted mussels (derived from Bodega Head, a relatively uncontaminated site) rather than relying on the sampling of wild populations, and all samples are deployed at the study sites concurrently, for periods from 3 to 4 months. This provides clear benefits in terms of the interpretation of spatial trends in the data, as a true "snapshot" of contaminant bioavailability (over the period of transplantation) is produced. The continuation of the programme from year to year, with transplantation occurring at the same time each year, also assists in providing data which are meaningful for the interpretation of temporal trends. Unfortunately, however, there has been a tendency in recent years to alter site locations (usually to permit attempts to monitor closer to hot-spots of contamination), and the temporal database has been degraded as a result (Martin and Richardson, 1991). Nevertheless, the Californian programme is probably the best-designed of all the US-based biomonitoring work undertaken to date.

It may be concluded that researchers wishing to commence new biomonitoring programmes for trace contaminants in aquatic ecosystems should clearly identify their precise objectives prior to continuing with more detailed aspects of the design of the programme. This is because the objectives essentially define not only the particular type of biomonitor to be utilised (see below), but also the sampling strategy and design. Much of the biomonitoring work undertaken to date has suffered from an inattention to this most basic of requirements, with consequent problems in the interpretation of data.

C. THE SELECTION OF A BIOMONITORING SPECIES

Once the overall objectives of a new biomonitoring programme have been defined, the researcher will wish to select a species to be employed in the investigations. The basic attributes required for a species to act as an efficient biomonitor have been discussed in Chapter 4, and will not be repeated in detail here. However, several of the more sophisticated aspects of species selection merit consideration here, and these include the following:

- The physicochemical form of the contaminant(s) likely to be present in the water bodies studied, especially the degree of association of the contaminants of interest with suspended particulate matter.
- The availability of wild populations of potential biomonitoring species in the study area, and their biology and ecology.
- The potential for transplantation of a desired biomonitor from locations within or outside the study area to all sites to be included in the monitoring effort, and the possible implications of the introduction of exotic species.
- The degree of spatial definition to be attained in the biomonitoring survey.

With respect to the first of these, it is important to note that the various types of biomonitors which are available for use in any given study respond to different portions of the contaminant load present in the aquatic biosphere. This is of particular importance with respect to the study of trace elements in aquatic ecosystems, as these contaminants may exist in significant amounts both in solution and in association with particulate material. For example, macroalgae are known to accumulate contaminants principally from solution rather than from suspension, whereas filter feeders such as bivalve molluscs take up contaminants from both these phases. Bivalves not only take up contaminants from their food, but may also strip those which are adsorbed to any ingested inorganic particulates.

This distinction has been emphasised in studies in the United Kingdom and Scandinavia, and a few examples are useful here to illustrate the distinct capabilities of different biomonitoring species. Bryan and Hummerstone (1977) noted that the patterns of silver contamination in different types of biomonitors from the Looe Estuary in the United Kingdom varied significantly, according to the capability of the organisms studied to respond to silver adsorbed to particulate matter (Fig. 78). Macroalgae and herbivorous limpets exhibited little variation in silver levels with site in the estuary, whilst cockles and other filter feeders contained greater concentrations of the element in upriver locations, indicating the presence of higher levels of silver adsorbed to particulates (and available only to filter feeders) in the upper estuary.

Ireland and Wootton (1977) produced similar data for herbivorous and carnivorous gastropods around the coasts of Wales, and Phillips (1979a)

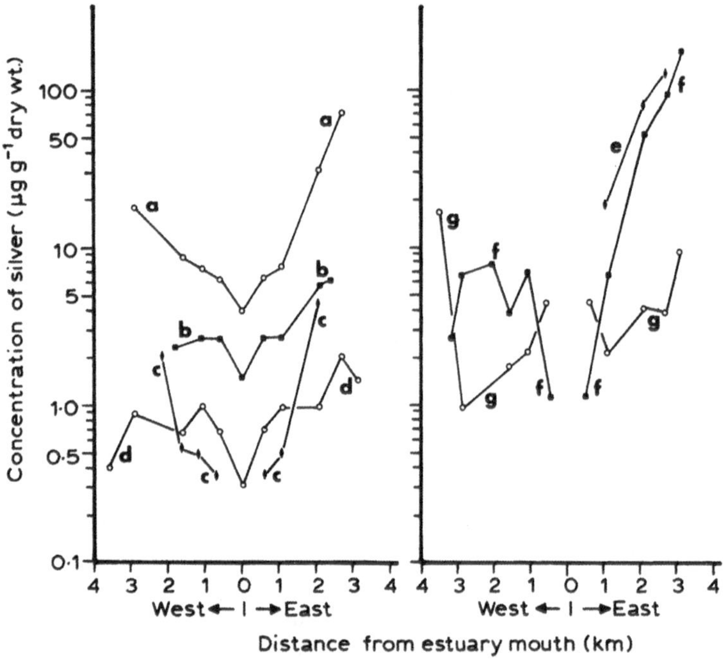

Fig. 78. Concentrations of silver ($\mu g\ g^{-1}$ dry weight) in the tissues of a range of biomonitors in the Looe Estuary, England. Species shown are (a) the gastropod *Littorina littorea*; (b) the limpet *Patella vulgata*; (c) the cockle *Cerastoderma edule*; (d) the brown alga *Fucus vesiculosus*; (e) the clam *Macoma balthica*; (f) the clam *Scrobicularia plana*; and (g) the polychaete *Nereis (Hediste) diversicolor*. After Bryan and Hummerstone (1977).

found that the profiles of trace metal contamination exhibited by the macroalga *Fucus vesiculosus* and the mussel *Mytilus edulis* in the Sound (Öresund) in Denmark differed, again due to the variations between the species in their response to metals adsorbed to particulates. Similar data have been reported more recently by other authors (e.g. Bryan and Gibbs, 1983; Bryan *et al.*, 1985; Langston, 1986), and Phillips (1979a, 1980) considered that the use of several types of biomonitors at each location could be of benefit in certain circumstances, to differentiate the bioavailability of metal loads adsorbed to particulates from that present in solution.

The distributions of wild populations of potential biomonitoring organisms often dominate decisions taken on the selection of species for use in monitoring surveys. Thus, some authors will simply sample the most abundant naturally-present species of alga, mollusc, crustacean or fish, assuming that these will act as efficient biomonitors of the contaminants of interest. This approach is totally inappropriate, and often gives rise to the use of organisms which either regulate their accumulation of particular contaminants (Phillips and Rainbow, 1988), or are sufficiently mobile to render interpretation of data on contaminant distributions impossible.

The approach which should be taken is exemplified by certain of the early studies on species which are now accepted as established biomonitors, and by some of the more recent investigations in areas where no established biomonitors are present, and the bioaccumulative capability of a range of species was studied prior to any decision being made on the most appropriate organism for routine biomonitoring.

With respect to the first of these, the acceptance of the mussel *Mytilus edulis* as a biomonitor developed through several studies in a wide variety of locations, these involving the testing of contaminant uptake kinetics and the comparison of spatial patterns of contaminant abundance in mussels with the distributions of known inputs of such contaminants. Using laboratory investigations, Schulz-Baldes (1972, 1973) established that the accumulation of lead by *M. edulis* occurred in simple proportion to the amount of lead in the ambient water. This was followed by careful studies of the distribution of lead in mussels from the Weser Estuary and German Bight, which exhibit a defined gradient of contamination by this metal (Schulz-Baldes, 1974). Phillips (1976a,b) undertook extensive investigations in both laboratory and field conditions of the net uptake of several trace elements by the same species. The effects of external variables such as salinity and temperature on the accumulation of metals by *M. edulis* were studied, and the patterns of contamination of mussels in Port Phillip and Western Port Bays in Australia were compared to licensing data on industrial discharges.

Such investigations laid the basis for a robust understanding of the capability of *M. edulis* to act as an accurate biomonitor of many trace elements, and later work extended this to additional contaminants, to other species of bivalve molluscs and to a range of new environments in the temperate zones. It is important to note that not all attempts to confirm the capability of the various organisms studied to act as efficient biomonitors were successful, even for bivalves, which are generally considered to be among the most useful species for such purposes. Thus, for example, the accumulation of copper by *M. edulis* is affected by several variables which cannot be accounted for in the design of sampling stra-

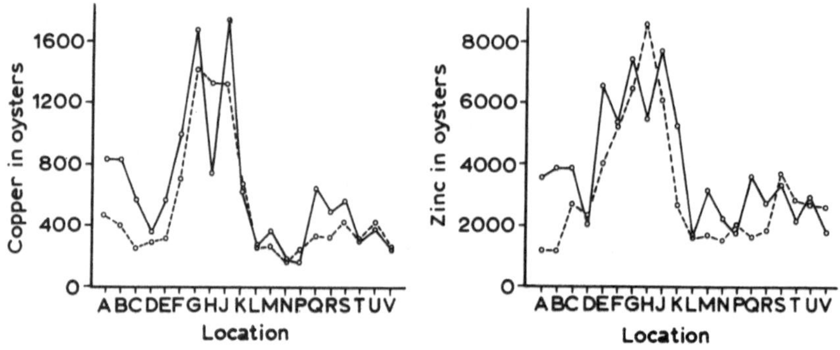

Fig. 79. Concentrations of copper and zinc ($\mu g\,g^{-1}$ dry weight) in whole soft parts of rock oysters of the genus *Saccostrea*, from coastal waters of Hong Kong. Locations A–E are in the west of the Territory; sites F–J are in the contaminated area of Victoria Harbour; sites K–V are in the east and south of the Territory. Dashed lines refer to a survey from March and April 1978, after Phillips (1979b); solid lines refer to samples taken in August 1978, after Phillips and Yim (1981).

tegies, and this species is therefore not considered to be a useful biomonitor of copper under certain conditions (Phillips, 1976a, 1990b; Davenport, 1977; Davenport and Manley, 1978; Manley, 1983; Redpath, 1985).

The development of biomonitoring techniques in subtropical and tropical zones has lagged significantly behind such studies in temperate waters (Phillips, 1991). While some early investigations of trace metal and organochlorine distributions were undertaken in areas such as the Gulf of Thailand (Huschenbeth and Harms, 1975), these covered a wide range of species and were of greater relevance to public health issues than to biomonitoring *per se*. However, studies in the late 1970s and 1980s in both Hong Kong and north-eastern Australia have provided potential candidates for contaminant biomonitoring in large areas of the tropical zone.

Hong Kong was ideal for such investigations, as a well-defined contamination gradient exists for both trace metals and organochlorines; this generally radiating outwards from the main urbanised area of Victoria Harbour to the relatively less polluted waters to the west and east of the Harbour (Phillips, 1989). The initial studies here involved rock oysters of the genus *Saccostrea*, and confirmed the elevated abundance of several trace elements in and around Victoria Harbour (Fig. 79; Phillips, 1979b; Phillips and Yim, 1981). However, while this species appeared to be a capable biomonitor of many metals, taxonomic uncertainties surrounding the genus as a whole suggested that its use over a wider geographical area might be more problematical, as the various different species of the genus *Saccostrea* co-exist in several parts of the Indo-Pacific, and are easily confused with each other. This led to further work in Hong Kong on the mytilid *Septifer virgatus* and the Pacific oyster *Crassostrea gigas* (Phillips and Yim, 1981; Phillips *et al.*, 1982), but neither of these was found to be ideal for biomonitoring studies in the region.

The problem was solved through additional investigations in both Hong Kong and Thailand, these involving the green-lipped mussel *Perna viridis* (Phillips, 1985b; Phillips and Muttarasin, 1985). *P. viridis* exhibits a wide distribution in the Indo-Pacific, and is also a significant dietary item in parts of Asia. It was shown to accurately reflect the distributions of most metals in Hong Kong waters (Table 25), although it partially regulates the content of zinc in its soft tissues (Phillips, 1985b; Chan, 1988; Phillips and Rainbow, 1988). Interestingly, the later studies by Phillips and Rainbow (1988) also identified several species of barnacle (*Balanus amphitrite, Capitulum mitella* and *Tetraclita squamosa*) to be of potential use in monitoring trace metals in subtropical and tropical environments. Despite their cosmopolitan distribution, barnacles have received insufficient attention as biomonitors, in either temperate or tropical waters (Rainbow, 1987).

Perna viridis was also shown by Phillips (1985b) to be a capable biomonitor of organochlorine pesticides and PCBs (Table 25), reflecting the considerable contamination of Hong Kong coastal waters by DDT and its metabolites, HCH, and (in one site in particular) PCBs. The extreme contamination of northern Junk Bay (to the east of Victoria Harbour) by PCBs was studied in detail, providing evidence both of continuing discharges of PCBs to Junk Bay and of the uptake of the highly toxic and persistent coplanar PCBs by mussels (Tanabe *et al.*, 1987b; Kannan *et al.*, 1989).

The studies of Burdon-Jones and co-workers also merit discussion here with respect to the development of biomonitors in tropical environments. These authors completed extensive investigations of trace elements in many species in the Great Barrier Reef area of Australia, emphasising bivalve molluscs and macroalgae in particular. Klumpp and Burdon-Jones (1982) studied the accumulation of metals in nine species of bivalves, using both wild populations and transplantation techniques to identify the biomonitoring capability of the various species. They considered that the most efficient biomonitor amongst the species investigated was the hairy

Table 25 Concentrations of trace elements and organochlorines in the whole soft parts of green-lipped mussels (*Perna viridis*) in Hong Kong waters. Sites 1 and 2 are in the west of the territory; sites 3–11 are in the urbanised area around Victoria Harbour; and sites 12–15 are in the north-east of the territory (after Phillips, 1985b)

Sample	*Location*	*Cadmium*	*Copper*	*Lead*	*Mercury*	*Zinc*	*HCB*	*HCH*	*PCBs*	*DDT*	*DDE*	*DDD*	*ΣDDT*
1	Chek Lap Kok	1.23	10.5	5.1	<0.11	128	<30	<30	<60	<30	<30	26	NC
2	Reef Island	1.44	10.2	3.1	<0.11	126	<30	211	<60	<30	<30	30	NC
3	Kennedy Town	0.29	11.5	12.6	<0.11	115	<30	<30	<60	745	106	213	1064
4	Mei Foo	0.31	30.3	18.0	<0.11	118	<30	92	<60	<30	86	221	NC
5	Queens Pier	0.19	15.7	9.4	<0.11	126	<30	61	<60	306	61	123	490
6	Kowloon Pier	0.21	22.8	8.3	<0.11	118	<30	<30	<60	439	<30	568	NC
7	Hung Hom	0.18	31.6	18.2	<0.11	164	<30	52	<60	<30	75	185	NC
8	Causeway Bay	0.07	15.6	7.8	<0.11	137	<30	88	1696	760	140	415	1315
9	North Point	0.18	38.1	15.5	0.12	149	<30	49	<60	219	67	213	499
10	Kwun Tong	0.29	278.5	19.3	0.14	129	<30	44	<60	56	69	131	256
11	Rennies Mill	1.43	16.0	60.5	<0.11	143	<30	<30	1904	466	667	910	2043
12	Sha Tin	0.55	29.4	7.5	<0.11	89	<30	<30	131	231	237	400	868
13	Tai Po Kau	0.38	20.1	2.7	<0.11	88	<30	<30	<60	54	60	223	337
14	Wu Kwai Sha	0.30	20.8	4.3	<0.11	77	<30	<30	<60	364	121	636	1121
15	Lai Chi Chong	0.59	8.5	1.4	<0.11	79	<30	<30	216	43	<30	56	NC

NC: not calculated (some components present below the limit of detection) mussel *Trichomya hirsuta*, although tissue levels of zinc were apparently partially regulated by this species.

Later work by Burdon-Jones and Denton (1984) extended to several species of giant clams (*Tridacna* spp.), which were found to accumulate large quantities of metals in their kidneys and showed considerable promise as tropical biomonitors. Extensive analyses of macroalgae in the same region of Queensland were also reported by this group (Burdon-Jones *et al.*, 1982; Denton and Burdon-Jones, 1986). It was concluded that most species of macroalgae were sensitive to the impacts of external factors such as salinity and temperature on trace metal accumulation, and that these effects reduced their usefulness as biomonitors. However, the green alga *Chlorodesmis fastigiata* was thought to be of potential for studies of metal distributions in tropical environments, as it was relatively unresponsive to the effects of external variables and appeared to faithfully reflect the ambient abundances of trace elements.

Such investigations are useful here to illustrate the care needed in the selection of a biomonitoring species for the study of contaminants in aquatic ecosystems. Where no previously-studied biomonitor exists, researchers should consider all known aspects of the biology and ecology of potential biomonitors, and should select species on the basis of their bioaccumulative capabilities and their adherence to the prerequisites for biomonitors noted in Chapter 4, rather than simply on their abundance in the study area. Where few previous data exist on contaminant accumulation by the species of interest, laboratory investigations of contaminant kinetics can provide useful information on the capacity of an organism to act as an efficient biomonitor, and these should be undertaken prior to the commencement of monitoring. Biological data are useful particularly in respect of the reproductive cycle of the species under consideration, as this may affect contaminant kinetics markedly and may thus influence the timing of sampling (see below).

Where wild populations of a species to be used as a biomonitor are not sufficiently widespread to support a monitoring programme, transplantation may be employed. The techniques for the transplantation of macroalgae, bivalves and other species are now well-developed (e.g. see Young *et al.*, 1976; Curran *et al.*, 1986; Green *et al.*, 1986), and should present no particular problems in most locations (although in certain parts of the world, protecting the deployed cages from disturbance by fishermen and recreational yachtsmen can present a challenge). Indeed, some authors have suggested that transplantation should be preferred to the use of wild populations in biomonitoring programmes (e.g. Ritz *et al.*, 1982), and this argument has merit in certain circumstances. Thus, it is possible that the use of hatchery-reared individuals may give rise to lower inherent variability in contaminant concentrations in populations of organisms such as bivalve molluscs (see below and Boyden and Phillips, 1981). In addition, the use of transplantation techniques artificially defines the time interval over which contaminant bioavailabities are measured by the chosen biomonitor (Ritz *et al.*, 1982; Phillips and Segar, 1986).

It should be noted here, however, that the transplantation of biota may in certain circumstances give rise to the unwanted introduction of exotic species (either those transplanted, or other organisms introduced inadvertently with the transplants). Perhaps the best example of this concerns the benthic fauna of San Francisco Bay, which is now dominated in terms of both species numbers and biomass by opportunistic species which were introduced from elsewhere (many from the east coast of the USA, with oysters; see Carlton, 1979).

The selection of species for biomonitoring surveys should also be undertaken with a clear view of the degree of spatial definition required from the monitoring programme, and this refers particularly to the issue of the mobility of the species selected. If a monitoring survey is to cover a small area in detail (e.g. an estuary), the selected biomonitor should be either sessile or sedentary in nature, such that it will reflect conditions at or near to the precise site of sampling. Finfish are clearly inappropriate for such studies, even where these are demersal and territorial in nature. However, if the region to be surveyed is extensive, rather more mobile species may be selected, such as demersal finfish (e.g. Johnels *et al.*, 1967; Dix *et al.*, 1976; Moilanen *et al.*, 1982; Schüler *et al.*, 1985), although a robust knowledge of their biology is required (e.g. their seasonal migrations for spawning or other reasons). An interesting example of the impacts of the latter is provided by the data of Smith and Cole (1970) on winter flounder (*Pseudopleuronectes americanus*) in the Weweantic Estuary in Massachusetts. Adult flounders from this location revealed different seasonal profiles of pesticides from those of juveniles, due to the migration of the sexually mature adults into and out of the estuary each year related to spawning, whilst the juvenile fish were resident year-round in the contaminated waters within the estuary (see Phillips, 1980). In broad-scale national surveys, even migratory or highly mobile species such as seals may provide useful data on regional differences in the contamination of water masses (see Chapter 6), although a knowledge of the routes and timing of migrations is essential if the monitoring data from such species are to be correctly interpreted.

This issue of the "spatial integration" of data provided by a biomonitor is fundamental to the success or otherwise of contaminant monitoring surveys. If organisms are selected which are unrepresentative of their area of sampling, all attempts at meaningful interpretation of analytical data will be frustrated.

D. THE TIMING AND FREQUENCY OF MONITORING

Having selected a species to act as a biomonitor, decisions are required on the precise timing and frequency of the monitoring surveys to be undertaken. A variety of factors must be considered here, the most important of these being:

- The available information on contaminant fluctuations in local waters, or (in the absence of such data) on the seasonal changes in factors which might influence the local abundance of contaminants.
- Biological changes occurring in the selected biomonitoring species with season (especially with respect to reproduction).
- The kinetics of uptake and depuration of the contaminants of interest in the selected biomonitoring species.

In some instances, pre-existing data will be available on contaminant abundance in natural waters, sediments or biomonitors in the study area, and these will obviously be of assistance in defining decisions on the most appropriate timing of monitoring surveys. Most natural waters exhibit seasonal changes in contaminant abundance (e.g. see Chapter 6), these occurring because of meteorological factors (rainfall and runoff being generally the most important), due to the episodic nature of contaminant sources, or for other reasons.

It must be recognised here that the most appropriate timing for monitoring surveys in any given area may vary for the different types of contaminants. Thus, for example, in industrialised estuaries which drain agricultural catchments, both trace metals and organochlorine pesticides may be of concern. It may be anticipated in such catchments that the seasonal peak in pesticide abundance in the receiving waters will depend on the interplay between the timing of pesticide applications and rainfall (Butler *et al.*, 1972; Phillips, 1980). However, if industrial sources of trace metals are predominant, the maximum element concentrations in receiving waters may occur at times of lowest rainfall (i.e. when least dilution is available from the relatively uncontaminated receiving waters from upstream). It is therefore important for researchers to consider the comparative magnitudes of individual contaminant sources to a study area, and plan their monitoring surveys accordingly. In certain cases, it must be accepted that the most appropriate timing for a survey of pesticide abundance will be distinct from that for the monitoring of trace elements, and discrete sampling should be undertaken in such instances for the study of each major class of contaminants (NAS, 1980; Phillips and Segar, 1986).

The most important biological aspect relating to biomonitors and their use to measure contaminant abundance is the annual cycle of growth and reproduction, and several examples are provided here to illustrate the potential impacts of this phenomenon on the data from monitoring surveys. Data on bivalve molluscs are emphasised here, but similar arguments may be made for many other species.

Phillips (1976a) concluded from studies of the seasonal variation of trace elements in the mussel *Mytilus edulis* that temporal fluctuations of metal concentrations were due in many cases to changes in tissue weights of the mussels sampled, these occurring principally because of the gametogenesis and spawning cycle. Later data from studies elsewhere in the temperate zone tended to support this hypothesis (e.g. Cossa *et al.*, 1979; Simpson, 1979; Boyden and Phillips, 1981), but not all populations of bivalves exhibit a well-defined seasonal change in tissue weights. Thus, certain bivalve populations (especially in the tropical zone, but also in some locations in temperate waters) exhibit an extended period of spawning, and the temporal changes in metal levels therefore tend to be less well defined in these populations (e.g. see Goldberg *et al.*, 1978).

Even in temperate waters, sub-populations of mussels present in a given location may vary significantly in condition. Seed (1975, 1976) presented extensive information on this aspect (Fig. 80), and it is evident from these and other data that completely synchronous spawning is a relatively rare event in many bivalve populations. This has considerable importance in monitoring surveys which seek to compare contaminant concentrations (or more strictly, contaminant bioavailability) at different locations using biomonitors, and it has been concluded that the most appropriate timing of sampling is the period over which least fluctuations occur in tissue weight or animal condition (NAS, 1980; Phillips, 1980). In many cases, this will be in the pre-spawning period, which is also often the preferred time for the sampling of biomonitors for organochlorine or hydrocarbon analyses (as lipid weights are highest at this time). However, individual populations and sub-populations vary with respect to their precise gametogenesis and spawning cycles (Fig. 80), and a knowledge of the

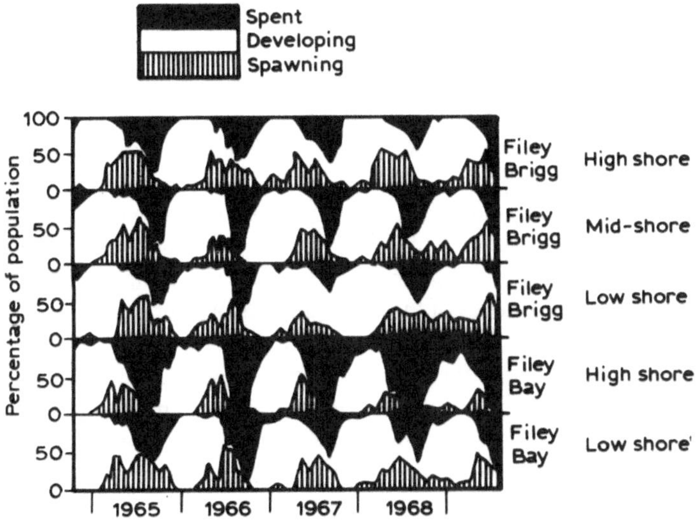

Fig. 80. Seasonal variations in the percentage distributions of spent, developing and spawning individuals in five sub-populations of the mussel *Mytilus edulis* in England between 1964 and 1969. After Seed (1975).

reproductive biology of the selected biomonitor at the various study locations is essential if the impacts of seasonal changes in animal condition on contaminant monitoring surveys are to be minimised.

It might also be noted here that some authors have attempted to circumvent this problem either by the use of sexually immature biomonitors (e.g. Cossa *et al.*, 1979) or through the stripping of gonads from the samples to be analysed (e.g. in the Californian State Mussel Watch; see Martin, 1985). The latter technique is not always simple to apply in the case of bivalve molluscs, however, as the gonads are intimately associated with other tissues in certain species (Bayne, 1976).

The most appropriate frequency of sampling in biomonitoring surveys is defined by two main considerations: the study objectives; and the kinetics of uptake and depuration of the contaminants of interest in the selected biomonitor. If the study seeks to provide a true "snapshot" of contaminant availability at a given time (and perhaps compare these data with similar analyses completed at a different time), the researcher is faced with few problems other than defining the most appropriate timing of sampling to provide the "snapshot". As noted above, this should be determined through a consideration of meteorological factors, known or inferred changes in contaminant abundance, and the biology of the selected biomonitor. This approach is typified by the Californian State Mussel Watch Program, which is designed to produce an annual picture of contaminant bioavailability at the various sites studied, providing an indication of the success or otherwise of regulatory initiatives (Martin, 1985; Martin and Richardson, 1991).

However, if the intention of a monitoring survey is to provide data on the average bioavailabilities of contaminants over the whole of a given period, greater problems arise. In this event, it is most important to amass information on the kinetics of the contaminants of interest in the biomonitor selected for study, as the rates of uptake and depuration of these contaminants will define the degree of time-integration of ambient conditions offered by the particular species employed (Phillips, 1980).

Many trace metals are depurated relatively slowly from biomonitors, with biological half-lives varying from months to years in certain cases (e.g. see NAS, 1980). However, as discussed in Chapter 6, some organochlorines and many hydrocarbons exhibit much more rapid kinetics of uptake and depuration in biomonitors, and the elucidation of the time-averaged bioavailabilities of these contaminants in the field can therefore be challenging (especially where their abundances are known to fluctuate episodically). In the case of these contaminants of short half-life in biomonitors, it is recommended that researchers consider the use of time-bulking techniques if a picture of average contamination over an extended period is desired (Phillips and Segar, 1986). The use of transplantation may also be considered in such situations, as it provides an initial starting point for the uptake of contaminants as well as defining the period over which the time-integration of ambient contamination conditions may extend (e.g. see Ritz *et al.*, 1982; Martin, 1985; Tanabe *et al.*, 1987b; Kannan *et al.*, 1989).

It is notable here that the sampling in biomonitoring surveys for organochlorines and hydrocarbons has varied considerably in frequency from study to study. Monthly sampling has been preferred by certain authors (e.g. Butler, 1971); quarterly sampling by others (Claisse, 1989), and longer periods extending up to a year have been employed in studies such as the US Mussel Watch Program and the more recent derivation of this programme run by the National Oceanic and Atmospheric Administration (Goldberg *et al.*, 1978; Lauenstein *et al.*, 1990; Sericano *et al.*, 1990). It is clear that the studies undertaken at longer intervals do not truly reflect ambient contamination conditions over the entire inter-survey period for contaminants of shorter half-life. For example, kinetic data from other studies (e.g. Tanabe *et al.*, 1987b) show that in the case of PCBs, the time-integration offered by bivalves extends from a few days (for PCBs of low chlorination) to several weeks (for higher-chlorinated homologues). The time-bulking of samples is the only method proposed to date to overcome this problem in surveys where a longer degree of time-integration of ambient contamination conditions is required (Phillips and Segar, 1986).

E. SITE SELECTION

Biomonitoring surveys may be designed in several fashions with respect to site selection, depending on their precise objectives (Phillips and Segar, 1986). Thus, many surveys of the spatial distribution of contaminants seek to identify so-called "hot-spots" or areas of particular contaminant abundance, whereas others prefer to delineate conditions over broader scales, in water masses. Indeed, these are not always mutually exclusive, and both aims may be satisfied by some spatial surveys, through the careful selection of sampling sites close to known or suspected sources of contaminants, and in reference or control locations elsewhere.

Similarly, investigations of temporal changes in contaminant abundance or bioavailability may also be designed using sites close to known contaminant sources (perhaps as a method of monitoring improvements in such discharges, allied to regulatory initiatives) or those in more distant locations (reflecting temporal changes in the contamination of water masses, rather than conditions at sites which are directly impacted by discharges).

Where water masses are of greatest interest and the selected biomonitor is sedentary or sessile in nature, the option of "space-bulking" of samples may be considered (Phillips and Segar, 1986). This involves the collection of biomonitoring samples at discrete sites within a larger area, but their bulking for analysis; it essentially provides an artificial method of smoothing out the more extreme fluctuations in contaminant bioavailabilities which may be encountered at only one or a few of the sites in a study area.

Most authors attempt to include both highly contaminated and reference or control sites in monitoring surveys, and some researchers design programmes to include gradients of suspected contamination. Besides providing comparative data on such sites, this has particular merit where the selected biomonitor has been little-studied previously, as useful data from reference locations will be provided on the "natural" fluctuations in contaminant levels in the biomonitor used, which may be due to factors such as changes in tissue weights or lipid levels with season. These may then be contrasted with the fluctuations seen at sites close to contaminant sources, where the biomonitor responds not only to changes caused by natural factors, but also to the additional contaminant sources and their temporal variations.

Very little detailed argument has been published concerning the rationale behind the selection of sites for biomonitoring surveys. However, certain factors are particularly worthy of consideration at this stage of the design of a monitoring programme, and these merit brief discussion here. The most important factors influencing site selection for biomonitoring surveys are generally as follows:

- In the case of coastal surveys, the presence of freshwater inflows in the study area, and the nature of the catchment drained by these.
- The existence and extent of urban areas in the region of study, and the locations of sewage outfalls.
- The siting and types of industries in or near the study area.
- The hydrodynamics of the region, particularly in estuarine locations.

It is evident that the study of land use characteristics and particularly of the types of agricultural practices and industries present in a region will provide much useful data on the contaminants which should be measured in any biomonitoring survey (see also Chapter 3). For example, it is possible in most parts of the world to access data on pesticide imports and/or utilisation rates, and these are often useful in defining the particular organochlorines likely to be present in riverine locations or estuaries. Similarly, particular types of industries have been documented to produce discharges containing specific trace metals and other contaminants, and this knowledge can be useful in decisions on both the siting of sampling points and the contaminants to be analysed in biomonitoring surveys.

The hydrodynamics of the water body of interest are particularly relevant to decisions on site selection in biomonitoring surveys, especially where known or suspected gradients in contamination are to be monitored. Where no data exist on the hydrodynamics of the study area, float or drogue tracking or even mathematical modelling studies of water movements may be considered useful, to support decisions on site selection and to provide background information to assist in the interpretation of analytical results.

F. OVERCOMING THE EFFECTS OF VARIABLES

By contrast to the dearth of data concerning site selection, a very large database exists describing the impacts of both biological and environmental variables on the uptake and depuration of contaminants by biomonitors. Much of this information is covered elsewhere in this book, and exhaustive detail is not provided here. However, the potential effects of variables which should be considered at the design stage of biomonitoring surveys merit particular mention here, and these are the biological variables of animal condition, size and sex.

The importance of animal condition has already been alluded to in the section above on the timing of monitoring surveys. The condition of a biomonitor with respect to both the growth and reproductive cycles is known to significantly affect its bioaccumulation of contaminants, and this should be taken into account in the design of surveys. As noted previously, in most types of surveys an attempt should be made to sample at the period of least change in organism condition, as this will tend to minimise the condition-based differences in contaminant concentrations between study locations. For bivalve molluscs, the most appropriate sampling time is often a few months prior to spawning; however, this varies between different species and populations, and a knowledge of the biology of the selected biomonitor in the chosen study area is vital if the correct decisions are to be made in this respect.

As noted in Chapters 5 and 6, contaminant concentrations in biomonitors may vary significantly with organism size, weight or age. The reasons for these variations are not entirely clear with respect to trace metals (see Boyden, 1974, 1977), particularly as the size-concentration relationships seen may vary from site to site and with time at any one location (Phillips, 1980; Strong and Luoma, 1981). By contrast, such variations as occur for organochlorines in biomonitors are nearly always associated with changes in lipid weights with size, weight or age of a species (Phillips, 1978, 1980), and direct correlations between size and organochlorine levels are the most common by far.

Whatever the reasons for such variations, they should be taken into account in the design of sampling strategies for biomonitoring surveys. The researcher has four options with respect to this variable:

- The potential impacts of organism size on the data from biomonitoring surveys may be ignored, and samples differing in size (age, tissue weight) may be accepted from the various monitoring locations studied.

- An attempt may be made to employ only those biomonitors which do not exhibit significant changes in contaminant concentrations with size.
- A normalisation procedure may be employed, to attempt to rationalise all data to a given "standard organism".
- A restricted range of organism sizes or tissue weights may be demanded for samples from all study locations.

The relative merits and practicalities of these approaches differ from case to case, depending on the particular biomonitor studied and the contaminants to be analysed. Thus, for example, the concentrations of many trace elements vary considerably with size in most bivalve mollusc populations (e.g. see Boyden, 1974, 1977; Phillips, 1980), whereas the variation in organochlorine or hydrocarbon concentrations with size in bivalves is generally much less and is often absent altogether (Phillips, 1980). By contrast, finfish species frequently exhibit size-dependent variations for all contaminant classes, these usually being due in the case of organochlorines to the increase in tissue lipids with size in many freshwater and marine fish (see Chapter 6 for examples).

In any event, the extensive database produced to the present concerning size-dependent variations of contaminant concentrations in biomonitors reveals that such variations are common, and may in certain instances be highly significant compared to inter-site differences in contaminant levels in biomonitoring species. This implies that the effects of organism size on the interpretation of data from biomonitoring surveys should not be discounted, and that effort should be expended to eliminate this variable at the time of sampling.

The second option above (i.e. the use of biomonitors which do not exhibit size-dependent variations in contaminant concentrations) is generally impractical, as such variations are encountered so frequently. It should be noted here that the absence of significant size-concentration regressions for any given contaminant in a biomonitor from particular locations should not be taken as evidence of the universal lack of such relationships for that species and contaminant. Thus, for example, the variations in trace metal concentrations with size in bivalve molluscs may differ between populations, or even within a population with time (Phillips, 1976a, 1980; NAS, 1980; Strong and Luoma, 1981). An example of this phenomenon is provided in Fig. 81, and this emphasises the

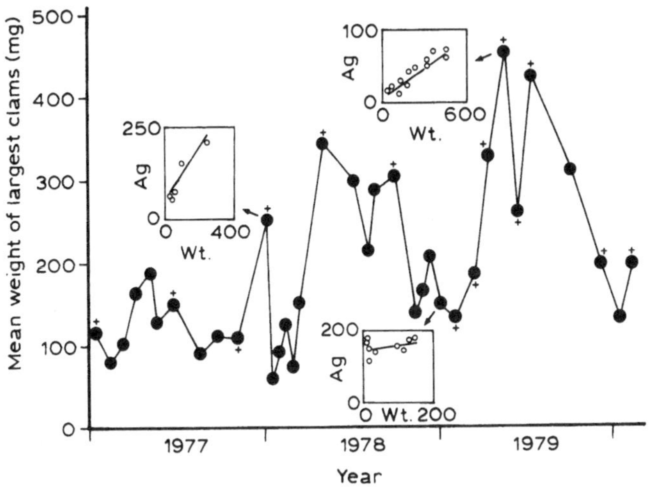

Fig. 81. Temporal variations in the mean dry weight (mg) of the largest size pool of clams (*Macoma balthica*) collected at a site in San Francisco Bay between 1977 and 1980. Points marked with (+) denote the presence of significant effects of tissue weight on silver concentrations in these samples. Examples of three relationships (expressed as silver concentrations in $\mu g\, g^{-1}$ dry weights against dry tissue weight in mg) are shown as insets. After Strong and Luoma (1981).

considerable effect which organism size may exert in biomonitoring surveys.

In view of the above arguments, the researcher is left with a choice between two alternatives in attempts to account for the size-related variable in biomonitoring investigations: the use of a normalisation procedure, or the sampling of organisms of a restricted size range at each study site.

Normalisation procedures were first employed by Johnels *et al.* (1967) in their classic studies of mercury in pike (*Esox lucius*) in Sweden. It was noted that the comparison of mercury bioavailabilities at different locations through the analysis of the element in axial muscle tissues of pike was affected by the size of fish sampled at each site. This was due to an increase in mercury concentrations with size in *E. lucius*, fish from the locations of greater contamination exhibiting a more rapid increase in element levels with length compared to those from relatively pristine sites (Fig. 82). It was impossible to sample pike of identical sizes from each location investigated, although a range of fish sizes was available at most

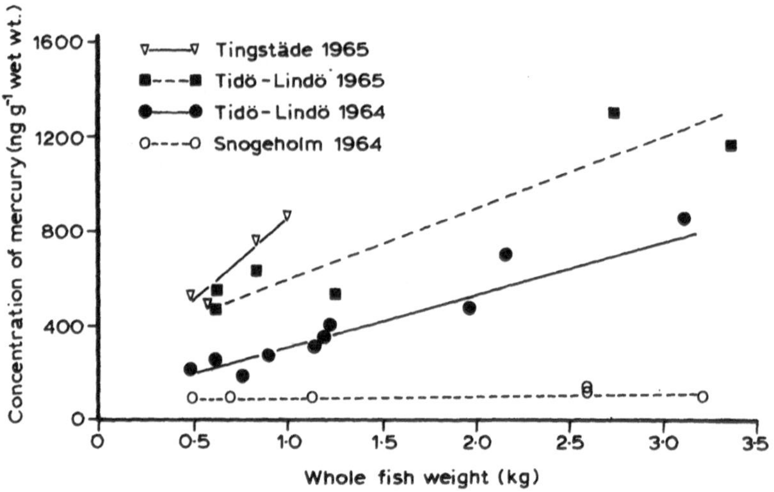

Fig. 82. Relationships between mercury concentration (ng g^{-1} wet weight) and whole fish weight for pike (*Esox lucius*) from various lakes in Sweden during 1964 or 1965. After Johnels *et al.* (1967).

study sites. Johnels *et al.* (1967) thus opted to analyse pike varying in size at each location, hence deriving a size-concentration regression slope for each site (Fig. 82). Thereafter, the relative levels of contamination (i.e. relative mercury bioavailabilities to pike) at each location could be compared by statistically deriving data for a notional "standard 1-kg pike" at each site.

A similar method was introduced by Phillips (1976a,b) to compare trace metal bioavailabilities to the mussel *Mytilus edulis* in Port Phillip Bay, Australia. In this instance, size normalisation was necessary for only certain of the samples taken from the field, as only these exhibited significant size-dependence of metal concentrations (see also NAS, 1980; Strong and Luoma, 1981; Cain and Luoma, 1985).

As noted previously, the changes in concentrations of organic contaminants with size (age; tissue weight) in biomonitors are most commonly due to increases (and occasionally, decreases; see Chapter 6) in lipid weights of organisms with size. Such changes in lipid contents with size or age may be related to fecundity in certain species (e.g. see Bayne, 1976) and may thus only be apparent at certain seasons, such as prior to spawning. Phillips (1978, 1980, 1986) has argued that the simplest and most appropriate approach to account for the impacts of organism size in these circumstances is to base data for contaminant concentrations on the lipid weights of samples, rather than on wet or dry tissue weights. This approach constitutes a form of normalisation for size (and for the related lipid changes), and offers considerable advantages to the interpretation of data for organochlorines in biomonitors (Phillips and Segar, 1986).

It should be noted, however, that size-related normalisation techniques rely in most cases on the analysis of a relatively large number of individual organisms (representing a considerable size range; see Boyden, 1974, 1977) from each study site. This is both expensive and time-consuming, and many authors have preferred to bulk samples for analysis to reduce the costs associated with biomonitoring programmes (see section G below for further discussion of sample bulking). In this event, the impacts of organism size on the resulting data can be reduced or eliminated only through the sampling of a restricted size range of the biomonitor employed, at each location included in a survey.

This has been the most common approach to date in attempts to account for size-related changes in contaminant concentrations in biomonitors, and has been employed in most of the national and international surveys of contaminants (e.g. see Goldberg *et al.*, 1978; Phillips, 1985b; Claisse, 1989). However, the size ranges employed in certain studies have been criticised as being too great (Phillips and Segar, 1986) and the sampling of a very restricted size range of organisms at multiple locations is sometimes impossible, due to natural inter-population variations in the availability of individuals of particular sizes. The Californian State Mussel Watch Program circumvents this dilemma by the transplantation of mussel samples pre-selected for size from a single area of derivation (Bodega Head) to each study site, and this approach has merit where natural populations of biomonitors exhibit restricted size ranges which vary between sites (e.g. see Martin, 1985; Smith *et al.*, 1986).

The final biological variable which deserves mention here is organism sex. Examples of sex-based variations in trace metal and organochlorine concentrations in biomonitors have been discussed in Chapters 5 and 6, and will thus not be reviewed in detail here. Compared to size-related impacts on contaminant levels in biomonitors, sex-based variations are rare. Nevertheless, certain examples have been documented, and these cover a range of species and contaminant types. As a result, the potential impacts of organism sex on the data resulting from biomonitoring investigations merit attention by researchers, and some type of sampling strategy is required to account for this variable. The strategy employed will vary from case to case depending upon circumstances, but will involve the exclusive sampling of either male or female individuals (e.g. see Burnett, 1971), or the deliberate stratification of samples for a given mix of sexes at each study site.

G. INHERENT VARIABILITY AND THE USE OF STATISTICS

The essential objective of most biomonitoring investigations is the elucidation of differences in the concentrations of contaminants between samples of biomonitors taken at different points in space and/or time. For this to be possible (and to be statistically meaningful), the intra-population variation in contaminant concentrations should be established at each study site, such that statistical tests may be undertaken to demonstrate that the concentrations of any particular contaminant are significantly different between sites (i.e. that site-to-site differences are greater than within-site variances).

It has long been established that contaminant concentrations vary significantly between individuals in any one population of biomonitoring organisms. Some of the sources of such variations have already been discussed (organism condition, including lipid content; size of individuals; organism sex), and others also exist (e.g. relative height of individuals on the shoreline; see Phillips, 1980). However, even when all known sources of variability in contaminant concentrations have been accounted for, a degree of residual variation remains between the levels of any one contaminant exhibited by different individuals in a population. This has been termed "inherent variability" (Boyden and Phillips, 1981), and its causes are essentially unknown, although it is thought that genetic differences between individuals may contribute to its existence.

In certain cases, this inherent variability may be of considerable magnitude. Thus, the concentrations of most contaminants in biomonitors tend to be log-normally distributed (e.g. see Giesy and Wiener, 1977), with some individuals exhibiting particularly high concentrations of a given contaminant compared to the population mean. An excellent example of this trend is provided by data for zinc in the mussel *Mytilus edulis* (Lobel *et al.*, 1982; Lobel and Wright, 1983; Lobel, 1986), where frequency distributions of element concentrations are highly positively skewed in some populations, due to the presence of so-called "superaccumulators" of zinc.

Gordon *et al.* (1980) evaluated the extent of trace metal variability in two populations of the mussel *Mytilus californianus* from the Southern

Californian Bight, in an attempt to define the most appropriate sampling strategy for biomonitoring programmes. They noted that intra-population variability in trace element concentrations differed between the two sites studied, and also varied in magnitude between different metals. This is a universal finding, even when all known sources of variability have been accounted for in the sampling of biomonitors. For example, cadmium exhibits a low intra-population variance in *M. edulis* compared to most other trace elements, and the intra-population variance generally increases for all metals in mussels as the ambient bioavailability of each metal increases and concentrations rise in the biomonitor (Phillips, 1976b).

Gordon *et al.* (1980) published relationships between sample sizes for *M. californianus* and detectable inter-population differences in mean trace metal concentrations (Fig. 83). They emphasised that these relationships varied between locations and with time at any one site, and concluded that initial tests of intra-population variability in contaminant concentrations should constitute an essential first step in biomonitoring programmes. In terms of sampling methods, Gordon *et al.* (1980) stated the following:

> ...Adequate estimates of population means and variances [for contaminant concentrations] allow investigators to ascertain optimum sampling strategies in future studies. Such studies will usually include a compromise between resolution of differences between sample means (e.g. control vs treatment) which should be maximized, and sample size which should be minimized (i.e. cost/benefit ratio). This strategy will apply to both pooled and unpooled sample designs.

Similar studies to define intra-population variability in trace element concentrations in biomonitors have been reported by Boyden and Phillips (1981) and Wright *et al.* (1985). These have concluded that sample sizes of 15–20 individuals from each study site generally provide an adequate estimate of the mean contaminant concentrations in a biomonitor. However, as noted by Gordon *et al.* (1980), inherent variability differs in extent with both the species and the contaminant concerned, and this general guideline cannot be considered to be universally applicable.

It should also be noted here that certain trace metals may exhibit high intra-population variances in biomonitors because of their intimate association with inorganic particulate material. Both aluminium and iron fall into this category, and several authors have reported that ingested sediment or particulates caught in gill filaments may significantly affect the measured whole-body concentrations of such elements in biomonitors (e.g. Flegal and Martin, 1977; NAS, 1980; Brumbaugh and Kane, 1985;

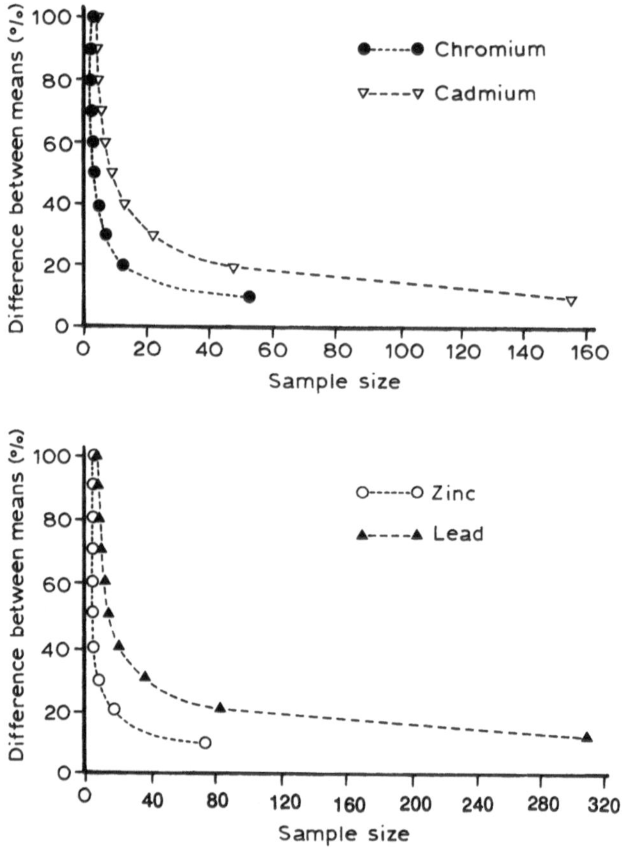

Fig. 83. Relationships between the numbers of individuals analysed and the percentage differences which may be detected between mean values at $p < 0.05$ for several metals in mussels (*Mytilus californianus*) from Coal Oil Point in California in January 1977. After Gordon *et al.* (1980).

Lobel *et al.*, 1991). It has been concluded on this basis that biomonitoring samples which may contain significant amounts of ingested or adsorbed inorganic particulates should be subjected to depuration for 36–48 h prior to their analysis for trace elements (NAS, 1980). However, precautions should certainly be taken to avoid any extraneous contamination of samples throughout this period. Depuration is not recommended for samples destined for analysis for organochlorines or hydrocarbons, as significant amounts of contaminants of shorter half-life in biomonitors may be excreted from their tissues during the depuration period (NAS, 1980).

Less information is available concerning the inherent variability of trace organic contaminants in biomonitors. As noted in Chapter 6, the principal cause of variation in organochlorine concentrations between individuals is related to differences in lipid contents. Thus, the intra-population variance in organochlorine levels is considerably reduced where analytical data are expressed on a lipid weight basis rather than on wet or dry weight bases (Phillips, 1978, 1980). Flores-Baez and Galindo-Bect (1989) have suggested on the basis of studies of DDT in mussels from Baja California that the sampling of 20 individuals from each population investigated provides an adequate estimate of mean DDT concentrations in *Mytilus edulis*. On this basis, it appears that the extent of individual variability in bivalve populations at least is broadly similar for the different classes of contaminants. However, the general paucity of data for organic contaminants in different populations is notable, and this topic merits further study.

It may be concluded here that researchers wishing to undertake biomonitoring surveys should match their detailed sampling design to the programme objectives (see Phillips and Segar, 1986), considering the need to pool samples and to optimise the sample sizes taken at each site on the basis of the approach suggested by Gordon *et al.* (1980). In many cases, preliminary studies of intra-population variability in contaminant concentrations will be necessary to define an optimum sampling strategy, and these will in any event be likely to improve the cost-benefit aspects of biomonitoring programmes. Certain investigations may also be amenable to the inclusion of more sophisticated statistical treatments, such as analysis of covariance or principal component analysis (e.g. see Popham and D'Auria, 1983; Phillips and Rainbow, 1988), but the need for this will again depend primarily on the precise objectives of each study.

H. SAMPLE ARCHIVAL

A few of the established biomonitoring programmes undertaken to date have specifically included the archival of samples. Thus, for example, the US Mussel Watch Program of the mid-1970s adopted this approach, to provide a "library" of frozen samples, available for later analysis with improved analytical techniques (Goldberg *et al.*, 1978). It was recognised that certain types of contaminants might suffer degradation in such samples over time (e.g. through the decay of radionuclides, or because of microbial activity and its potential for affecting the concentrations of certain organic contaminants). However, the archival of samples in either frozen or perhaps freeze-dried forms merits consideration in the design of new biomonitoring programmes, particularly as past experience suggests that many new contaminants of concern are likely to be discovered in the future.

I. SUMMARY

There can be no doubt that insufficient information is presently available on the abundances and bioavailabilities of trace contaminants in aquatic ecosystems. The robust nature of the biomonitoring approach and its advantages over other methods for the investigation of trace pollutants in aquatic environments (see Chapter 4) imply that many new studies (in both the temperate and tropical zones) will rely either completely or in part on the use of biomonitors.

The most important aspect of the design of new biomonitoring programmes is the establishment of their precise and unambiguous objectives (Phillips and Segar, 1986). These will most frequently include the delineation of spatial and/or temporal fluctuations in contaminant bioavailabilities in aquatic ecosystems, but other aims may also exist. Such objectives are important not only in their own right, but also because they define several aspects of the design of the biomonitoring programme to be employed.

The selection of a species to be employed as a biomonitor requires attention to the basic prerequisites which have been documented for such organisms (see Chapter 4), and also to certain more sophisticated considerations. The latter include: the physicochemical form of the contaminants to be studied (which assists in defining the type of biomonitor chosen); the availability of wild populations of potential biomonitors and the possibilities for transplantation; and the degree of spatial definition to be attained in the biomonitoring survey. A range of potential biomonitors is available for use in most locations, and developmental studies in both temperate and (more recently) tropical ecosystems can be relied upon to provide basic information and assistance in the selection of the most appropriate biomonitoring species. The transplantation of biomonitors may provide advantages under certain circumstances, and should be considered for possible use in new programmes. It is of critical importance to recognise that the species selected defines the degree of spatial definition provided by a biomonitoring survey, and data on the biology of the selected organism are generally required to assist in both the design of the programme and the interpretation of the analytical results.

Decisions on the timing and frequency of monitoring should be based upon considerations of factors which influence temporal variations in contaminant levels in local waters; on biological changes which occur in the species selected as a biomonitor; and on the kinetics of the contaminants of interest in the chosen species. The most important biological aspect is the reproductive state of the selected biomonitor and its seasonal changes, as this may affect its accumulation and retention of various classes of trace contaminants. From a biological standpoint, the most appropriate timing for a monitoring survey is the period in which least changes in organism condition occur, as differences between populations in this factor will tend to be minimal at that time. In the case of bivalve molluscs, this equates to the pre-spawning period in most populations (NAS, 1980).

The monitoring frequency to be employed should be defined by considerations of the study objectives and the kinetics of contaminant uptake and depuration in the biomonitor selected. Decisions are required here on whether the monitoring should provide a true "snapshot" of the ambient bioavailability of contaminants, or should measure pollution conditions over a longer period. If the latter is preferred, samples may be subjected to "time-bulking" (Phillips and Segar, 1986) to artificially lengthen the time-integration of contaminants offered by any particular biomonitoring species.

Decisions on site selection also depend on the precise study objectives, particularly in terms of the spatial definition required in the monitoring survey. In all cases, however, the use of both contaminated and "reference" or "control" sites is recommended, as data from the latter on the natural contaminant levels and their temporal changes in the selected biomonitor will assist in the interpretation of data from samples taken at the more contaminated locations. The presence of local known or suspected sources of contaminants, the existence of freshwater inflows in coastal regions, and the hydrodynamics of the study area are all of relevance to site selection for biomonitoring surveys.

The most important biological variables which must be taken into account in any biomonitoring investigation are organism condition, the sizes of the individuals sampled at each study site, and organism sex. Care should be taken at sampling to reduce or eliminate the potential effects of these variables, through appropriate decisions on the timing of sampling and through the design of a reproducible sampling strategy which may be followed at each study site.

Even where all known sources of contaminant variability within biomonitoring populations have been accounted for, some residual variations remain between individuals in their precise concentrations of contaminants. This "inherent variability" (Boyden and Phillips, 1981) has implications for the sampling strategy employed, and for the statistics used to demonstrate the existence of significant spatial and/or temporal differences between samples in the bioavailabilities of contaminants. The approach proposed by Gordon *et al.* (1980) is recommended, and researchers are urged to undertake preliminary investigations of the intra-population variances in the concentrations of contaminants in their selected biomonitoring species.

Finally, it is recommended that consideration be given to the establishment of a sample archival facility, in which frozen or freeze-dried samples may be stored for possible later analysis employing novel analytical techniques.

Chapter 9

Monitoring the Effects of Contaminants

A. INTRODUCTION

The biomonitoring of trace aquatic contaminants provides information on the temporal and geographical variation in the concentrations of those contaminants which are bioavailable to one or more chosen biomonitors. Valuable though such information may be, evidence for an elevated bioavailability of a particular contaminant at a given location does not necessarily imply that a toxic effect is created by the contaminant, simply because it is present in higher availability at that site than elsewhere. It is therefore often considered necessary to seek other evidence of toxic effects, as an adjunct to the monitoring of contaminant abundance and bioavailability. It should be noted that such "effects-based monitoring" is distinct in its objectives from the biomonitoring which attempts simply to establish the spatial and temporal abundances and bioavailabilities of contaminants, and programmes to study the toxic effects of contaminants should therefore be designed with their own specific objectives in mind (Phillips and Segar, 1986).

Evidence for the toxic effects of contaminants can be manifested at a variety of biological levels: (i) at that of the organism, such as in changes of morphology, physiology, biochemistry, cytochemistry or behaviour; (ii) at the level of the population with respect to, for example, changes in reproduction or recruitment rates; (iii) at the level of community structure, involving the interplay between several populations of different species; or (iv) in combinations of the above categories.

Bayne *et al.* (1985) listed criteria for assessing the suitability of a particular physiological response to be used as a measure of the condition of an animal when it is subjected to environmental stress and pollution. These criteria can be extended to any response to be used to monitor the biological effects of a contaminant. The criteria are as follows:

- A quantitative or predictable relationship should exist between the response and the contaminant dose.
- The response should have ecological significance and be shown (or convincingly argued) to be related to an adverse effect on the growth, reproduction or survival of the individual, the population and ultimately the well-being of the community.
- Sufficient sensitivity to the contaminant should exist to provide a large scope for response throughout the range of exposure, from optimal to lethal conditions.
- The response should reflect an integrated steady state condition that does not alter significantly with short-term fluctuations in contaminant availability (e.g. with tidal or diurnal variation).
- The reponse should be measurable with precision, and with a high signal-to-noise ratio. Thus, the response (signal) should be easily detectable above the natural variability (noise).
- The reponse should be easy to measure in the laboratory and/or field, without necessitating the use of very expensive equipment, complicated procedures or high running costs.

B. EFFECTS MONITORING AT THE ORGANISM AND TISSUE LEVELS

Responses at the level of the organism and its component tissues are divided (somewhat arbitrarily) into those relating to morphology, physiology and behaviour. Physiological responses are taken to include changes in growth and in the rates of metabolic processes, as well as biochemical and cytochemical effects which inevitably blend into each other. These various responses are dealt with in turn in the following sections.

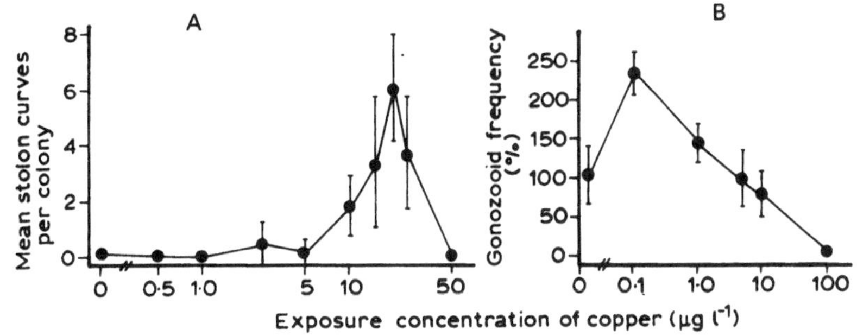

Fig. 84. (A) Stolon curving frequencies of colonies of the hydroid *Campanularia flexuosa* after exposure to different concentrations of copper for 14 days. Bars indicate standard errors of the mean. (B) Gonozooid frequencies in colonies of *C. flexuosa* after exposure to different concentrations of copper for 11 days. The ratio of gonozooids and buds thereof to the total number of colony members is expressed as a percentage of this same ratio in control samples. Bars indicate standard errors of the mean. After Bayne *et al.* (1985).

(i) Morphological Responses

Hydroids (the sedentary polyp stages of Cnidaria) have been of particularly frequent use in studies of altered morphology in response to aquatic contamination (e.g. Stebbing, 1976; Karbe *et al.*, 1984; Bayne *et al.*, 1985). The colonial marine hydroid *Campanularia flexuosa* (Bayne *et al.*, 1985) responds to unfavourable conditions by curling of the stolons during growth, deviating from the typical form of more or less linear radiation (Stebbing, 1979). Curvature of the stolons increases with elevated concentrations of toxins (Fig. 84(A)), and occurs typically in an anti-clockwise fashion. A second more sensitive morphological response of *C. flexuosa* to increasing concentrations of a toxin is the increased formation of gonozooids (Fig. 84(B)), the reproductive polyps which (atypically for a hydroid) emit gametes directly, without releasing the more usual dispersive medusoid stage.

These two morphological responses (together with changes in the growth rates of colonies; see below) can be used in bioassays of water quality. The problem of identifying the causal agents of the effects noted in water of poor quality may be approached by chemical manipulation of a water sample. For example, the responses of *C. flexuosa* can be measured before and after the sample has been passed through columns containing a Chelex-100 ion exchange resin (to remove metals) or an XAD resin (to remove polychlorinated hydrocarbons), with or without the subsequent

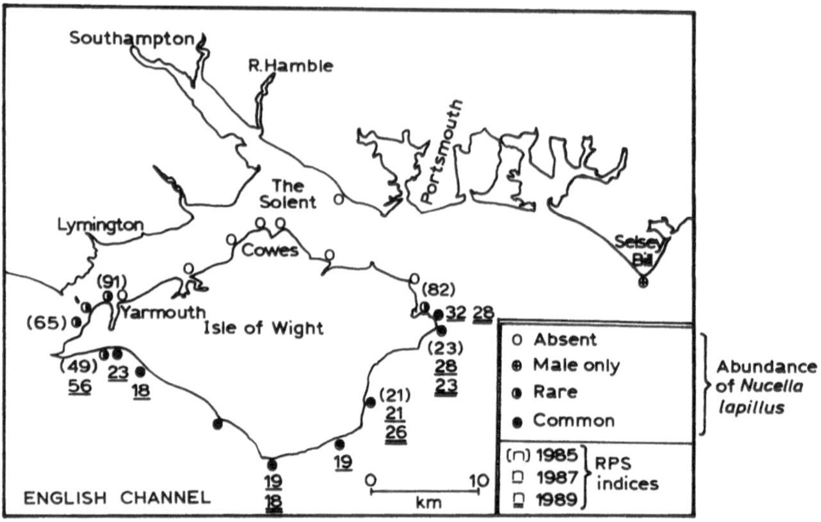

Fig. 85. Abundance of the dogwhelk *Nucella lapillus* around the Isle of Wight, England, and indices of relative penis size (RPS, a measure of imposex) in 1985, 1987 and 1989 at various sites. Sites at which *N. lapillus* was rare or absent had supported successful populations prior to the onset of the use of tributyltin anti-fouling compounds. After Langston *et al.* (1990).

addition of a known concentration of toxin (Stebbing, 1979; Bayne *et al.*, 1985).

A notorious morphological response to a toxin is that of stenoglossan gastropods such as the dogwhelk *Nucella lapillus* to tributyl tin (TBT), involving imposex — the imposition of male characters including a penis and vas deferens onto females (Langston *et al.*, 1990). Imposex causes loss of fertility and ultimately mortality of female gastropods, and the consequent eventual elimination of local populations such as has been noted for *N. lapillus* at sites on the Isle of Wight, England (Langston *et al.*, 1990; see Fig. 85). The intensity of imposex in a population of this type can be expressed on a Relative Penis Size (RPS) index, calculated as ([mean female penis length^3/mean male penis length^3] × 100). Gibbs *et al.* (1987) have assigned six stages to the development of imposex in a female dogwhelk, and this assists in measuring the intensity of the impacts due to TBT in any given population. The six stages are as follows:

(1) The growth of a proximal section of vas deferens close to the genital papilla.

Table 26 Summary of the effects of exposure to tributyl tin on the reproductive system of the gastropod *Nucella lapillus* (after Gibbs *et al.*, 1988)

TBT in water (ng Sn litre^{-1})	*RPS index (%)*	*VDS index*	*Effect on reproductive system*
<0.5	<5	<4	Breeding normal; development of penis and vas deferens
1–2	40+	4–5(+)	Breeding capacity retained by some females; others sterilised by oviduct blockage; aborted capsules in capsule gland
3–5	90+	5(+)	Virtually all females sterilised; oogenesis apparently normal
10+	90+	5	Oogenesis suppressed; oocytes resorbed; spermatogenesis initiated
20	90+	5	Testis developed to variable extent; vesicula seminalis with ripe sperm in most-affected animals
100	90+	5	Sperm-ingesting gland undeveloped in some 'females'

(2) The commencement of penis development, with formation of a ridge behind the right tentacle.
(3) The formation of a small penis; development of the distal section of vas deferens from the base of this penis.
(4) The joining of sections of the vas deferens; at this stage, the penis size in females approaches that of males.
(5) The proliferation of the vas deferens, which overgrows the genital papilla and sterilizes the female.
(6) The accumulation in the capsule gland of compressed aborted egg capsules that cannot be expelled, forming a brown mass.

Mortality occurs in severely affected females, apparently as a result of the rupture of the distorted stage 6 capsule gland. The Vas Deferens Sequence (VDS) index is employed to describe the average stage of imposex for a population. When the VDS index exceeds 4, the presence of sterile females in a population reduces its reproductive capacity (Table 26).

Concentrations of TBT in solution in seawater as low as 1 to 10 ng litre^{-1} cause imposex in *Nucella lapillus* (Langston *et al.*, 1990; see Table 26), and similar levels cause shell abnormalities in the Pacific oyster, *Crassostrea gigas*, which was originally introduced into European waters for culture (Langston, 1990). Indeed, the occurrence of shell abnormalities in cultured stocks of *C. gigas* was one of the first indications of the high toxicity of TBT in the field (Thain and Waldock, 1986). While other metals such as cadmium and copper can cause shell deformities in bivalve molluscs (e.g. see Sunila and Lindstrom, 1985), the concentrations required in these cases are very much greater than those of TBT, and are environmentally unrealistic.

Trace metals can also cause gill and skeletal deformities in fish (Hughes and Perry, 1976; Bengtsson, 1979; Bengtsson and Larsson, 1986). Elements in solution may be responsible for the induction of fish lesions, although evidence for the latter is usually correlational and indirect (Bengtsson, 1979; Langston, 1990). However, the results of Yamashita *et al.* (1990) suggest that pigment cell disorders in the skin of fish (chromatophoroma in the croaker *Nibea mitsukurii* and skin pigment cell hyperplasia in the sea catfish *Plotosus anguillaris*) may be useful biomarkers for carcinogens in coastal waters.

(ii) Physiological Responses

Physiological responses of organisms to aquatic contaminants take place at the level of the whole organism (e.g. changes in growth and respiration rates); in organs and tissues (e.g. alterations in heart beat and body condition indices); and at the cellular level (both biochemically, as exemplified by enzyme activities, and cytochemically, as shown by changes in lysosomal hydrolase latencies). Divisions of this nature are necessarily artificial as responses often act at more than one level. Nevertheless, such subdivisions are useful here for presentational purposes.

Depledge (1989b) has attempted to place the detection of the early effects of marine pollutants using physiological indicators onto a rational basis, thereby improving our ability to detect significant effects of pollutants at an early stage and to define acceptable levels of pollution. He emphasised the ideas of Hatch (1962) that a distinction be made between departures from normal physiological and behavioural responses signifying health (the impairment scale in Fig. 86) and the consequences of this departure from health (the disability scale in Fig. 86). The impairment scale is more sensitive to the effects of pollutants than is the disability scale. Thus, monitoring of the impairment of behavioural and physiological responses will provide early warning of the onset of disabilities (Depledge, 1989b). Monitoring of the disability scale (e.g. through measuring the presence of disease) may not be sensitive enough to supply a warning until the pollutant has pushed the organism beyond the limit of compensation. Thus, monitoring of physiological effects asks "what levels of pollutants

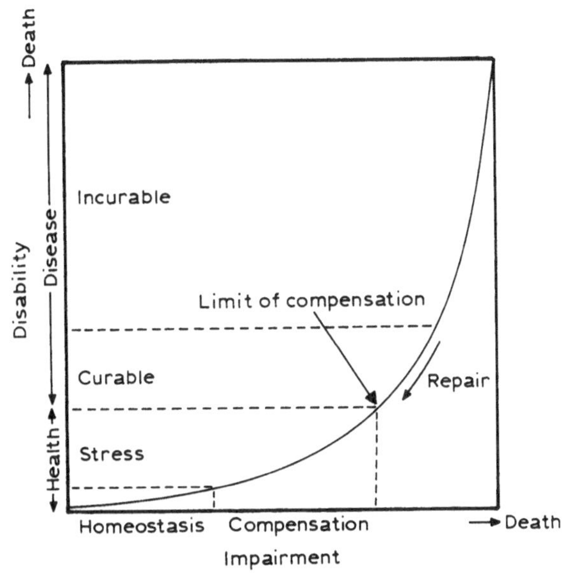

Fig. 86. Disability and impairment as indicators of the effects of contaminants. After Depledge (1989b).

produce effects that are likely to lead to biological damage", as opposed to "what levels of pollutants lead to biological damage" (Depledge, 1989b).

Growth Rates

There is much evidence in the literature of aquatic contaminants measurably affecting growth rates of aquatic organisms. Langston (1990) lists a selection of this evidence relevant to the effects of trace metals, emphasising phytoplankton which are among the most sensitive indicators of metal pollution. Given the caveat that the toxicity of dissolved trace metals is determined by the concentration of free metal ions rather than their total concentrations in solution (including both free ions and complexed metals; see Chapter 5), many estimates of metal levels causing growth inhibition of phytoplankton are of the same order as those found in moderately polluted estuaries (Langston, 1990). It would appear, therefore, that studies on the physiological responses of phytoplankton (e.g. through measurement of ^{14}C incorporation) are of considerable benefit in defining the impacts of contaminants in aquatic ecosystems.

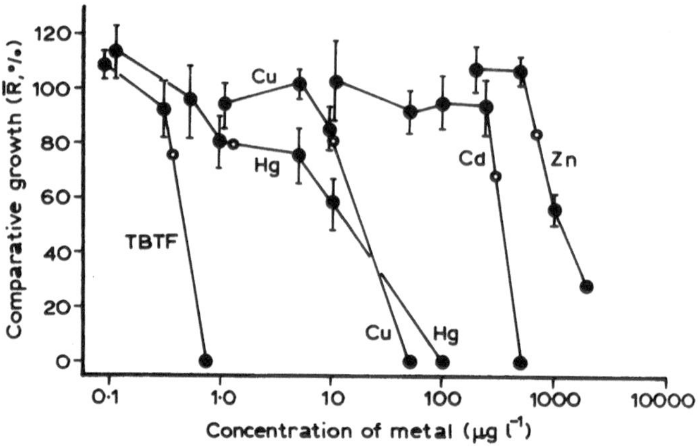

Fig. 87. Relationships between trace element concentrations and response, showing the effects of several contaminants on the growth of the colonial marine hydroid *Campanularia flexuosa* in exposure periods of 11 days. Growth is measured by comparison to controls, as noted in the text: open circles show thresholds for each metal. TBTF is tributyltin fluoride. After Bayne *et al.* (1985).

The colonial marine hydroid *Campanularia flexuosa* responds to the presence of a toxin by a change in cumulative colonial growth rate. This may be conveniently expressed in laboratory studies as the mean specific colonial growth rate measured as a percentage of the growth rate of the control (Stebbing, 1976; Bayne *et al.*, 1985). Given a null hypothesis that the specific growth rate does not differ from that of the control, error bars on either side of the experimental means can be plotted as multiples of the pooled standard error to indicate whether there is significant deviation from the growth rate of the controls. This has also been shown to provide a sensitive indication of the impacts of contaminants (Fig. 87).

Lande (1977) reported extensive data on the growth of the mussel *Mytilus edulis* in Trondheimsfjorden, central Norway. Mussels growing in the waters contaminated by copper and zinc near local mining industry exhibited a reduction in growth rate, in comparison with mussels from 'unpolluted' parts of the fjord, as well as increased mortalities and a partial loss of their ability to anchor themselves to the substrate with the byssus. Strömgren (1982) studied the short-term (10–22 days) effect of several trace metals on the increase in shell length of *Mytilus edulis* in running, unfiltered seawater at Trondheim. Significant reductions in shell growth were found at added concentrations of 0.3 μg Hg litre^{-1}, 3 μg Cu litre^{-1},

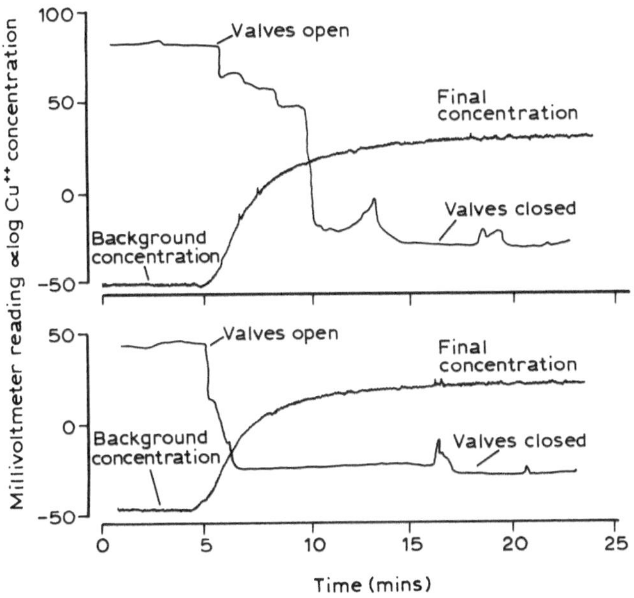

Fig. 88. Shell valve movements in mussels (*Mytilus edulis*) exposed to increasing concentrations of copper. After Davenport and Manley (1978).

10 μg Zn litre^{-1} and 10 μg Cd litre^{-1} (but see Davenport and Redpath, 1984). Concentrations of lead and nickel of up to 200 μg litre^{-1} had no effect.

At added levels of copper above 5 μg litre^{-1}, the mussels showed an apparently permanent closing of the valves, a process which has also been detailed by Davenport and Manley (1978; see Fig. 88). Strömgren (1982) concluded that the decrease in shell growth associated with exposure to copper may be caused by a combination of starvation stress due to valve closure, and direct toxic effects on shell growth. Similarly, Redpath (1985) noted significant inhibition of shell growth of *Mytilus edulis* at added concentrations of greater than 2 μg litre^{-1} of copper, with substantial variability in the growth rates of individual mussels; valve closure was not, however, found to occur in these studies. Analysis of growth bands on the shells of bivalves (e.g. see Richardson, 1989) has the potential to detect temporal and local differences in the shell growth rates of bivalve populations that might be attributable to the differential concentration of aquatic contaminants.

Physiological Processes

The effects of contaminants on the physiology of aquatic organisms are detectable at the level of processes contributing to the overall growth rate, such as respiratory and feeding processes.

The inspection of fish placed in flowing wastewater has long been used as a continuous monitor of effluent toxicity, many of the responses monitored being concerned with respiration (Cairns and van der Schalie, 1980). Very high dissolved concentrations of metals (1–10 mg litre^{-1}) are known to disrupt circulatory and respiratory activity in the shore crab *Carcinus maenas* (Depledge, 1984a), and increasing concentrations of copper and zinc (up to 1000 μg litre^{-1}) have been found to progressively reduce the oxygen consumption of the freshwater decapod *Macrobrachium carcinus* (Correa, 1987). Depledge (1984b) has shown that the exposure of *C. maenas* to concentrated solutions of the water-soluble fraction of Fortes crude oil (20%) and a dispersant (10%) increases the cardiac activity and oxygen consumption of the crabs, and disrupts normal feeding behaviour.

Other physiological processes are also affected by the presence of aquatic contaminants. Trace metals such as mercury and copper upset ionic and osmotic regulation in the shore crab *Carcinus maenas* (Bjerregaard and Vislie, 1985, 1986). Yearling coho salmon (*Oncorhynchus kisutch*) exposed to sublethal concentrations of copper in freshwater before their transfer to seawater exhibited decreases in microsomal Na^+; K^+ activated ATPase activity in the gills, probably associated with a loss of osmoregulatory ability. This effect led in turn to increased mortalities on exposure of the fish to seawater (Lorz and McPherson, 1976).

Widdows *et al.* (1982) investigated the physiological responses of the mussel *Mytilus edulis* in chronic exposures to low, more environmentally realistic concentrations of hydrocarbons. As shown in Fig. 89, the exposed mussels exhibited a higher rate of oxygen consumption than the controls. There was no significant difference in any treatment between oxygen consumption rates of the mussels during feeding and non-feeding periods. In the same series of experiments, Widdows *et al.* (1982) also measured other physiological processes, including clearance rates and ammonia excretion. Mussels exposed to hydrocarbons in the absence of food exhibited reduced average clearance rates, due to a reduction during the period of hydrocarbon dosing without food. Mussels exposed to hydrocarbons and algal food simultaneously maintained similar clearance rates to the control animals, which showed slightly higher clearance rates during the feeding periods. The rate of ammonia excretion (Fig. 89) was significantly

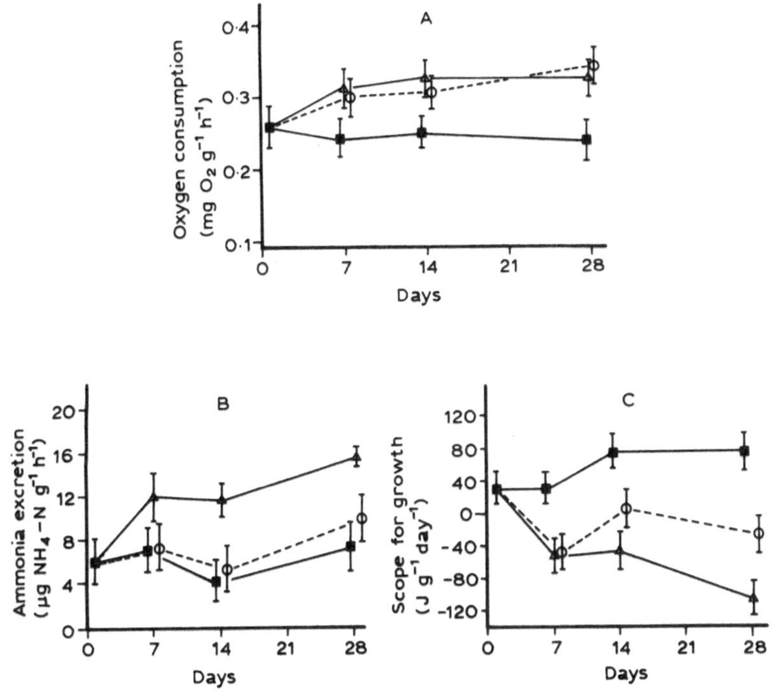

Fig. 89. Effects of exposure of mussels (*Mytilus edulis*) to the water-accommodated fraction of North Sea crude oil at 36 μg litre^{-1} on: (A) oxygen consumption; (B) excretion of ammonia; (C) scope for growth. ■ Control mussels; ○, mussels exposed to oil without feeding; △, mussels exposed to oil with feeding. After Widdows *et al.* (1982).

raised in the mussels exposed to hydrocarbons in the feeding phase, but not in those exposed in the non-feeding phase.

The apparent variability in the physiological responses of organisms to the presence of aquatic contaminants is difficult to interpret in isolation. However, it is possible to integrate the total effects of the different physiological responses in 'Scope for Growth' measurements. These were introduced initially in studies of fish bioenergetics by Warren and Davis (1967) but have since been extended into investigations of the energetics of invertebrates, especially the mussel *Mytilus edulis* (e.g. see Widdows and Bayne, 1971; Widdows *et al.*, 1982; Bayne *et al.*, 1985, Widdows, 1985; Widdows and Johnson, 1988).

Scope for Growth

Living organisms produce matter for both growth and reproduction, this resulting from the difference between energy intake and energy output. For an animal:

$$P = A-(R+U)$$

where: P is the energy incorporated into body growth and gamete production, A is the energy absorbed from food, R is the energy respired and U is the energy excreted.

It is possible to measure the physiological processes on the right-hand side of the above equation and express them in energy equivalents (e.g. joules per hour). The equation then gives an index of the energy available for growth and reproduction; hence the phrase "scope for growth" (Bayne *et al.*, 1985). Scope for growth is in effect an integration of all the responses of an organism to its environment, and may be altered by both natural and anthropogenic stresses. Examples of natural stressors affecting scope for growth include temperature, salinity, and dissolved oxygen. Perhaps most important among anthropogenic stressors is the presence of contaminants. Scope for growth can be positive when energy is available for growth and gamete production, and negative when the animal is using body reserves for maintenance metabolism.

Measurement of the physiological processes needed in the calculation of scope for growth requires transfer of the organism into the laboratory, although the use of a mobile laboratory can reduce time and distance from a field site before measurements are made (e.g. Widdows *et al.*, 1984). In the case of the mussel *Mytilus edulis* as an example:

- Respiration (R) is measured as the oxygen consumption rate (ml h^{-1}), with an allowance for body size (Bayne *et al.*, 1985). The rate of oxygen consumption standardised for a mussel of 1 g dry flesh weight is converted to a caloric equivalent (Widdows and Johnson, 1988).
- Energy absorbed from food (A) is calculated from the rate of filtration and the absorption efficiency (Bayne *et al.*, 1985). Rates of filtration are measured in the laboratory, for example using a Coulter Counter (e.g. see Widdows and Johnson, 1988), and are again calculated for a standardised mussel. Filtration rates can be converted to energy units, knowing the suspended concentration of food and the caloric content of the food (Bayne *et al.*, 1985). Absorption efficiency may be calculated according to Conover (1966) using ash-free dry weight:total dry weight ratios of the food (F), in this case the suspended material, and of the faeces (E). Thus:

$$\text{Absorption Efficiency} = \frac{(F-E)}{(1-E)F}$$

- Ammonia excretion (U) may also be measured in the laboratory, by estimating the dissolved ammonia concentrations released into experimental chambers in a known period (Widdows and Johnson, 1988).

Figure 89 shows scope for growth data for the mussels in the experiment of Widdows *et al.* (1982) discussed above. Figure 90 provides an example of the relationship between scope for growth and environmental pollution, this particular case involving a gradient of total hydrocarbons and nickel in Narragansett Bay (Widdows *et al.*, 1981). Widdows and Johnson (1988) provided further evidence of the decline of scope for growth in mussels along a pollution gradient in Langesundfjord, Norway. Scope for growth was also reduced with increasing exposure to copper and diesel oil in a mesocosm experiment (Table 27). Other examples of the use of scope for growth in relation to environmental quality assessments are summarised by Widdows (1985), emphasising the role of transplantation experiments in the design of such studies.

The measurement of scope for growth of mussels as an assessment of the relative suitabilities of sites to support invertebrate growth requires considerable laboratory backup, with the need for a great deal of replication to allow for intrinsic (e.g. size, age) and extrinsic (e.g. season, salinity, temperature) variables. Moreover, Jorgensen (1990) claims that scope for growth is a sensitive measure of growth only under conditions far below optimal, since scope for growth increases hyperbolically with assimilation.

A second index can also be calculated from the energy equation. The net growth efficiency (K) is the energy available for growth $[A-(R+U)]$ as a proportion of the energy absorbed from food (A), and is a measure of the efficiency with which food is converted into body tissue (Widdows, 1985). Thus:

$$K = \frac{A-(R+U)}{A}$$

A reduction in K (see the example in Table 27) indicates a stressed condition in an animal, since more of the energy absorbed from food is being used for metabolic maintenance as opposed to growth. Figure 90 shows how the net growth efficiency of mussels was also correlated with increasing pollution in Narragansett Bay (Widdows *et al.*, 1981).

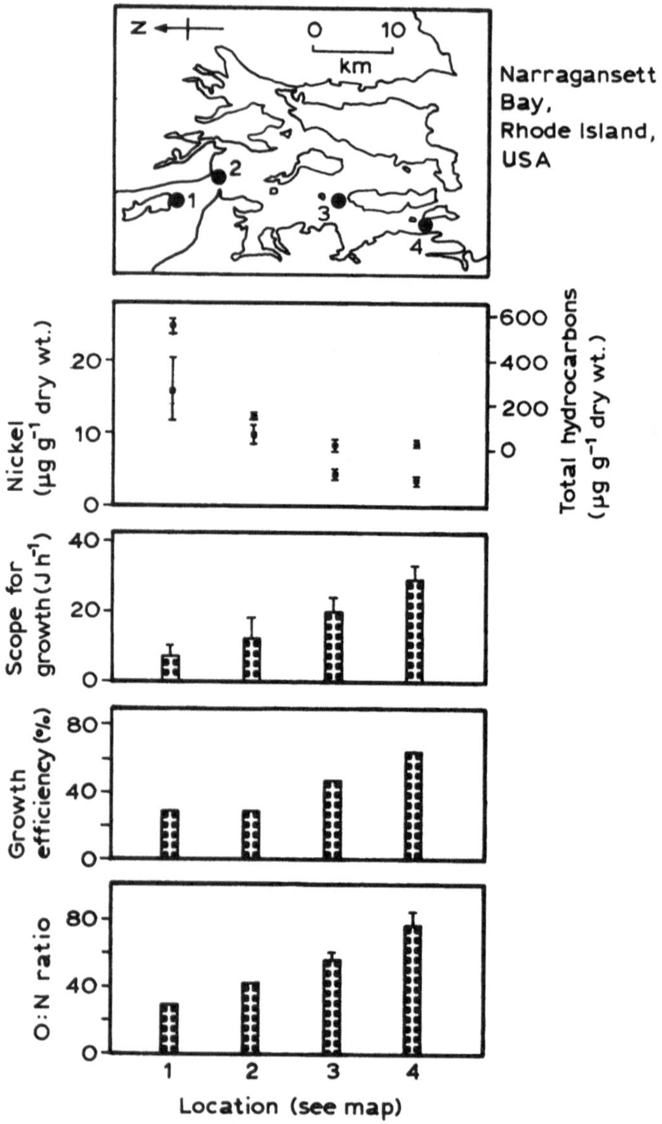

Fig. 90. Concentrations of nickel (solid circles) and total hydrocarbons (open circles; $\mu g\ g^{-1}$ dry weight) in the whole soft parts of mussels (*Mytilus edulis*) transplanted along a gradient of pollution in Narragansett Bay, USA (see map), and changes in scope for growth of the mussels, growth efficiency, and oxygen:nitrogen ratios. After Widdows (1985).

Table 27 Components of the energy budget and scope for growth (J g^{-1} h^{-1}) in mussels (*Mytilus edulis*) exposed to four concentrations of copper and hydrocarbons (control; low; medium; high) in a mesocosm experiment. Differences from controls denoted by *($p < 0.05$) or **($p < 0.01$). (After Widdows and Johnson, 1988)

Source	*Energy consumed*, C	*Energy absorbed*, A	*Energy respired*, R	*Energy excreted*, U	*Scope for growth, SFG*	*Growth efficiency*
Basin						
C	19.46 ± 0.65	7.78 ± 0.26	4.57 ± 0.38	0.32 ± 0.02	2.91 ± 0.52	0.37
L	18.50 ± 1.10	7.40 ± 0.44	4.61 ± 0.25	0.28 ± 0.02	2.79 ± 0.48	0.38
M	16.05 ± 0.79	6.41 ± 0.31	5.03 ± 0.39	0.37 ± 0.03	1.02 ± 0.46*	0.16
H	3.62 ± 0.34	0.00	1.53 ± 0.40	0.01 ± 0.03	−1.54 ± 0.42**	—

Condition Index

A crude index of the condition of bivalves is provided by calculation of the proportion of the internal shell volume which is occupied by the body tissues (Bayne *et al.*, 1985). Any increase or decrease in the condition index depends on the balance between the rates of food assimilation and of catabolism (metabolic breakdown of complex organic molecules with liberation of energy). The condition index therefore responds to anthropogenic stress, but also to the use of metabolic reserves when accumulating gametes. In *Mytilus edulis*, gamete development is reflected by changes in the percentage contribution of the mantle (the major site of the gonads in mussels) to body weight (e.g. Widdows *et al.*, 1984). The relative size of a tissue or organ can be calculated as a body component index, which expresses the dry weight of the tissue as a proportion of the total dry weight of the soft tissues. The relative size of the digestive gland of mussels (or its equivalent for other invertebrates) is often relevant, since these organs constitute a major store of nutrients and will respond to changes in the energy demands of the organism.

The use by an organism of nutrient reserves to meet extra metabolic requirements in conditions of stress affects the balance between the catabolism of carbohydrate, protein and lipid. The ratio between oxygen consumed and nitrogen excreted (the O:N molar ratio) provides an index of the relative use of protein in energy metabolism (Bayne *et al.*, 1985). A high rate of protein catabolism relative to that of carbohydrates and lipids (which is considered to be a "fallback position" in times of stress) results in a low O:N ratio. Conversely, a high O:N ratio is produced by the dominance of lipid and carbohydrate catabolism over that of proteins, and this is considered to reflect a more healthy condition. When using the O:N ratio, care must be taken to allow for factors causing natural variability

Table 28 The taurine: glycine ratio, and the sum of threonine and serine (μmol g^{-1} dry weight) in whole tissues less digestive gland of field populations of mussels (*Mytilus edulis*) from environments differing in their degree of contamination. Data shown as means ± standard errors (after Bayne *et al.*, 1985)

Population	*Environment*	*T:G ratio*	*(thr + ser)*
Mumbles	clean-coastal	1.9 ± 0.03	106 ± 8
Teignmouth	domestic sewage	1.9 ± 0.1	99 ± 2
Minehead	domestic sewage	2.0 ± 0.3	50 ± 3
Lynher	hydrocarbons	2.5 ± 0.2	64 ± 4
Houb of Scatsa	peat bed run-off	2.9 ± 0.5	26 ± 4
Atlantic College	industrial effluent	3.0 ± 0.1	27 ± 2
Swansea Dock	hydrocarbons, metals	3.2 ± 0.4	24 ± 3
Swale	paper mill effluent	3.5 ± 0.3	35 ± 3

in this ratio. These include both intrinsic and extrinsic parameters, such as seasonal/gametogenic cycles, food availability, and temperature (Bayne *et al.*, 1985). The application of the O:N ratio in mussels to a field situation (Widdows *et al.*, 1981) is illustrated in Fig. 90.

Biochemical Responses

Physiological responses by organisms to the presence of aquatic contaminants blend into those that can be described as biochemical, the O:N ratio described above being a case in point. The term "biochemical response" is restricted here to denote a significant change in concentration of a molecular component of cells or tissues, or of extracellular fluids such as blood serum.

As discussed briefly by George (1990), contaminants will induce changes in serum concentrations of cholesterol, cortisol, glucose, lactate, pyruvate and protein in fish. However, many other factors (e.g. catching stress, season, sex, temperature, and reproductive cycles; see Wedermeyer and Yasutake, 1977) must be accounted for before such changes can be used diagnostically for the recognition of an effect of a contaminant. Many of the complex responses observed are due to perturbation in the hypothalamo-pituitary axis, and are probably of little diagnostic use (George, 1990).

The relative concentrations of amino acids in the tissues have been proposed as indicators of "stress", for example in bivalve molluscs (Bayne *et al.*, 1985). The use of the molar ratio of taurine to glycine (T:G) in tissues has been evaluated as an index of this type by Bayne *et al.* (1985; see Table 28). These authors concluded that the T:G ratio might be applicable for studies of stress in several species of marine bivalve molluscs, although the absolute values found differ between species.

The combined concentration of threonine and serine (thr + ser) in whole tissues of mussels (*Mytilus edulis*) also appears to be a sensitive index of stress. This concentration is reduced in mussels from sites considered polluted (Table 28) and is more sensitive to contamination than is the T:G ratio, although large seasonal variations occur in the former parameter. Bayne *et al.* (1985) envisaged the use of the T:G ratio and the combined threonine and serine concentration together in monitoring programmes of the biological effects of contaminants, these employing the mussel *Mytilus edulis*.

An estimate of the amount of metabolic energy potentially available for the cells of contaminant-exposed organisms is provided by the adenylate energy charge (AEC) proposed by Atkinson (1977). This is calculated according to the following equation:

$$\mathrm{AEC} = \frac{\mathrm{ATP} + \frac{1}{2}\mathrm{ADP}}{\mathrm{ATP} + \mathrm{ADP} + \mathrm{AMP}}$$

In this equation, ATP (adenosine triphosphate), ADP (adenosine diphosphate) and AMP (adenosine monophosphate) are expressed in molar concentrations. AEC exhibits a value between 0 (when all adenine nucleotide is in the form of AMP with no high energy phosphate groups available) and 1 (all adenine nucleotide is ATP and the adenylate energy system is completely full). The use of AEC has limitations as a biochemical index as discussed by Bayne *et al.* (1985), although it may be useful for the early and rapid detection of stress in polluted environments. For example, Zaroogian *et al.* (1982) showed a correspondence between the possible effects of metal exposure of bivalve molluscs on oxidative phosphorylation and the decrease of the AEC value.

Enzyme Activities

Aquatic contaminants affect the activities of enzymes in exposed organisms. One particular enzyme system that has received much attention in this regard is the microsomal Mixed Function Oxidase (MFO) system. Hydrophobic contaminants are removed from potential sites of toxic action in exposed organisms by partitioning into lipid pools, as discussed in Chapter 6. Any metabolic conversion of such contaminants requires the presence of enzyme systems which convert these apolar compounds into more polar metabolites, the latter being relatively easily excreted. Such enzyme systems are apparently present universally in

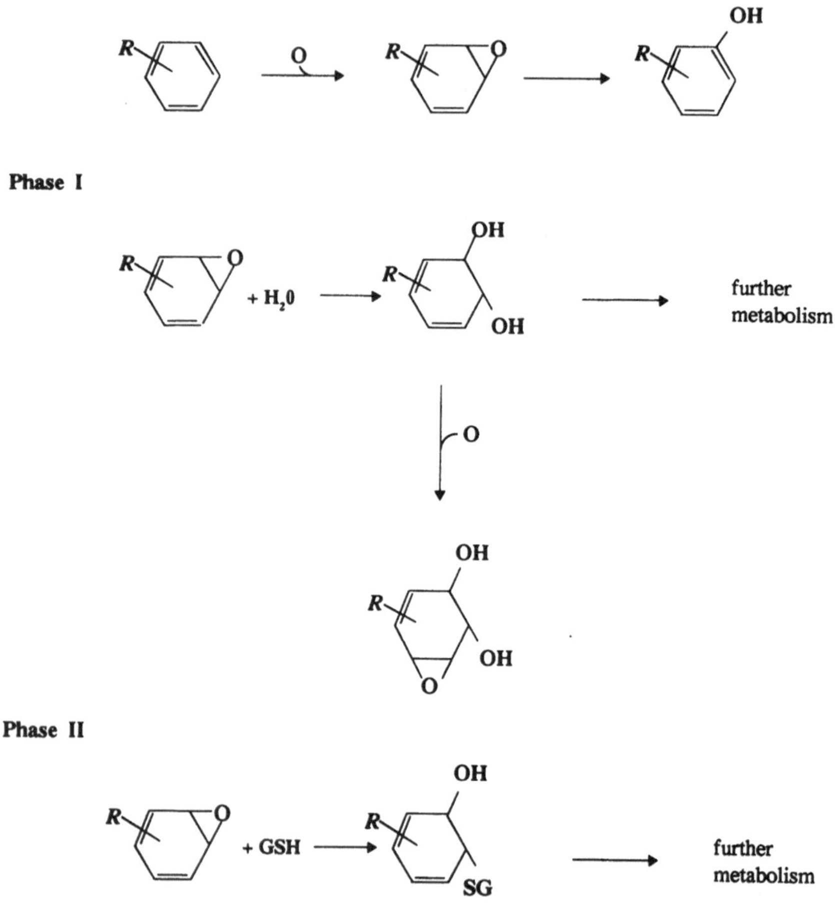

Fig. 91. Phase I and II reactions involved in the metabolism of hydrophobic contaminants. After Bayne *et al.* (1985).

organisms, although their activities vary markedly between different species.

The metabolism of hydrophobic contaminants can be classified into Phase I (biotransformation) and Phase II (conjugation) reactions (Livingstone, 1985; Addison, 1988). The first phase involves oxidation through various monooxygenase reactions, including epoxidation, hydroxylation and dealkylation (Fig. 91) catalysed by the MFO enzyme system. The MFO system involves an iron-containing protein (cytochrome P-450), and a flavoprotein which is generally known as NADPH-cytochrome P-450 reductase (Bayne *et al.*, 1985). Phase I enzymes include ethoxyresorufin O-de-ethylase (EROD) and benzo(*a*)pyrene hydroxylase (B(*a*)PH or BPH; see Addison, 1988).

In Phase II reactions, the products or intermediates of Phase I metabolism may be converted to dihydrodiols using epoxide hydralases (EH), and/or conjugated with glutathione or glucuronic acid. The latter reactions employ glutathione-S-transferases (GST) and UDP-glucuronyl transferases respectively (Fig. 91; see Livingstone, 1985; Addison, 1988). Further monooxygenation of some products may also occur, as may conjugation with other chemicals including sulphate, and the subsequent metabolism of conjugates (e.g. conversion of glutathione conjugates to mercapturic acids; see Livingstone, 1985). The eventual result of these processes is the conversion of hydrophobic contaminants to metabolites which are hydrophilic and more readily excreted.

A crucial feature of the MFO system is that many of the enzymes involved are inducible in nature, i.e. their activities may be increased by prior exposure of an organism to contaminants such as petroleum hydrocarbons and PCBs (Livingstone, 1985). Increases in the activities of cytochrome-P450 and NADPH-cytochrome P-450 reductase on such exposure to contaminants have been widely documented, although BPH may not be of a highly inducible nature. Increases are also observed in the MFO-associated neotetrazolium reductase (NTR) activity of digestive gland cells of mussels exposed to contaminants such as phenanthrene (Moore *et al.*, 1984; Livingstone, 1985).

Figure 92 provides examples of the increases in MFO enzyme activities in mussels (*Mytilus edulis*) and periwinkles (*Littorina littorea*) experimentally exposed to mixtures of contaminants (Livingstone, 1988; Suteau *et al.*, 1988). Suteau *et al.* (1988) considered that BPH and EH activities could be employed to monitor organic pollution, but Livingstone (1988) concluded that use of MFO enzyme activities in bivalve molluscs as biochemical indices of pollution remained limited by a lack of understanding of the basic nature and function of the molluscan MFO system.

The experiments discussed above were carried out in concert with several other studies; most notably, MFO enzyme activities in fish were also investigated (Addison, 1988; Addison and Edwards, 1988). The latter authors considered that EROD measurements in liver of the flounder *Platichthys flesus* gave the clearest and most sensitive response to expected pollutant gradients in the field samples. Long and Buchman (1990) found similar patterns of response of the liver MFO system of the related

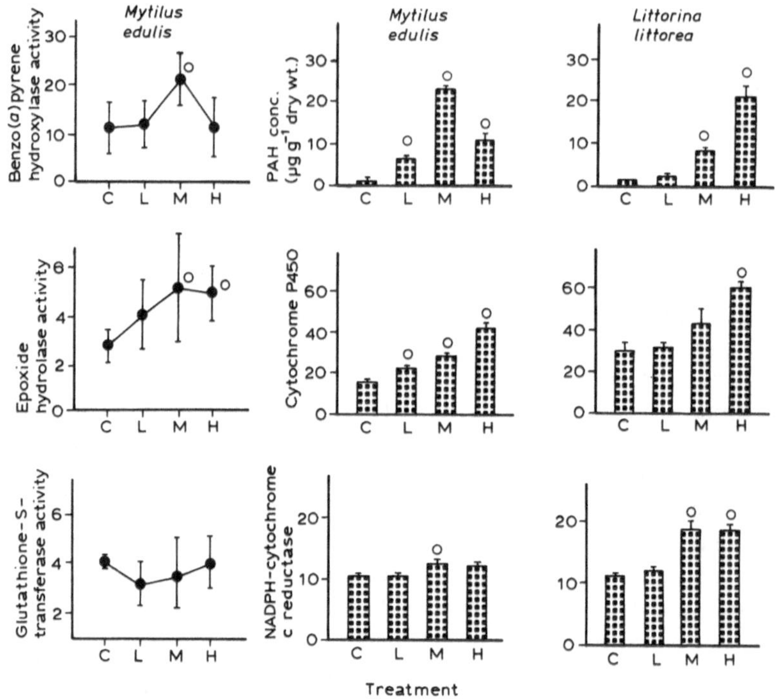

Fig. 92. Responses of various microsomal enzymes in mussels (*Mytilus edulis*) and gastropods (*Littorina littorea*) in a mesocosm-based exposure to four dosage levels (control, C; low, L; medium, M; high, H) of contamination by copper and PAHs. Mean values ± standard errors are shown, and significant differences from controls are denoted by (○). After Livingstone (1988) and Suteau *et al.* (1988).

flounder *Platichthys stellatus* from Californian waters to exposure to organic contaminants, with between-site discriminatory power being highest for EROD activity.

Some studies of this type have also been undertaken in fresh waters. For example, Jiminez *et al.* (1990) studied enzymes of the liver MFO system in the redbreast sunfish, *Lepomis auritus*, from Tennessee streams. Fish from a stream impacted by an industrial effluent exhibited significantly higher EROD activities than those from a reference stream, although fish from the most severely impacted stream did not display the highest EROD levels. This apparent anomaly was possibly a result of damage to the livers of the animals at the most exposed site, reducing the capacity of the fish to respond (Jiminez *et al.*, 1990). As in the marine system, more information is needed on the response of the MFO system in fish to many environmental, physiological and toxicological factors in order to properly interpret its potential role in biomonitoring.

It should also be noted here that, by contrast to the effect of lipid-soluble organic contaminants on the MFO system, certain trace metals may actually inhibit MFO activity (see Viarengo, 1989).

Brief comments are also required here on the impacts of contaminants on other enzyme systems in aquatic biota. These include the effects of lead on enzymes of various types in salmonid and other fish species (Hodson *et al.*, 1977, 1978; George, 1990), and the use of acetylcholinesterase activity to monitor the impacts of organophosphate and carbamate pesticides on crustaceans and fish (Galgani and Bocquené, 1988; Bocquené *et al.*, 1990). However, Bayne *et al.* (1985) noted that while much information of this type exists, none of the enzymes or their responses in fish and aquatic invertebrates are sufficiently well understood to be employed as reliable sublethal indices of contaminant toxicity. Thus, the most useful monitoring techniques developed to date are considered to involve the enzymes of the MFO system, and even in these instances, further developmental work is required.

Genotoxic Responses

In addition to their effects on enzymes, contaminants may cause other changes at the cytological level, and at least some of these appear to occur at the genetic level. For example, Hinton and Laurén (1990) described microscopic structural alterations in the livers of fish suffering chronic toxicity, including hepatocyte coagulative necrosis, spongiosis hepatis and neoplasia. Cormier and Racine (1990) have reported a very high incidence of hepatocellular carcinoma in Atlantic tomcod, *Microgadus tomcod*, from the Hudson River, New York, with lower incidences in other less industrially impacted rivers. Granulocytomas (non-neoplastic inflammatory responses to environmental contaminants) occur in the digestive gland and mantle tissue of the mussel *Mytilus edulis* (Bayne *et al.*, 1985). The incidence of the granulocytomas correlates with levels of anthropogenic stress and is thought to be indicative of a general loss of condition in mussels.

Haemopoetic neoplasms (involving infiltration, invasion or replacement of normal cells by atypical cells which undergo active mitosis) have been described in several bivalves including *M. edulis* (Lowe and Moore, 1978; Bayne *et al.*, 1985), and their incidence may be linked to the presence of contaminants. Mussels exposed to the water-accommodated fraction of crude oil exhibited a significant reduction in the mean cell height of the digestive tubule epithelium, with a concomitant change in shape of the tubule lumen from convoluted to rounded in transverse section (Widdows *et al.*, 1982). Long-term exposure of *Mytilus edulis* to copper (5 μg Cu litre^{-1} for 18 months) also caused histopathological changes in the digestive tubule epithelium, which consisted of non-ciliated cuboidal cells as opposed to ciliated columnar cells (Calabrese *et al.*, 1984).

Bright and Ellis (1989) described histopathological changes in the digestive gland, ctenidia and ventral foot epithelium of the tellinid bivalve *Macoma carlottensis* sampled close to a previous discharge site of copper mine tailings in British Columbia, Canada. Bivalves nearer the discharge had progressively more extensive lesions in these tissues, including vacuolation and increased fragmentation of digestive gland cells, sloughing of the foot ventral epithelium and swelling of the subfilamentar tissue of the ctenidium.

Aquatic contaminants may also cause chromosomal and other nuclear aberrations in aquatic organisms. Such effects have, for example, been reported in the starry flounder *Platichthys stellatus*, in the polychaete *Neanthes arenaceodentata* and in the mussel *Mytilus edulis* (Dixon and Clarke, 1982; Bayne *et al.*, 1985; Spies *et al.*, 1990). Indeed, chromosomal damage (and related cellular effects) induced in different life history stages of the marine polychaete *Pomatoceros triqueter* can be used in laboratory bioassays of genetically harmful agents in seawater (Bayne *et al.*, 1985; see Table 29). Contaminants such as metals may cause abnormal development during embryological stages and larval development, for example in bivalve embryos and crab larvae (Martin *et al.*, 1981).

Shugart (1990) has provided a brief overview of biochemical approaches taken to monitor DNA damage by environmental contaminants. Such responses include primary DNA damage by the formation of adducts, involving the covalent bonding of compounds with DNA. Such changes may lead to the chromosomal aberrations referred to above. The potential biomonitoring role of the relative presence of DNA adducts in aquatic organisms is not yet conclusively proven, however, as DNA in the digestive gland of the mussel *Mytilus galloprovincialis* contains "natural adducts" which are apparently unrelated to the presence of pollutants (Kurelec *et al.*, 1990).

Table 29 Life stages of the marine polychaete *Pomatoceros triqueter* which may be employed to test for genetically harmful agents in seawater (after Bayne *et al.*, 1985)

Material exposed to toxicant	*Maximum exposure time*	*Stage at which scored*	*Observed effects*
Gonad (adult organism)	months	sperm	malformations of head
		oocytes	chromosome damage, e.g. translocation heterozygotes, aneuploids
Gametes	1.5 h (sperm) 6.0 h (oocytes)	fertilisation	reduced rate of fertilisation
Fertilised oocytes	1 h	same or early cleavage	chromosomal and cellular abnormalities, e.g. abnormal polar bodies and vesicle, septate ooplasm
Early stage embryos (<8 cells)	7 h	same	chromosomal aberrations, i.e. numerical and structural abnormalities; developmental effects, e.g. premature loss of synchrony, malformations

Metallothioneins

The exposure of aquatic organisms to increased concentrations of certain trace metals will induce a response in a specific biochemical index — that of the tissue concentration of metallothioneins.

Metallothioneins are a group of low molecular weight proteins (6000–7000 Da), found in solution in the cytosol of particular cells. They are characterised by their affinity for particular heavy metals (cadmium, copper, mercury, silver and zinc), their stability to heat, a virtual lack of aromatic amino acids and histidine, and by an unusually high content (30–35 mol %) of the sulphur-containing amino acid cysteine (Viarengo, 1985; George, 1990). Metallothioneins usually bind 6 to 7 metal atoms per mole by clusters of thiolate bonds, with more than one metal generally present. Cadmium, mercury, or excess copper will displace zinc from zinc/copper-containing metallothioneins which are usually present in the cytosol, and excess metal ions will induce the synthesis of new metallothioneins. Thus, metals accumulated within cells will be bound by metallothioneins and thereby detoxified (see Chapter 5). Toxic effects will only occur if the rate of metal influx into the cell exceeds the rate of metallothionein synthesis and/or the maximum level of these proteins produced by the cell (Viarengo, 1985). The toxic metal ions may then "spill over" from the metallothionein pool, to bind with other subcellular components, with potentially toxic effects.

Metallothioneins are ubiquitous in eukaryotes and display a very highly conserved structural homology (George, 1990). Since they are present in the tissues of uncontaminated organisms, they are believed to play a physiological role, probably in the control of essential Zn and Cu metabolism (Brady, 1982). Metallothioneins can be used as an indicator of exposure to certain trace metals because, whilst low amounts of these proteins are present in tissues under normal conditions, sublethal exposure to the five elements noted above induces metallothionein synthesis, causing their levels to increase by as much as 30-fold (see George, 1990).

Organs such as the liver and kidney of fish or the digestive gland or kidney (or equivalents) of invertebrates contain particularly high levels of metallothioneins. Exposure to metals will not only cause changes in the total amounts of metallothioneins, but will also affect their relative metal content, since metals do not bind to metallothioneins with equal affinity, and differential displacement and induction will occur. The general order of affinity of metallothioneins for metals is as follows: mercury > copper > cadmium > zinc.

The induction of metallothioneins has been amply demonstrated in laboratory studies (e.g. George *et al.*, 1979; Viarengo *et al.*, 1984; Roesijadi, 1986; Langston *et al.*, 1989; see Fig. 93). In the field, Viarengo *et al.* (1982) found that the concentration of metallothionein-like proteins binding copper in the digestive gland of mussels (*Mytilus galloprovincialis*) was three times higher in samples from a metal-polluted environment than in those from a clean environment. Langston *et al.* (1989) found cadmium was predominantly bound to metallothionein in periwinkles (*Littorina littorea*) from uncontaminated sites, but the amounts of cadmium bound in this fashion increased markedly in samples from cadmium-contaminated sites.

There are, however, several caveats over the use of metallothionein levels as an indicator of metal exposure, and it is necessary to understand the characteristics of this response in the particular test organism used (George, 1990). Such characteristics include the dose/response relation-

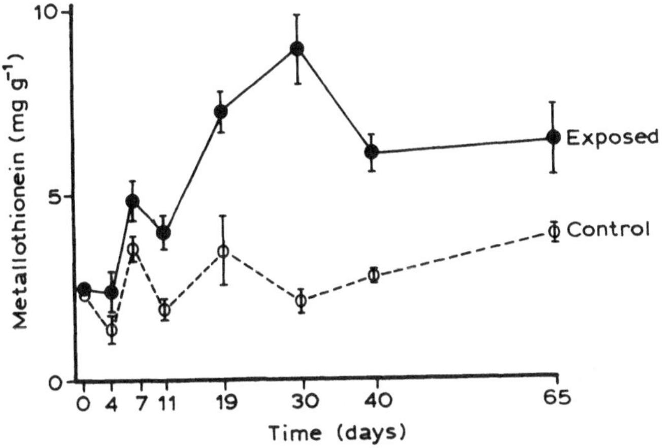

Fig. 93. Concentrations of metallothioneins in control mussels (*Mytilus edulis*) and in mussels exposed to cadmium at 400 μg litre^{-1} for 65 days. After Langston *et al.* (1989).

ships, the longevity and magnitude of the response (i.e. induction and degradation kinetics for the metallothioneins involved), and the effects of several parameters which affect metallothionein production in organisms. The latter include seasonal variations, reproductive and moult cycles, and the possible roles of other stressors including temperature, the presence of organic contaminants, steroid hormone levels, tissue injury, and other factors (Benson *et al.*, 1990; George, 1990; Petering *et al.*, 1990; Sanders, 1990). Furthermore, metal accumulation strategies differ between organisms (see Chapter 5), with resulting differential dependence on the use of metallothioneins for the detoxification of metals. For example, the mussel *Mytilus edulis* binds cadmium in metallothionein-like proteins, induction being quantitatively related to cadmium exposure. By contrast, British populations of the tellinid bivalve *Macoma balthica* fail to produce metallothionein-like proteins even in conditions of extreme cadmium contamination (Langston *et al.*, 1989).

Intracellular Metal Distributions

The intracellular distributions of trace elements may provide information on the presence of a toxic effect of exposure to metals, indicating the existence of "spillover" of excess metal from any existing detoxification process. Thus, changes in the distributions of metals among lysosomes, proteins of high and low molecular weights, and specific metal-binding ligands of high affinity have been investigated both in laboratory and field exposures (see the review by George, 1990). An example is provided by the studies of Viarengo *et al.* (1988) on metal distributions in the digestive glands of mussels (*Mytilus edulis)* along a pollution gradient in Langesundfjord, Norway. These authors showed that the concentration of copper bound to thioneins was essentially the same in mussels sampled at reference and polluted sites, but the levels of copper bound to other cytosolic proteins significantly increased at the polluted field sites.

This "spillover" of metals from detoxification pathways into other biochemical pools may affect the health of the organism involved. For example, on exposure to copper, larvae of the crab *Rhithropanopeus harrisii* are able to limit the non-specific cellular binding of the element over a range of three orders of magnitude of cupric ion activity. However, this cellular regulation breaks down at very high exposure levels, and copper accumulates in the pool of cell metabolites of very low molecular weights, with a consequent disturbance of larval growth (Sanders *et al.*, 1983; Sanders and Jenkins, 1984). Similarly, an accumulation of cadmium in the very low molecular weight pool of cell metabolites of cadmium-exposed polychaetes (*Neanthes arenaceodentata*) was associated with loss of reproductive potential and other signs of stress, including sluggish behaviour, tremors, reduced tube building and reduced growth (Jenkins and Sanders, 1986; Jenkins and Mason, 1988).

Cytochemical Responses

The effects of contaminants on the structure and/or function of cell organelles have been classified by Slater (1978) and Moore (1985) into four main categories, as follows:

- Depletion or stimulation of metabolites or co-enzymes. This may be sufficient to produce a morphologically evident lesion in the cell, for example by altering the intracellular redox state.
- Inhibition or stimulation of enzymes and other specific proteins. This has been discussed above at the biochemical level, with respect to the effects of organic contaminants on the MFO system and the induction of metallothioneins in aquatic biota by their exposure to trace metals.
- Activation of an organic contaminant to produce a more toxic molecular species. For example, active metabolites produced by the MFO system (e.g. see Livingstone *et al.*, 1990) may damage DNA, proteins or membranes.

- Membrane disturbances. Examples of membrane damage include changes in cellular compartmentalisation (e.g. injury to cell organelles), lipid peroxidation, and changes in membrane fluidity.

Some of these effects have already been considered above, as biochemical responses. Others are better described as cytochemical, for example those involving lysosomal membrane stability, lysosomal enlargement and lipofuscin accumulation.

Many contaminants interact with and alter the function of cell membranes and the membranes of subcellular organelles. Any damage to the membranes of lysosomes will release normally latent hydrolases into the cell cytoplasm, with resulting autolytic cell damage. The induction of free enzyme activity (as opposed to latent activity) which may result from the destabilisation of the lysosomal membrane can be measured histochemically as the labilisation period. This corresponds to the length of time of preincubation of a tissue section needed for maximum staining intensity for the lysosomal hydrolases: arylsulphatase, β-glucuronidase or N-acetyl-β-hexosaminidase. A reduction in this labilisation period (reduced hydrolase latency) for lysosomal hydrolases has been used as an indicator of pollutant-induced lysosomal membrane damage, particularly in molluscs such as the mussel *Mytilus edulis* and the gastropod *Littorina littorea* (Moore, 1980, 1985, 1988a,b; Moore *et al.*, 1987a,b), and the hydroid *Campanularia flexuosa* (Stebbing, 1976).

Moore *et al.* (1987b) carried out a cytochemical investigation of the digestive gland cells of *Littorina littorea* near Sullom Voe oil terminal in Shetland, Scotland (Fig. 94). Gastropods from near the oil terminal had reduced latencies of lysosomal β-glucuronidase, increased total β-glucuronidase activities (Table 30) and enlarged lysosomes, in comparison with the same species from a nearby reference site which was less polluted. Similarly, Moore (1988a) used acid labilisation characteristics of latent lysosomal hydrolases to determine lysosomal stability in digestive gland cells of *Mytilus edulis* and *Littorina littorea* exposed to a diesel oil and copper mixture experimentally, and in the same species along an expected field pollution gradient in Langesundfjord (Norway). Lysosomal membrane stability was reduced in both molluscs with increased contamination in the field (Fig. 95), but the experimental data were difficult to interpret. Lipofuscin content and the number of tertiary lysosomes increased in contaminated field mussels (Moore, 1988a).

Moore (1988b) concluded that lysosomal enlargement, increased lysosomal fragility, lipid accumulation and lipofuscin accumulation are all good descriptors of pathological effects in the digestive gland cells of

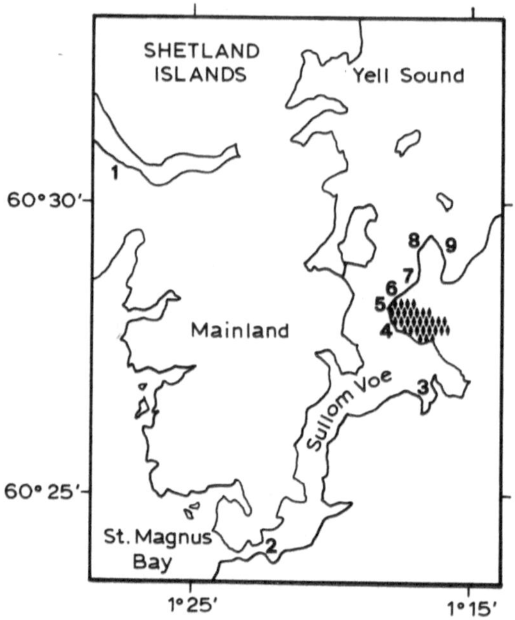

Fig. 94. Locations of populations of the gastropod *Littorina littorea* employed for studies of the impacts of oil on microsomal enzymes in the digestive gland (see Table 30). Sites marked are as follows: (1) Ronas Voe, the reference site; (2) Mavis Grind, a heavily-contaminated site; (3) Outer Houb of Scatsta; (4) Tanker Jetty; (5) Kames Tidal Pool; (6) The Kames; (7) Brei Wick; (8) Skaw Taing; and (9) Swarta Taing. The Sullom Voe Oil Terminal is also shown, as a hatched area. After Moore *et al.* (1987b).

mussels exposed to contaminants in the field. These pathological alterations at the cellular level were interpreted as indicating the presence of increased autophagocytosis with increased catabolism of macromolecules, leading ultimately to cellular atrophy (Moore, 1988b).

Moore (1988a) was also able to measure histochemically the activity of NADPH-ferrihemoprotein reductase in these mussels. These results correlated well with the biochemical measurements of Livingstone (1988) and Suteau *et al.* (1988) discussed above. Reductions in scope for growth of these contaminated mussels (Widdows and Johnson, 1988) have been reported above, indicating that the cytochemical responses described were present in mussels that were suffering physiologically from the toxic effects of contaminants. More directly, Bayne *et al.* (1979) showed how reduction of the labilisation period of the lysosomal hydrolase N-acetyl-β-hexosami-

Table 30 Latencies of lysosomal β-glucuronidase and relative activities of this enzyme in the digestive gland of gastropods, *Littorina littorea*, from sites in Sullom Voe exhibiting different degrees of oil contamination. See Fig. 94 for sample sites. (After Moore *et al.*, 1987b)

Sample site	*Mean (range) labilisation period of β-glucoronidase (min)*	*Mean relative activities (% of Ronas Voe values)*
Ronas Voe	24 (20, 25)	100.0 ± 2.0
Mavis Grind	6 (5, 10)**	153.2 ± 12.6**
Outer Houb of Scatsta	18 (15, 20)*	88.4 ± 1.8**
Tanker Jetty	9 (5, 10)**	149.1 ± 8.0**
Kames Tidal Pool	5.4 (2, 10)**	167.4 ± 8.3**
The Kames	14 (10, 15)**	121.8 ± 7.6**
Brei Wick	9 (5, 10)**	121.4 ± 3.7**
Skaw Taing	13 (10, 15)**	144.3 ± 9.5**
Swarta Taing	9 (5, 10)**	107.0 ± 4.3

Note: * Mann-Whitney U-test, $p \leqslant 0.05$; ** Mann-Whitney U-test, $p \leqslant 0.01$.

nidase from digestive gland cells is correlated with decreasing scope for growth in *Mytilus edulis* in contaminated conditions (Fig. 96).

Trace elements also accumulate intracellularly (often in tertiary lysosomes) in aquatic organisms exposed to high metal concentrations, par-

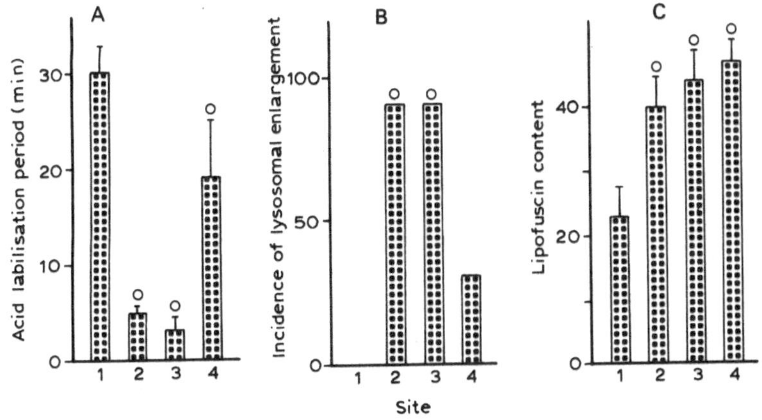

Fig. 95. Responses of lysosomes in the digestive glands of mussels (*Mytilus edulis*) from field sites in Langesundfjord, Norway. (A) Lysosomal membrane stability. (B) Incidence of lysosomal enlargement. (C) Lysosomal content of lipofuscin. Differences from the reference site (site 1) are denoted by (○). After Moore (1988a).

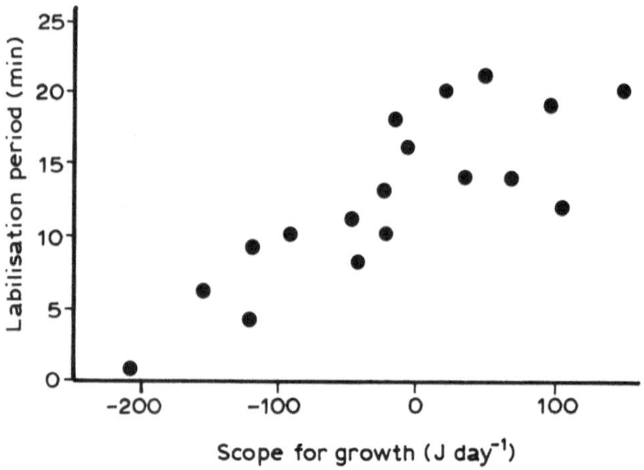

Fig. 96. Relationship between the labilisation period of the lysosomal enzyme *N*-acetyl-β-hexosaminidase and scope for growth in mussels (*Mytilus edulis*) from four different populations. After Bayne *et al.* (1979).

ticularly those employing an accumulation strategy of strong net accumulation (see Chapter 5). Such accumulations may be visualised using electron microscopy and sometimes light microscopy; however, they do not in themselves indicate that a toxic effect is present. Viarengo (1989) has presented an excellent review of the effects of heavy metal exposure at the cellular level.

(iii) Behavioural Reponses

The behavioural responses of aquatic organisms to the presence of aquatic contaminants are manifold. Behaviour patterns shown to be modified by pollution include sensory responses (phototaxis, geotaxis, chemotaxis, chemoreception, temperature preferences, tactile inhibition, and lateral line sensitivity); rhythmic activities (daily, tidal, moult, and reproductive activities); motor activity (avoidance, attraction, shelter seeking including substrate attachment and burrowing, equilibrium, and swimming activities); motivation and learning phenomena (feeding and avoidance); inter-individual responses (migration, aggregation, aggression, and predation vulnerability); and respiration (Eisler, 1979; Capuzzo and Kester, 1987).

Such behavioural responses are of use in early warning biological monitoring systems of effluent quality (see Chapter 1 and Cairns and van der Schalie, 1980), but are of limited value in assessments of the toxic effects of contaminants in the field (cf. Depledge and Andersen, 1990). They may, however, constitute the primary cause of certain types of changes in community structure (see below). Similarly, behavioural responses will affect some of the other indices already discussed, for example the valve closure displayed by the mussel *Mytilus edulis* on exposure to copper (Davenport and Manley, 1978) will affect filtration rates, and subsequently scope for growth.

C. EFFECTS MONITORING AT THE POPULATION LEVEL

Responses at the population level to the presence of aquatic contaminants are exemplified by changes in the numbers of individuals, the reproductive output, or the recruitment rates of affected organisms. In addition, contaminant exposure may alter gene and genotype frequencies in a population, through the selection of tolerant individuals. Such responses in populations inevitably lead on to alterations at the level of community structure.

Any reduction in the number of individuals in a population of a sedentary organism that results from aquatic contamination is clearly a lethal effect (Mance, 1987). However, some species increase in numbers in the absence of competitors in polluted areas. Such niceties of macrobenthic succession have been discussed at length by Pearson and Rosenberg (1978), with respect to organic enrichment of the marine environment. Gray (1979) briefly summarised some of the relevant data for chemically-polluted marine macrobenthic communities. Certain species are considered typical of contaminant-tolerant benthic populations; these include the polychaetes *Capitella capitata* and *Polydora ciliata*, and the oligochaetes *Peloscolex benendeni* and *Paranais littoralis* (Gray, 1976, 1979). Thus, for example, the numbers of *Capitella capitata* (later followed by the other polychaetes *Polydora ligni* and *Prionospio cirrifera)* increased dramatically in sediments affected by a spill of fuel oil at West Falmouth, USA (Gray, 1979). Such species are considered opportunistic (r strategists), with high rates of population increase and mortality and large population sizes in the absence of competitors (Gray, 1979).

Kingston (1987) presented data for the numbers of macrobenthic species affected by contaminated discharges of oil-based drilling fluids from North Sea oilfield platforms. The spatial distribution of benthic macrofauna at increasing distances from the platforms conformed generally with the division of adaptive strategies to pollution recognised by Gray (1979). Firstly, r strategists (e.g. *Capitella capitata*) are able to mature and reproduce rapidly, and are therefore quick to colonise and dominate disturbed areas. Secondly, T strategists (e.g. *Pholoe minuta, Prionospio cirrifera*), although less fecund than r strategists, are able to tolerate contaminated ("stressful") conditions and therefore outcompete other species. Finally, K strategists, controlled by normal forces of inter-specific competition, are unable to compete where there is environmental stress and therefore decline in numbers on approaching the contamination source (Gray, 1979; Kingston, 1987).

Effects on fecundity and reproductive rates are sublethal in nature. As discussed above, the hydroid *Campanularia flexuosa* produces more gonozooids in the presence of very low concentrations of toxins (Stebbing, 1979), the result of which is increased sexual reproduction, with subsequent larval dispersion and potential recruitment to populations elsewhere. However, reduced fecundity is a more common sublethal response to an exposure to aquatic contaminants, and this may occur at concentrations considerably lower than those which are acutely toxic to adults (Langston, 1990). An extreme example of reduced fecundity, that of the effect of tributyltin on the dogwhelk *Nucella lapillus*, has been discussed earlier.

Reish (1978) showed significant suppression of reproductive rates (measured by numbers of eggs or offspring produced) of the polychaetes *Neanthes arenaceodentata, Capitella capitata, Ctenodrilus serratus* and *Ophryotrochus diadema* by sublethal concentrations of various trace metals, with sensitivities varying between the species. Further examples of this type are provided by Langston (1990). Fecundity may be affected by contaminants through effects on fertilisation and subsequently on embryological development; indeed, echinoderm gametes and embryos are commonly used in bioassays of water quality using fertilisation, cleavage and gastrulation as markers (Kobayashi, 1984). The gametes and embryos of fish are also sensitive to contaminants (Langston, 1990).

Nevertheless, it is not clear that reduced reproductive output necessarily has an impact on the abundances of populations nor on their contributions to food webs (Underwood and Peterson, 1988). The high mortality rate of larvae of benthic invertebrates (Mileikovsky, 1971), together with a density-dependent regulation of the numbers of recruits (Connell, 1985), may render negligible the impact of reduced reproductive output on overall population sizes. Furthermore, many benthic invertebrates have dispersive planktonic larval phases and recruit from stock in areas away from those which are contaminated (Underwood and Peterson, 1988).

In the case of recruitment, Langston (1990) considered that this phase of the life cycle is less sensitive to sublethal contamination (at least by trace elements) than is fecundity, and therefore recruitment is not affected except at the most severely metal-contaminated sites. Restronguet Creek in Cornwall, England provides an example of such a site, with very high concentrations of many elements, including arsenic, cadmium, copper, iron, lead, manganese and zinc (Bryan and Gibbs, 1983). The fauna of the Creek are severely impoverished in bivalve molluscs, probably as a result of the sensitivity of embryonic and larval bivalves to copper and zinc, with consequent effects on population recruitment (Bryan and Gibbs, 1983). Elsewhere, McGreer (1982) considered that the degree of contamination of the substratum of a mudflat receiving sewage effluent in the Fraser River estuary, British Columbia, determined the distribution of the tellinid bivalve *Macoma balthica*, by affecting the settlement and survival of larvae and juveniles.

Aquatic contaminants such as cadmium and mercury are able to affect gene and genotype (allozyme) frequencies in exposed populations of molluscs and crustaceans (Nevo *et al.*, 1978; Lavie and Nevo, 1986; Ben-Shlomo and Nevo, 1988). Since the changes in allozyme frequencies are sensitive to and vary with the concentration and type of pollutant (Lavie and Nevo, 1986), such changes have the potential to be used as a biological monitoring system to measure the levels of specific pollutants (Beardmore, 1980; Beardmore *et al.*, 1980). For example, laboratory exposure of five species of marine gastropods to lethal cadmium concentrations gave rise to genetic selection of genotypes homozygous for the enzyme phosphoglucose isomerase, with a significantly higher proportion of heterozygotes among the dead animals (Lavie and Nevo, 1986). Similarly, Patarnello *et al.* (1991) provided evidence of selection due to metal exposure in the lagoon of Venice, Italy, of three polymorphic allozymes in a population of the barnacle *Balanus amphitrite*; this was again expressed as a significant reduction of genetic polymorphism (see also Nevo *et al.*, 1978).

Thus, changes in particular allozyme frequencies in a population of marine organisms are to be expected if contaminants are present in sufficient concentrations (and at sufficient bioavailabilities) to exert a selective effect. These changes may occur either by differential lethality or possibly through differential sublethal effects on local reproduction and recruitment. Changes in genotype frequencies of local populations can therefore be interpreted as evidence for the existence of a toxic effect from a local contaminant.

D. EFFECTS MONITORING AT THE COMMUNITY STRUCTURE LEVEL

Changes in the numbers of individuals in populations or the complete loss or addition of species will clearly give rise to alterations in the local community structure, and many studies have attempted to correlate such effects with the presence of environmental contamination. The communities considered include planktonic and benthic macrofaunal and meiofaunal assemblages of organisms.

Much of the work on contaminant-induced changes in planktonic community structure has involved containment studies or mesocosms, using natural plankton assemblages (Langston, 1990). The exposure of such communities to trace metals has led typically to changes in dominance patterns of phytoplankton, but with such variable results that it is difficult to predict resistant species (Langston, 1990). Nevertheless, one consistent effect seems to be the replacement of centric diatoms by pennate species (e.g. Thomas and Siebert, 1977; Sanders and Vermersch, 1982; Langston, 1990). Contaminants may alter the species compositions of communities through biological effects, including altered competitive abilities or predation pressure, rather than by exerting direct differential toxicities. Thus, changes in the species composition of phytoplankton communities may result from alterations in zooplankton grazing pressures. Similarly, changes in phytoplankton diversity resulting from contaminant effects may alter zooplankton diversity (Sanders, 1986). A feature of planktonic communities is that they are relatively short-lived (with rapid recovery rates) in comparison to macrobenthic associations, and by definition, plankton may be moved away passively by water currents from the site of contaminant action. Planktonic communities therefore offer limited scope for the detection of the toxic effects of aquatic contamination.

With respect to benthic communities, the studies of Lande (1977) on the effects of metal pollution in Trondheimsfjorden, Norway, may be cited. These concerned the local littoral and sublittoral assemblages of macrofauna and macroflora. The sessile organisms adjacent to mining areas were considerably reduced in quantity (measured as total number of species, mean number of total individuals m^{-2}) and diversity, the latter being measured using Simpson's and Margalef's indices. Reference has already been made above to Restronguet Creek in Cornwall, England, which receives extremely high concentrations of several trace elements in acid solution from the Carnon River (Bryan and Gibbs, 1983; Bryan *et al.*, 1987). A comparison of Restronguet Creek faunal assemblages with those of

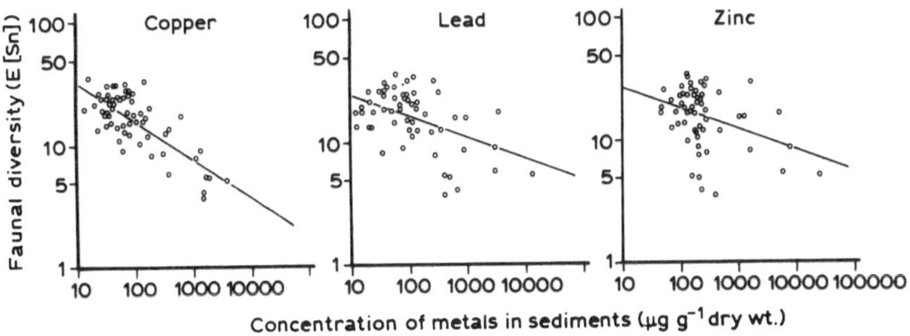

Fig. 97. Correlations between faunal diversity and the concentrations of copper, lead and zinc ($\mu g\ g^{-1}$ dry weight) in the sediments of Norwegian fjords. Correlation coefficients and significance are as follows: copper, $R = -0.76$, $p \leqslant 0.001$; lead, $R = -0.49$, $p \leqslant 0.001$; zinc, $R = -0.37$, $p \leqslant 0.01$. After Rygg (1985).

similar creeks in the area indicates that the former supports only a sparse fauna. Most notably, bivalves are absent except for the tellinid *Scrobicularia plana* on the margins of Restronguet Creek (Bryan and Gibbs, 1983). Bivalve larvae and juveniles are evidentially unable to withstand the toxic conditions. However, the flora and fauna are not as obviously affected in general as would be predicted from laboratory-based toxicity data, probably because of the presence of metal-tolerant populations (Bryan and Gibbs, 1983).

Rygg (1985) used correlation and regression analyses on data from Norwegian fjords to show that species diversity (after Hurlbert, 1971) in benthic faunal communities was negatively correlated with metal contamination of the sediments (Fig. 97). These correlations were strong with copper, moderate with lead, and weak with zinc. Of the 50 most frequently occurring species, 20 occurred at significantly lower numbers at the stations exhibiting greatest concentrations of copper, different groupings of species being identified by correlation plots between copper levels in the sediments and the occurrence of selected species (Rygg, 1985).

Kingston (1987) measured indices of diversity and equitability in an investigation of the effects of hydrocarbons on benthic macrofaunal communities near North Sea oil platforms. Diversity indices are affected by two community components: species richness (the total number of species in the community); and equitability (the evenness with which the individuals in the community are distributed among the species). A commonly used diversity index in such studies is the Shannon-Wiener index, essen-

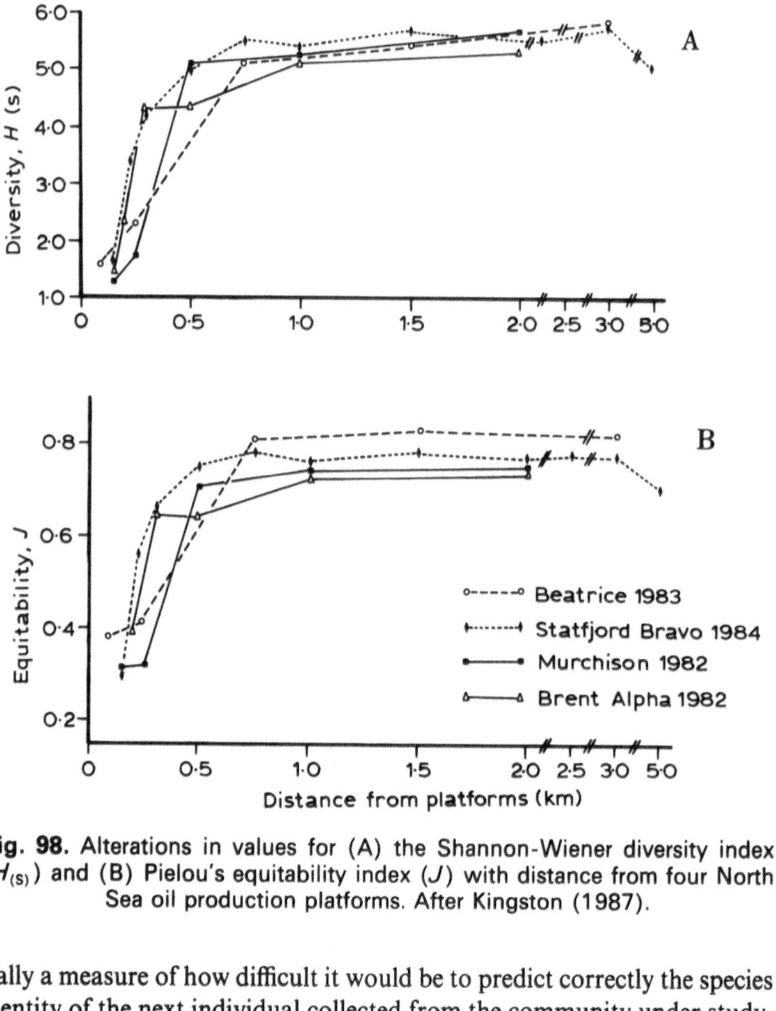

Fig. 98. Alterations in values for (A) the Shannon-Wiener diversity index ($H_{(s)}$) and (B) Pielou's equitability index (J) with distance from four North Sea oil production platforms. After Kingston (1987).

tially a measure of how difficult it would be to predict correctly the species identity of the next individual collected from the community under study. The theory behind the use of diversity indices is based on the premise that communities with a high diversity result from less environmental stress than those with low diversity. The validity of this premise is certainly debatable (Gray, 1979; Gray and Pearson, 1982), but in studies such as those of Kingston (1987) involving local gross pollution, such measures nevertheless appear quite useful (Fig. 98(A)). The equitability component of diversity is also frequently used in such studies (Fig. 98(B)), in order to indicate the degree to which a species abundance distribution may be dominated by a proportion of its members. For many North Sea oil platforms, there is a dramatic drop in the observed diversity index of benthic macrofauna within 750 m of the platform, indicating a gross change in community structure (Kingston, 1987).

The community parameters described have no regard to individual species identity. In fact, there is remarkable similarity between different platforms studied in the North Sea in respect of the species which are most abundant close to the platforms, and the top five species account for between 89 and 99% of the total number of individuals from those stations (Kingston, 1987). These species are r strategists, and groups of T strategists and K strategists (see above) are found in turn with distance away from the platforms (Gray, 1979; Kingston, 1987).

In a review of benthic macrofaunal data, Gray (1979) concluded that diversity indices are relatively insensitive to pollution-induced changes (see also Gray and Pearson, 1982), and he championed the use of log-normal distribution models to assess perturbation effects on communities. The distribution of individuals among species in a large sample taken from a community is expected to fit a log-normal distribution, a straight line being derived for a plot on probability paper of cumulative percentage species against species grouped in geometric classes. Changes in community structure brought about by pollution cause a departure from a log-normal model, and the resulting deviation from the straight-line correlation can be tested statistically; a measure not possible in the case of diversity indices (Gray, 1979).

If the number of species is plotted on the y axis instead of cumulative percent species, a log-normal distribution produces a bell-shaped curve. Moreover, any log-normally distributed groups of species contributing to the whole distribution will separate out (Gray and Pearson, 1982). Thus, it becomes possible to recognise the presence of the three indicator groups of species (r strategists, T strategists and K strategists), or variations thereof. Such groups will also move along the horizontal axis of the plot as stations are compared in space or time, according to changes in contamination (Gray and Pearson, 1982). Groups of species fitting different categories (e.g. those "sensitive to pollution") can be identified between regions, without resorting to the use of the theoretically less acceptable concept of universal indicators.

In a statistically precocious study, Ward and Young (1982) examined the effects of trace metal contamination in sediments on the community structure of the epibenthic seagrass fauna near a lead smelter in South Australia. One pattern recognisable in the faunal distribution of 20 common species (mostly of fish) described decreased frequencies correlated with the concentration of metals in the sediments. Analysis of variance was used to compare the species richness (number of species per sample, square root-transformed to provide homogeneity of variance) of four sites along each of three transects varying in metal contamination, on three separate occasions. This treatment showed that species richness varied between transects, as well as between sites and times, but its effect was not uniform at all sites, nor at all times within these sites (Ward and Young, 1982). The classification of species compositions was investigated using a combination of classification and ordination procedures. Such statistical procedures are now common in analyses of changes in benthic community structure (Warwick and Clarke, 1991).

Several statistical techniques were employed to describe sublittoral macrobenthic community structures at six stations along the copper/hydrocarbon pollution gradient at Langesundfjord, Norway (Gray *et al.*, 1988; Warwick, 1988). The techniques used included multivariate methods to discriminate between sites based on their faunal attributes (classification, ordination and discrimination tests), and univariate methods to determine levels of disturbance or "stress" at given sites (abundance and biomass of individuals, diversity index, and species richness and evenness).

Multivariate analyses produced generally similar results, dividing the sites studied into three groups, with differences in water depth masking possible pollution effects (Gray *et al.*, 1988; Warwick, 1988). There was a high degree of redundancy in the data, and transformed biomass data (grouped at family and even phylum level) produced an appreciable improvement in separation of replicates at certain sites (Warwick, 1988). This conclusion suggests that such studies might be carried out successfully, even with limited taxonomic expertise. Univariate stress measures in combination ranked the sites in order of increasing disturbance, but again correlated poorly with measured levels of contaminants in sediments (Gray *et al.*, 1988).

Warwick and Clarke (1991) have produced a detailed comparison of statistical methods for analysing changes in benthic community structure, dividing them into univariate, graphical/distributional, and multivariate techniques. The detailed arguments are beyond the scope of the present chapter, but in summary, Warwick and Clarke (1991) favoured the use of multivariate ordinations, making four general conclusions, as follows:

- The similarity between stations based on univariate or graphical/distributional properties is usually different from their clustering in multivariate analyses.
- Species-dependent (multivariate) methods are much more sensitive than species-independent (univariate and graphical/distributional) methods in discriminating between stations.
- When more than one component of the fauna has been studied, univariate and graphical/distributional methods may give different results for different components, whereas multivariate methods tend to give the same results for each component studied.
- The key environmental variables responsible for community change may be identified by matching multivariate ordinations from subsets of environmental data to an ordination of faunistic data.

Although the above discussion has cited examples relating only to macrofauna, analysis of the community structures of meiofauna (sieved between 63 μm and either 0.5 or 1.0 mm) is becoming increasingly important. Macrofauna have the advantage of relative longevity, thereby reflecting environmental conditions integrated over long periods (years in most cases). In addition, a robust taxonomic expertise exists, as do well-developed methodologies for sampling and processing (Warwick, 1988). However, Heip *et al.* (1988) have provided convincing arguments for analysing meiofaunal communities, including the less labour-intensive sampling involved and the shorter generation times of meiofaunal species. The latter may be expected to lead to faster potential response times to pollution incidents (Warwick, 1988). Examples of meiobenthic studies include those of Bartlett and Hennig (1987), Heip *et al.* (1988) and Sundelin and Elmgren (1991).

E. TOLERANCE AS A MEASURE OF CONTAMINANT EFFECTS

The selection of contaminant-tolerant populations in particularly heavily impacted aquatic habitats has been alluded to above. Bryan and Gibbs (1983) and Bryan *et al.* (1987) have identified metal-tolerant populations of seaweeds and invertebrates in Restronguet Creek, Cornwall. For example, the brown seaweed *Fucus vesiculosus* from the Creek is tolerant to copper but not zinc (Fig. 99), and the local populations of the polychaete *Nereis (Hediste) diversicolor* and the crab *Carcinus maenas* are insensitive to both copper and zinc (Bryan and Gibbs, 1983). Such tolerances may be a result of the selection of a metal-insensitive population

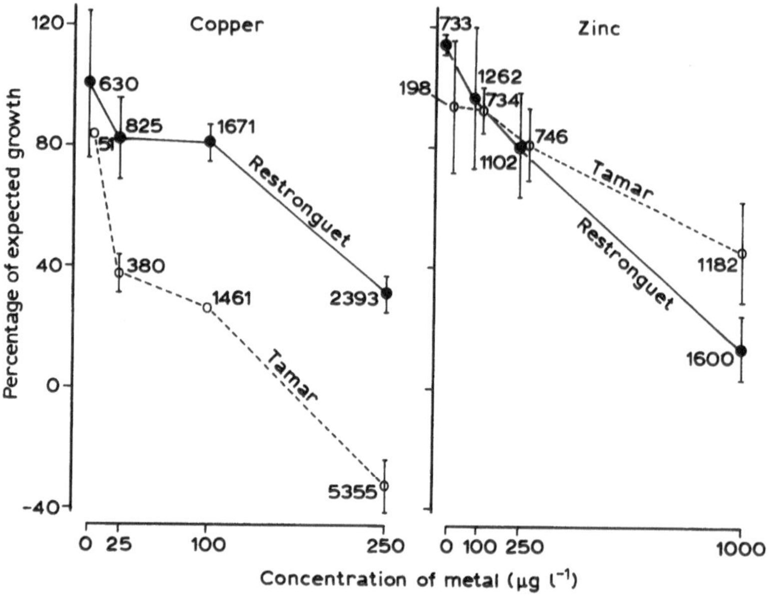

Fig. 99. Comparative effects of copper and zinc on the growth of the brown macroalga *Fucus vesiculosus* from Restronguet Creek and the Tamar Estuary, England. The effects of metals were assessed by comparing the measured growth during metal exposure to that predicted from the pre-exposure period. Figures denote the accumulated tissue concentrations of metals ($\mu g\ g^{-1}$ dry weight); vertical lines denote the range in growth. After Bryan and Gibbs (1983).

breeding within the Creek (e.g. *Nereis diversicolor*), or of the selection of individual organisms migrating into the Creek from elsewhere (this may be the case for *Carcinus maenas*). In fresh water, populations of the isopod crustacean *Asellus meridianus* are tolerant to both copper and lead in the copper-contaminated River Hayle in Cornwall, but are tolerant to lead only in the lead-contaminated River Gannel (Brown, 1976, 1977, 1978).

As reviewed by Klerks and Weis (1987), a genetic basis exists to the tolerance of individuals in some populations which are exposed to contamination in the field. As noted above, exposure to contaminants can change genotypic frequencies in a population, and such changes will reflect the selection of more tolerant individuals. Luoma (1977b) argued that where local populations exhibit an unusual tolerance to a toxicant, this provides direct evidence that the bioavailability of the toxicant in the environment of the resistant population is sufficient to elicit a detrimental biological effect. The presence of a toxicant-resistant population of one species in a habitat further suggests that other species could be affected by the toxicant (Luoma, 1977b).

Klerks and Levington (1989) provided an apt example, involving the very rapid evolution of genetically determined resistance to cadmium- and nickel-contaminated sediment in Foundry Cove (New York) by the oligochaete *Limnodrilus hoffmeisteri*. Grant *et al.* (1989) followed up the suggestion of Luoma (1977b) directly, using the inherited metal tolerance of the polychaete *Nereis diversicolor* in Restronguet Creek (see Bryan and Gibbs, 1983) to map the ecological impact of trace metal contamination in the Creek. Account should be taken in such work of the possible distorting effect of a downstream drift or migration of tolerant individuals beyond the geographical extent of ecologically significant contamination, unless this is outweighed by local intraspecific competition (Grant *et al.*, 1989).

Both these examples concern tolerance recognisable by changes in lethal responses to toxicants, and both are associated with sediments of extraordinarily high contamination by trace elements. There remains a need to identify field examples of the evolution of tolerance to contaminants realising sublethal responses. Such insensitivity to sublethal exposure is likely to be of wider ecological significance, and therefore of more use as a biological marker of the toxic effects of an environmental contaminant.

F. EFFECTS MONITORING USING COMBINATIONS OF APPROACHES

Although the detection of biological responses to the toxic action of contaminants has been divided here into the levels of the organism and its tissues, the population, and the community, not all published studies (to their credit) fall into an individual discrete category.

Many of the studies on the mussel *Mytilus edulis* have combined morphological and physiological parameters, including scope for growth, biochemical and cytochemical indices (e.g. Moore *et al.*, 1984, 1987a; Bayne *et al.*, 1985). Lack and Johnson (1985) made use of a wide range of indices in assessing the biological effects of sewage sludge disposed of at a licensed site off Plymouth, England. These included scope for growth, lysosomal stability, mixed function oxygenase activity, metal-protein complexing in caged mussels *Mytilus edulis*, and the bioassay of water samples using the hydroid *Campanularia flexuosa*. Martin (1985) described the combined use of scope for growth and laboratory bioassays in the California State Mussel Watch Program. In assessing the impacts of tributyl tin in estuaries and coastal regions, Langston *et al.* (1990) used chemical measures in sediments, biomonitoring (with the bivalve *Scrobicularia plana*), and morphological data (imposex in *Nucella lapillus*).

Chapman and Long (1983) emphasised that it is not sufficient to document marine pollution only in terms of chemical concentrations of contaminants, but that bioassays should be a central and integral part of any marine pollution assessment. Long and Chapman (1985) expanded this proposal to put forward the concept of a "Sediment Quality Triad" to characterise the contamination of sediments in aquatic environments, and the impacts deriving from this. Essentially this approach calls for the use of three types of studies in combination: (i) the analysis of trace contaminants in sediments; (ii) bioassays of sediment toxicity in lethal and sublethal laboratory tests; and (iii) investigations of the infaunal community structure. Long and Chapman (1985) showed that a strong overall correspondence existed between these three components in Puget Sound (Fig. 100). The Sediment Quality Triad approach has much to recommend it, particularly if modern multivariate statistical analyses are used to recognise changes in infaunal community structure. In practice, its use is supplemented by biomonitoring studies undertaken within the framework of the National Mussel Watch Program in the USA, and in many senses, the Triad is in fact based upon four sets of data rather than three.

An example of this approach had in fact already been published before the designation of the term "Sediment Quality Triad". Bryan and Gibbs (1983), in their study of Restronguet Creek, monitored the concentrations of relevant trace metals in water, sediment and resident biota; tested the toxicities of local water and sediments; and also investigated differences in the distribution of local fauna and flora. Furthermore, Bryan and Gibbs (1983) met the criterion of Luoma (1977b), confirming the presence of metal-tolerant populations of algae and invertebrates in Restronguet Creek.

There can be little doubt that the identification of meaningful biological effects of an aquatic contaminant requires the recognition and measurement of responses at all levels: those of the organism, the population and the community. Table 31 summarises the present state of understanding of the worth of specific biological responses to this end (Howells *et al.*, 1990). A modified version of the Sediment Quality Triad remains attractive, with the incorporation of biomonitors and modern multivariate approaches to analysis of community structure. However, there is also a

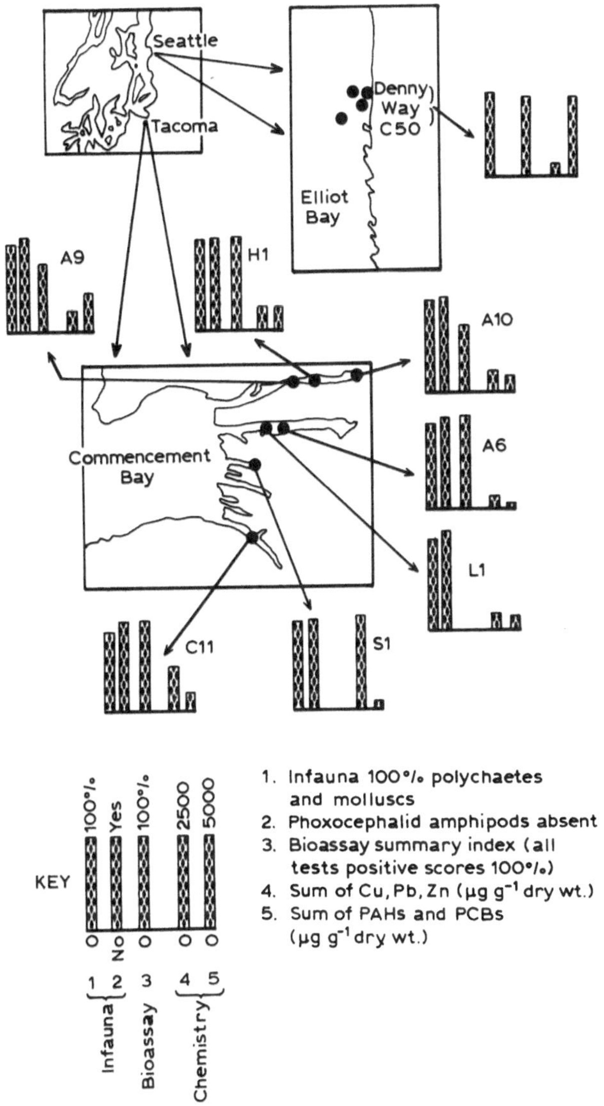

Fig. 100. Correspondence among measures employed in the Sediment Quality Triad (see the key at bottom) for samples from Commencement Bay and Elliot Bay, USA. After Long and Chapman (1985).

Table 31 An assessment table for the monitoring of the effects of contaminants on aquatic organisms (after Howells *et al.*, 1990)

Level of biological organisation	*Response characteristics*[a]						*Methods of measurement*				
	Specificity of response	*Sensitivity of response*	*Persistence*	*Dose-response*	*Recovery*	*Ecological significance*	*Sampling validity*	*Specificity*	*Quality control/ quality assurance*	*Statistical testability*	*Potential for monitoring*
Cellular/Tissue											
MFO	▲◑	△○	▲◑	▲•	?	△○	▲	▲	▲	▲	▲
Metallothionein	●	●	●	●	?	◑	●	●	●	●	●
Histopathology	○	◑	◑	○	?	◑	◑	◑	◑	◑	○
Genetic change	•	•	•	•	•	•	•	•	•	•	
Individual											
Scope for growth	○	●	•	●	●	●	●	●	●	●	●
Reproduction	•	•	•	•	•	•	•	•	•	•	
Survival	•	•	•	•	•	•	•	•	•	•	
Behaviour	•	•	•	•	•	•	•	•	•	•	
Migration	•	•	•	•	•	•	•	•	•	•	
Population											
Age/size structure	•	•	•	•	•	•	•	•	•	•	
Abundance	•	•	•	•	•	•	•	•	•	•	
Recruitment	•	•	•	•	•	•	•	•	•	•	
Migration	•	•	•	•	•	•	•	•	•	•	
Community											
Species/abundance/ biomass	□	◧	■	□	■	■	■	◧	■	◧	■
Individuals/species	□	■	■	□	■	■	■	◧	■	◧	◧
Size structure	□	◧	■	□	■	■	◧	◧	◧	□	◧
Diversity	□	◧	■	□	■	■	■	□	◧	■	■
Production/respiration	□	□	◧	□	◧	■	□	□	□	□	□

○ *Mytilus edulis*, △ *Platichthys flesus*, □ community responses (● strong, ◑ moderate, ○ weak, • not measured).
[a] Specificity: Is the response specific to one or a few contaminants only? Sensitivity: Is the response sensitive to one or more contaminants? Persistence: Does the response persist with the continuous dosing? Dose response: Have dose-response relationships been established for one or more contaminants? Recovery: Is the response able to measure recovery? Ecological significance: Is the response measured known to be ecologically significant?

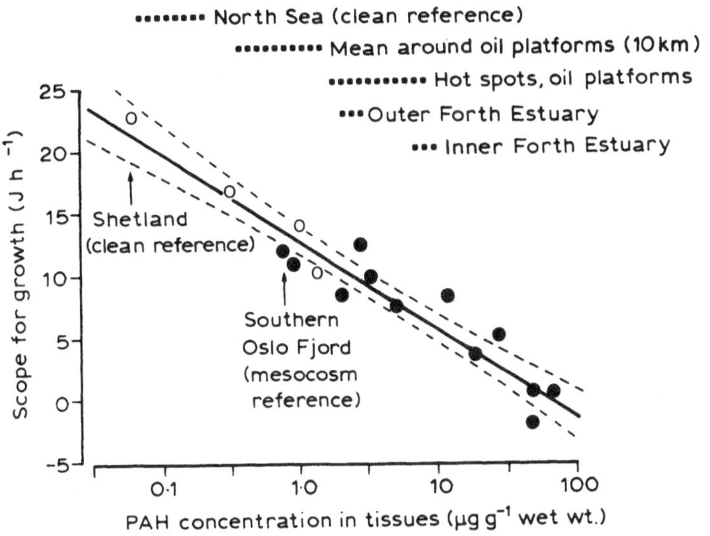

Fig. 101. Relationship between scope for growth (for a standardised animal and conditions) and the concentrations of two-ringed and three-ringed PAHs ($\mu g\ g^{-1}$ wet weight) in tissues of the mussel *Mytilus edulis*. ●, Data from experimental exposures to oil in Norway; ○, data from field studies in Sullom Voe, United Kingdom. Horizontal bars represent estimates of the tissue concentrations of hydrocarbons in mussels, based upon measurements in seawater at various sites. After Moore *et al.* (1987a).

need to incorporate the confirmation of the inherited tolerance of exposed populations (most usefully at sublethal exposures), in order to delimit the geographical extent of biologically significant contamination.

A need remains to further identify the relevance of stress measures to both individual organisms and to populations and communities of aquatic biota. There is a need for the identification of more specific indices, such that observed biological responses can be traced to individual contaminants. Moreover, a requirement exists to elucidate the importance of impacts at the population level to the overall ecology of affected aquatic habitats. McCarthy (1990) has provided a useful overview of such requirements in the design of an environmental monitoring programme involving such indices.

Given the discussion earlier on metal accumulation strategies (see Chapter 5), there is no prospect in most cases of being able to relate accumulated metal concentrations of biomonitors directly to quantifiable toxic effects. However, such a possibility is not ruled out in the case of lipophilic contaminants. An example is shown in Fig. 101, which describes a significant regression between scope for growth and accumulated concentrations of two- and three-ringed aromatic hydrocarbons in the soft tissues of the mussel *Mytilus edulis* around the North Sea. Further research is required on such matters, in order that the robust basic information on the distributions and abundances of contaminants provided through biomonitoring studies may be matched by an equally reliable measure of the toxic impacts of such contaminants on living organisms.

Chapter 10

The Present Status and the Future

This final chapter provides a brief overview of the present status of the use of biomonitors to define spatial and temporal differences in contaminant availabilities in aquatic ecosystems, and considers the future growth of this area.

The very considerable increase in the use of biomonitoring techniques over the last two decades demonstrates both the utility and versatility of these methods. The early studies of temperate species (dominated heavily by work on bivalve molluscs, with particular emphasis on cosmopolitan species such as the mussel *Mytilus edulis* and oysters of the genus *Crassostrea*) have slowly been extended into subtropical and tropical environments during this period. While problems exist to some extent in the latter regions due to the paucity of potential biomonitoring species with particularly widespread geographical distributions, these have to some extent been overcome, both through the use of mussels of the genus *Perna* (see Phillips, 1984) and by the development of new species as biomonitors. Over the same period, investigations have also further clarified the utility of different types of organisms as biomonitors, ranging from algae, through invertebrate animals, to fish and marine mammals. Many species have a place in biomonitoring, but careful and logical decisions are required in the design of monitoring programmes if the technique is to be employed to full advantage (see Chapter 8 in particular).

However, there can be no doubt whatsoever that our knowledge of trace contaminant distributions in the subtropics and tropics remains extremely rudimentary at present, due to the overall paucity of studies of this kind (Phillips, 1991). This implies that many rivers and coastlines in these areas of the world are at significant risk from pollution, particularly given the increases in population and industrial development in the developing nations of the subtropics and tropics.

This problem appears soluble through two methods, both of which should be accorded high priority. Firstly, legislators and scientists in the developing nations must be educated in the utility and use of these techniques, such that they may be introduced to provide the much-needed data on the health of aquatic ecosystems. Secondly, financial assistance will clearly be required from the developed nations. The latter can be justified on many grounds, not least the fact that some of the contaminants involved are no respectors of political or geographical divisions between nations (or even hemispheres), being transported internationally and/or globally through the atmosphere and other routes (e.g. see Fig. 50).

Contaminants such as DDT and PCBs exhibit sufficient mobility in this fashion that the imposition of controls on their use in only certain parts of the world will have relatively little impact on their overall distributions or effects. Thus, although a "southward tilt" may be expected in the distributions of such contaminants pursuant to the imposition of restrictions on their utilisation in the northern hemisphere (Goldberg, 1975), the overall global problem created by their impacts has not been significantly reduced by these controls. Indeed, it is possible that the adverse effects of certain contaminants in the tropical regions may be even greater than those experienced in the temperate zones, for a variety of reasons (Phillips, 1991).

It should also be noted that recent data on these same contaminants in temperate regions suggest that adverse impacts are continuing, at least in certain instances. Perhaps the best example of this involves the PCBs, which are still being released from a reservoir created historically in all the western nations (Tanabe, 1988; see Table 16). Data from the studies of Reijnders (1986) and others provide convincing evidence of continuing impacts of PCBs on marine mammal populations, and there is also cause for concern over their effects on fish reproduction (e.g. Cross and Hose, 1988; Spies and Rice, 1988; Spies *et al.*, 1990).

Concerns should be raised further through the events of the last two decades surrounding the introduction of tributyltin compounds for marine anti-fouling purposes. It is now clear that these compounds exert sufficiently acute "side effects" on particularly susceptible species such as oysters and gastropods (e.g. see Fig. 85) that their widespread use should be deemed unacceptable (as evidenced by the restrictions and bans on their utilisation introduced over the last five years in many western nations).

Against this background, it is sobering to reflect further upon the quotation from Jernelöv (1974), provided towards the end of Chapter 3. There can be little doubt that present monitoring programmes are over-constrained with respect to the compounds included, and that additional contaminants of concern in aquatic environments will emerge as our knowledge improves and analytical techniques of greater sensitivity are introduced (see also Tanabe *et al.*, 1989a). The fact that PCBs were recognised as a contaminant of global concern only in 1966, some 37 years after their initial use by industry, suggests that many more "skeletons" may exist in the aquatic "cupboard" (Jensen, 1972; Phillips, 1986).

The implication arising from each of these considerations is clear. Aquatic environments must be monitored for trace contaminants, and this monitoring should involve both the quantification of the amounts of contaminants in such environments, and some measure of their adverse effects. The thesis developed in this text suggests that biomonitoring techniques should be considered the method of choice for such studies, for the reasons enumerated in Chapter 4 and elsewhere.

However, much remains to be achieved. While the present state of development of biomonitoring techniques to quantify the abundances and distributions of trace metals and organochlorines in aquatic ecosystems may be considered to be reasonably robust, the methods employed can no doubt be further improved. It is also clear that insufficient studies of this type have been undertaken on hydrocarbon compounds, and this arena remains constrained to some extent by analytical methodologies, which require further development (see Chapter 6).

The other major area requiring improvement involves "effects-related biomonitoring", as covered briefly here in Chapter 9. This field can be legitimately considered to be rather younger than other areas of biomonitoring, and many of the techniques remain at present in the stage of research and development. Such work is undoubtedly of great importance, as the quantification of contaminant distributions in aquatic environments is of limited application unless these data can be related (directly or indirectly) to the impacts of such contaminants on biological resources.

We may conclude that much progress has been made in a relatively short period, and that biomonitoring techniques should now be considered the method of choice in many instances for studies of trace contaminants in aquatic ecosystems. However, further development of biomonitoring techniques is required in many spheres. Our vision remains that methods based upon biomonitoring will provide the backbone for the global protection of aquatic environments. The challenge continues, such that this vision may be fully realised.

References

Addison, R.F. (1988). Biochemical effects of a pollutant gradient — summary. *Mar. Ecol. Prog. Ser.*, **46**, 75–77.

Addison, R.F. and P.F. Brodie (1977). Organochlorine residues in maternal blubber, milk, and pup blubber from grey seals (*Halichoerus grypus*) from Sable Island, Nova Scotia. *J. Fish. Res. Board Can.*, **34**, 937–941.

Addison, R.F. and A.J. Edwards (1988). Hepatic microsomal mono-oxygenase activity in flounder *Platichthys flesus* from polluted sites in Langesundfjord and from mesocosms experimentally dosed with diesel oil and copper. *Mar. Ecol. Prog. Ser.*, **46**, 51–54.

Addison, R.F. and T.G. Smith (1974). Organochlorine residue levels in Arctic ringed seals: variation with age and sex. *Oikos.*, **25**, 335–337.

Addison, R.F., S.R. Kerr, J. Dale and D.E. Sergeant (1973). Variation of organochlorine residue levels with age in Gulf of St. Lawrence harp seals (*Pagophilus groenlandicus*). *J. Fish. Res. Board Can.*, **30**, 595–600.

Addison, R.F., P.F. Brodie, M.E. Zinck and D.E. Sergeant (1984). DDT has declined more than PCBs in eastern Canadian seals during the 1970s. *Environ. Sci. Technol.*, **18**, 935–937.

Aguilar, A. (1983). Organochlorine pollution in sperm whales, *Physeter macrocephalus*, from the temperate waters of the eastern North Atlantic. *Mar. Pollut. Bull.*, **14**, 349–352.

Ahmed, M. (1975). Speciation in living oysters. *Adv. Mar. Biol.*, **13**, 357–397.

Ahsanullah, M. and T.M. Florence (1984). Toxicity of copper to the marine amphipod *Allorchestes compressa* in the presence of water- and lipid-soluble ligands. *Mar. Biol.*, **84**, 41–45.

Akielaszek, J.J. and T.A. Haines (1981). Mercury in the muscle tissue of fish from three northern Maine lakes. *Bull. Environ. Contam. Toxicol.*, **27**, 201–208.

Al-Omar, M.A., F.H. Abdul-Jalil, N.H. Al-Ogaily, S.J. Tawfiq and M.A. Al-Bassomy (1986). A follow-up study of maternal milk contamination with organochlorine insecticide residues. *Environ. Pollut. Ser. A.*, **42**, 79–91.

Amiard, J.C., C. Amiard-Triquet, B. Berthet and C. Métayer (1986). Contribution to the ecotoxicological study of cadmium, lead, copper and zinc in the mussel *Mytilus edulis*. I. Field study. *Mar. Biol.*, **90**, 425–431.

Amiard, J.C., C. Amiard-Triquet, B. Berthet and C. Métayer (1987). Comparative study of the patterns of bioaccumulation of essential (Cu, Zn) and non-essential (Cd, Pb) trace metals in various estuarine and coastal organisms. *J. Exp. Mar. Biol. Ecol.*, **106**, 73–89.

Amiard-Triquet, C., C. Métayer and J.C. Amiard (1987). Études *in situ* et expérimentales de l'écotoxicologie de quatre métaux (Cd, Pb, Cu, Zn) chez des algues et des mollusques gastéropodes brouteurs. *Water Air Soil Pollut.*, **34**, 11–30.

Amico, V., G. Impellizzeri, G. Oriente, M. Piattelli, S. Sciuto and C. Tringali (1979a). Levels of chlorinated hydrocarbons in marine animals from the central Mediterranean. *Mar. Pollut. Bull.*, **10**, 282–284.

Amico, V., G. Oriente, M. Piattelli and C. Tringali (1979b). Concentrations of PCBs, BHCs and DDTs residues in seaweeds of the east coast of Sicily. *Mar. Pollut. Bull.*, **10**, 177–179.

Anas, R.E. and A.J. Wilson (1970a). Organochlorine pesticides in fur seals. *Pestic. Monit. J.*, **3**, 198–200.

Anas, R.E. and A.J. Wilson (1970b). Organochlorine pesticides in nursing fur seal pups. *Pestic. Monit. J.*,, **4**, 114–116.

Anderson, P.D. and P.A. Spear (1980). Copper pharmacokinetics in fish gills — I. Kinetics in pumpkinseed sunfish, *Lepomis gibbosus*, of different body sizes. *Water Res.*, **14**, 1101–1105.

Anderson, R.B. and W.H. Everhart (1966). Concentrations of DDT in landlocked salmon (*Salmo salar*) at Sebago Lake, Maine. *Trans. Amer. Fish. Soc.*, **95**, 160–164.

Anderson, R.B. and O.C. Fenderson (1970). An analysis of variation of insecticide residues in landlocked Atlantic salmon (*Salmo salar*). *J. Fish. Res. Board Can.*, **27**, 1–11.

Anon. (1966). Report of a new chemical hazard. *New Scientist*. **32**, 612.

Anon. (1991a). Sources of dioxins are re-evaluated. *Environ. Impact.*, **V (9)**, 1–4.

Anon. (1991b). $1 billion Exxon settlement collapse. *Mar. Pollut. Bull.*, **22**, 320.

Armstrong, F.A.J. and A. Lutz (1977a). *Lake Huron, 1974: PCB, chlorinated insecticides, heavy metals and radioactivity in offshore fish*. Fisheries and Marine Service, Environment Canada: Technical report no. 692.

Armstrong, F.A.J. and A. Lutz (1977b). *Lake Superior, 1974: PCB, chlorinated insecticides, heavy metals and radioactivity in offshore fish*. Fisheries and Marine Service, Environment Canada: Technical report no. 693.

Armstrong, F.A.J. and D.P. Scott (1979). Decrease in mercury content of fishes in Ball Lake, Ontario, since imposition of controls on mercury discharges. *J. Fish. Res. Board Can.*, **36**, 670–672.

Atkinson, D.E. (1977). *Cellular energy metabolism and its regulation*. Academic Press, New York.

Aulio, K. and M. Salin (1982). Enrichment of copper, zinc, manganese, and iron in five species of pondweeds (*Potamogeton* spp.). *Bull. Environ. Contam. Toxicol.*, **29**, 320–325.

Ayling, G.M. (1974). Uptake of cadmium, zinc, copper, lead and chromium in the Pacific oyster, *Crassostrea gigas*, grown in the Tamar River, Tasmania. *Water Res.*, **8**, 729–738.

Ayres, P.A. (1975). Mussel poisoning in Britain with special reference to Paralytic Shellfish Poisoning. *Environ. Hlth.* July 1975: 261–265.

Bache, C.A., J.W. Serum, W.D. Youngs and D.J. Lisk (1972). Polychlorinated biphenyl residues: accumulation in Cayuga Lake trout with age. *Science.*, **177**, 1191–1192.

Badri, M.A. and S.R. Aston (1983). Observations on heavy metal geochemical associations in polluted and non-polluted estuarine sediments. *Environ. Pollut. Ser. B.*, **6**, 181–193.

Badsha, K.S. and C.R. Goldspink (1982). Preliminary observations on the heavy metal content of four species of freshwater fish in NW England. *J. Fish. Biol.*, **21**, 251–267.

Bahner, L.H., A.J. Wilson, J.M. Sheppard, J.M. Patrick, L.R. Goodman and G.E. Walsh (1977). Kepone bioconcentration, accumulation, loss, and transfer through estuarine food chains. *Chesapeake Sci.*, **18**, 299–308.

Baker, E.K. and P.T. Harris (1991). Copper, lead, and zinc distribution in the sediments of the Fly River Delta and Torres Strait. *Mar. Pollut. Bull.*, **22**, 614–618.

Bakir, F., S.F. Damluji, L. Amin-Zaki, M. Murtadha, A. Khalidi, N.Y. Al-Rawi, S. Tikriti, H.I. Dhahir, T.W. Clarkson, J.C. Smith and R.A. Doherty (1973). Methylmercury poisoning in Iraq. An interuniversity report. *Science.*, **181**, 230–241.

Baldi, F. and R. Bargagli (1982). Chemical leaching and specific surface area measurements of marine sediments in the evaluation of mercury contamination near cinnabar deposits. *Mar. Environ. Res.*, **6**, 69–82.

Ballschmiter, K., H. Buchert, S. Bihler and M. Zell (1981). Baseline studies of the global pollution. IV. The pattern of pollution by organo-chlorine compounds in the North Atlantic as accumulated by fish. *Fresenius Z. Anal. Chem.*, **306**, 323–339.

Barcellos, C., C.E. Rezende and W.C. Pfeiffer (1991). Zn and Cd production and pollution in a Brazilian coastal region. *Mar. Pollut. Bull.*, **22**, 558–561.

Barnett, B.E. and C.R. Ashcroft (1985). Heavy metals in *Fucus vesiculosus* in the Humber estuary. *Environ. Pollut., Ser. B.* **9**, 193–213.

Barrington, E.J.W. (1967). *Invertebrate structure and function.* Nelson, London.

Bartlett, P.D. and H.F.-K.D. Hennig (1987). Pollution studies in False Bay, South Africa: chemical versus meiofaunal indicators. In *Oceanic processes in marine pollution, Vol. 1. Biological processes and wastes in the ocean*, ed. J.M. Capuzzo and D.R. Kester, R.E. Krieger Publ. Co., Malabar, Florida, pp. 231–240.

Bastürk, O., M. Dogan, I. Salihoglu and T.I. Balkas (1980). DDT, DDE, and PCB residues in fish, crustaceans and sediments from the eastern Mediterranean coast of Turkey. *Mar. Pollut. Bull.*, **11**, 191–195.

Bates, J. (1991). The Paris Convention for the prevention of marine pollution from land-based sources. *Land Manage. Environ. Law Rep.* **3**, 146–149.,

Bayne, B.L. (Ed.) (1976). *Marine mussels: their ecology and physiology*. Cambridge University Press, Cambridge.

Bayne, B.L., M.N. Moore, J. Widdows, D.R. Livingstone and P. Salkeld (1979). Measurements of the responses of individuals to environmental stress and pollution: studies with bivalve molluscs. *Phil. Trans. Roy. Soc. Lond. B.*, **286**, 563–581.

Bayne, B.L., D.A. Brown, K. Burns, D.R. Dixon, A. Ivanovici, D.R. Livingstone, D.M. Lowe, M.N. Moore, A.R.D. Stebbing and J. Widdows (1985). *The effects of stress and pollution on marine animals*. Praeger Publishers, New York.

Beanlands, G.E. and Z. Si (1988). Coping with Environmental Impact Assessment in the developing world. In *Pollution in the urban environment: POLMET 88*, ed. P. Hills, R. Keen, K.C. Lam, C.T. Leung, M.A. Oswell, M. Stokes and E. Turner, Vincent Blue Copy Company, Hong Kong. Vol. 1, pp. 12–17.

Beardmore, J.A. (1980). Genetical considerations in monitoring effects of pollution. *Rapp. P.-V. Réun. Cons. Int. Explor. Mer.*, **179**, 258–266.

Beardmore, J.A., C.J. Barker, B. Battaglia, J.F. Payne and A. Rosenfield (1980). The use of genetical approaches to monitoring biological effects of pollution. *Rapp. P.-V. Réun. Cons. Int. Explor. Mer.*, **179**, 299–305.

Bedford, J.W., E.W. Roelofs and M.J. Zabik (1968). The freshwater mussel as a biological monitor of pesticide concentrations in a lotic environment. *Limnol. Oceanogr.*, **13**, 118–126.

Bengtsson, B.-E. (1979). Biological variables, especially skeletal deformities in fish, for monitoring marine pollution. *Phil. Trans. Roy. Soc. Lond. B.*, **286**, 457–464.

Bengtsson, B.-E. (1980). Long-term effects of PCB (Clophen A50) on growth, reproduction and swimming performance in the minnow, *Phoxinus phoxinus*. *Water Res.*, **14**, 681–687.

Bengstsson, B.-E. and A. Larsson (1986). Vertebral deformities and physiological effects in fourhorn sculpin (*Myxocephalus quadricornis*) after long-term exposure to a simulated heavy metal-containing effluent. *Aquatic Toxicol.*, **9**, 215–229.

Ben-Shlomo, R. and E. Nevo (1988). Isozyme polymorphism as monitoring of marine environments: the interactive effect of cadmium and mercury pollution on the shrimp, *Palaemon elegans*. *Mar. Pollut. Bull.*, **19**, 314–317.

Benson, W.H., K.N. Baer and C.F. Watson (1990). Metallothionein as a biomarker of environmental metal contamination: species-dependent effects. In *Biomarkers of environmental contamination*, ed. J.F. McCarthy and L.R. Shugart, Lewis Publishers, CRC Press, Boca Raton, Florida, pp. 255–265.

Bertine, K.K. and E.D. Goldberg (1977). History of heavy metal pollution in Southern California coastal zone — reprise. *Environ. Sci. Technol.*, **11**, 297–299.

Bevelander, G. and H. Nakahara (1966). Correlation of lysosomal activity and ingestion by mantle epithelium. *Biol. Bull. Woods Hole, Mass.*, **131**, 76–82.

Bias, R. and L. Karbe (1985). Bioaccumulation and partitioning of cadmium within the freshwater mussel *Dreissena polymorpha* Pallas. *Int. Revue. Ges. Hydrobiol.*, **70**, 113–125.

Bierman, V.J. and W.R. Swain (1982). Mass balance modeling of DDT dynamics in Lakes Michigan and Superior. *Environ. Sci. Technol.*, **16**, 572–579.

Biggar, J.W., L.D. Doneen and R.L. Riggs (1966). *Soil interaction with organically*

polluted water. Summary Report, Department of Water Science and Engineering, University of California, Davis, California.
Birch, P.B., G.G. Forbes and N.J. Schofield (1986). Monitoring effects of catchment management practices on phosphorus loads into the eutrophic Peel-Harvey Estuary, Western Australia. *Water Sci. Technol.*, **18**, 53–61.
Bjerregaard, P. (1982). Accumulation of cadmium and selenium and their mutual interaction in the shore crab *Carcinus maenas* (L). *Aquatic Toxicol.*, **2**, 113–125.
Bjerregaard, P. and T. Vislie (1985). Effects of mercury on ion and osmoregulation in the shore crab *Carcinus maenas* (L.). *Comp. Biochem. Physiol.*, **82C**, 227–230.
Bjerregaard, P. and T. Vislie (1986). Effect of copper on ion- and osmoregulation in the shore crab *Carcinus maenas* (L.). *Mar. Biol.*, **91**, 69–76.
Bjerregaard, P., S. Topcuoglu, N.S. Fisher and S.W. Fowler (1985). Biokinetics of americium and plutonium in the mussel *Mytilus edulis*. *Mar. Ecol. Prog. Ser.*, **21**, 99–111.
Björklund, I., H. Borg and K. Johansson (1984). Mercury in Swedish lakes — its regional distribution and causes. *Ambio.*, **13**, 118–121.
Boalch, R., S. Chan and D. Taylor (1981). Seasonal variation in the trace metal content of *Mytilus edulis*. *Mar. Pollut. Bull.*, **12**, 276–280.
Bocquené, G., F. Galgani and P. Truquet (1990). Characterization and assay conditions for use of AChE activity from several marine species in pollution monitoring. *Mar. Environ. Res.*, **30**, 75–89.
Boehm, P.D. and J.W. Farrington (1984). Aspects of the polycyclic aromatic hydrocarbon geochemistry of recent sediments in the Georges Bank region. *Environ. Sci. Technol.*, **18**, 840–845.
Boon, J.P. and J.C. Duinker (1985). Kinetics of polychlorinated biphenyl (PCB) components in juvenile sole (*Solea solea*) in relation to concentrations in water and to lipid metabolism under conditions of starvation. *Aquatic Toxicol.*, **7**, 119–134.
Bopp, R.F., H.J. Simpson, C.R. Olsen and N. Kostyk (1981). Polychlorinated biphenyls in sediments of the tidal Hudson River, New York. *Environ. Sci. Technol.*, **15**, 210–216.
Boryslawskyj, M., A.C. Garrood and M.J. Morphy (1985). Spatial and temporal patterns of dieldrin pollution in the Holme catchment, West Yorkshire, England. *Environ. Pollut. Ser. B.*, **10**, 129–139.
Boudou, A. and F. Ribeyre (1984). Influence de la durée d'éxposition sur la bioaccumulation par voie directe de deux dérivés du mercure par *Salmo gairdneri* (Alevins) et rélation 'poids des organismes - concentration en mercure'. *Water Res.*, **18**, 81–86.
Boudou, A., F. Ribeyre, A. Delarche and R. Marty (1980). Bioaccumulation et bioamplification des dérivés du mercure par un consommateur de troisième ordre: *Salmo gairdneri* — incidences du facteur température. *Water Res.*, **14**, 61–65.
Bowman, M.C., F. Acree and M.K. Corbett (1960). Solubility of ^{14}C-DDT in water. *J. Agric. Fd. Chem.*, **8**, 406–434.
Boyden, C.R. (1974). Trace element content and body size in molluscs. *Nature, Lond.*, **251**, 311–314.
Boyden, C.R. (1977). Effect of size upon metal content of shellfish. *J. Mar. Biol. Ass. U.K.*, **57**, 675–714.

Boyden, C.R. and D.J.H. Phillips (1981). Seasonal variation and inherent variability of trace elements in oysters and their implications for indicator studies. *Mar. Ecol. Prog. Ser.*, **5**, 29–40.

Brady, F.O. (1982). The physiological function of metallothionein. *Trends Biochem. Sci.*, **7**, 143–145.

Branson, D.R., G.E. Blau, H.C. Alexander and W.B. Neely (1975). Bioconcentration of 2,2′,4,4′-tetrachlorobiphenyl in rainbow trout as measured by an accelerated test. *Trans. Amer. Fish. Soc.*, **104**, 785–792.

Bright, D.A. and D.V. Ellis (1989). Aspects of histology in *Macoma carlottensis* (Bivalvia: Tellinidae) and *in situ* histopathology related to mine-tailings discharge. *J. mar. biol. Ass. U.K.*, **69**, 447–464.

Brodtmann, N.V. (1970). Studies on the assimilation of 1,1,1-trichloro-2,2-*bis* (*p*-chlorophenyl) ethane (DDT) by *Crassostrea virginica* Gmelin. *Bull. Environ. Contam. Toxicol.*, **5**, 455–462.

Brown, B.E. (1976). Observations on the tolerance of the isopod *Asellus meridianus* Rac. to copper and lead. *Water Res.*, **10**, 555–559.

Brown, B.E. (1977). Uptake of copper and lead by a metal-tolerant isopod *Asellus meridianus* Rac. *Freshwater Biology.*, **7**, 235–244.

Brown, B.E. (1978). Lead detoxification by a copper-tolerant isopod. *Nature, Lond.*, **276**, 388–390.

Brown, B.E. (1982). The form and function of metal-containing 'granules' in invertebrate tissues. *Biol. Rev.*, **57**, 621–667.

Brown, B.E. and M.C. Holley (1982). Metal levels associated with tin dredging and smelting and their effect upon intertidal reef flats at Ko Phuket, Thailand. *Coral Reefs.*, **1**, 131–137.

Brown, M.P., M.B. Werner, R.J. Sloan and K.W. Simpson (1985). Polychlorinated biphenyls in the Hudson River. *Environ. Sci. Technol.*, **19**, 656–661.

Bruggeman, W.A., L.B.J.M. Martron, D. Kooiman and O. Hutzinger (1981). Accumulation and elimination kinetics of di-, tri-, and tetra chlorobiphenyls by goldfish after dietary and aqueous exposure. *Chemosphere.*, **10**, 811–832.

Bruland, K.W. (1983). Trace elements in seawater. In *Chemical oceanography, Vol. 8*, ed. J.P. Riley and R. Chester, Academic Press, London, pp. 157–220.

Bruland, K.W., G.A. Knauer and J.H. Martin (1978a). Zinc in north-east Pacific water. *Nature, Lond.*, **271**, 741–743.

Bruland, K.W., G.A. Knauer and J.H. Martin (1978b). Cadmium in northeast Pacific waters. *Limnol. Oceanogr.*, **23**, 618–625.

Brumbaugh, W.G. and D.A. Kane (1985). Variability of aluminum concentrations in organs and whole bodies of smallmouth bass (*Micropterus salmoides*). *Environ. Sci. Technol.*, **19**, 828–831.

Brunn, H.and D. Manz (1982). Contamination of native fish stock by hexachlorobenzene and polychlorinated biphenyl residues. *Bull. Environ. Contam. Toxicol.*, **28**, 599–604.

Bryan, G.W. (1964). Zinc regulation in the lobster *Homarus vulgaris*. I. Tissue zinc and copper concentrations. *J. mar. biol. Ass. U.K.*, **44**, 549–563.

Bryan, G.W. (1966). The metabolism of Zn and ^{65}Zn in crabs, lobsters and freshwater crayfish. In *Radioecological concentration processes*, ed. B. Åberg and F.P. Hungate, Pergamon Press, Oxford, pp. 1005–1016.

Bryan, G.W. (1968). Concentrations of zinc and copper in the tissues of decapod crustaceans. *J. mar. biol. Ass. U.K.*, **48**, 303–321.
Bryan, G.W. (1973). The occurrence and seasonal variation of trace metals in the scallops, *Pecten maximus* (L.) and *Chlamys opercularis* (L.). *J. mar. biol. Ass. U.K.*, **53**, 145–166.
Bryan, G.W. (1976). Heavy metal contamination in the sea. In *Marine pollution*, ed. R. Johnston, Academic Press, London. pp. 185–302.
Bryan, G.W. (1979). Bioaccumulation of marine pollutants. *Phil. Trans. Roy. Soc. Lond. B.*, **286**, 483–505.
Bryan, G.W. (1985). Bioavailability and effects of heavy metals in marine deposits. In *Wastes in the ocean, Vol. 6: Near shore waste disposal*, ed. B.J. Ketchum, J. Capuzzo, W. Burt, I. Duedall, P. Park and D. Kester, John Wiley and Sons, New York, pp. 41–79.
Bryan, G.W. and P.E. Gibbs (1979). Zinc — a major inorganic component of nereid polychaete jaws. *J. mar. biol. Ass. U.K.*, **59**, 969–973.
Bryan, G.W. and P.E. Gibbs (1980). Metals in nereid polychaetes: the contribution of metals in the jaws to the total body burden. *J. mar. biol. Ass. U.K.*, **60**, 641–654.
Bryan, G.W. and P.E. Gibbs (1983). Heavy metals in the Fal Estuary, Cornwall: a study of long-term contamination by mining waste and its effects on estuarine organisms. *Occ. Publ. mar. biol. Ass. U.K.*, **2**, 1–112.
Bryan, G.W. and L.G. Hummerstone (1973a). Brown seaweed as an indicator of heavy metals in estuaries in south-west England. *J. mar. biol. Ass. U.K.*, **53**, 705–720.
Bryan, G.W. and L.G. Hummerstone (1973b). Adaptation of the polychaete *Nereis diversicolor* to estuarine sediments containing high concentrations of zinc and cadmium. *J. mar. biol. Ass. U.K.*, **53**, 839–872.
Bryan, G.W. and L.G. Hummerstone (1977). Indicators of heavy-metal contamination in the Looe Estuary (Cornwall) with particular regard to silver and lead. *J. mar. biol. Ass. U.K.*, **57**, 75–92.
Bryan, G.W. and H. Uysal (1978). Heavy metals in the burrowing bivalve *Scrobicularia plana* from the Tamar estuary in relation to environmental levels. *J. mar. biol. Ass. U.K.*, **58**, 89–108.
Bryan, G.W. and E. Ward (1965). The absorption and loss of radioactive and non-radioactive manganese by the lobster, *Homarus vulgaris*. *J. mar. biol. Ass. U.K.*, **45**, 65–95.
Bryan, G.W., G.W. Potts and G.R. Forster (1977). Heavy metals in the gastropod mollusc *Haliotis tuberculata* (L.). *J. mar. biol. Ass. U.K.*, **57**, 379–390.
Bryan, G.W., W.J. Langston and L.G. Hummerstone (1980). The use of biological indicators of heavy-metal contamination in estuaries; with special reference to an assessment of the biological availability of metals in estuarine sediments from south-west Britain. *Occ. Publ. mar. biol. Ass. U.K.*, **1**, 1–73.
Bryan, G.W., W.J. Langston, L.G. Hummerstone, G.R. Burt and Y.B. Ho (1983). An assessment of the gastropod, *Littorina littorea*, as an indicator of heavy-metal contamination in United Kingdom estuaries. *J. mar. biol. Ass. U.K.*, **63**, 327–345.
Bryan, G.W., W.J. Langston, L.G. Hummerstone and G.R. Burt (1985). A guide

to the assessment of heavy metal contamination in estuaries using biological indicators. *Occ. Publ. mar. biol. Ass. U.K.*, **4**, 1–92.

Bryan, G.W., L.G. Hummerstone and E. Ward (1986). Zinc regulation in the lobster *Homarus gammarus:* importance of different pathways of absorption and excretion. *J. mar. biol. Ass. U.K.*, **66**, 175–199.

Bryan, G.W., P.E. Gibbs, L.G. Hummerstone and G.R. Burt (1987). Copper, zinc, and organotin as long term factors governing the distribution of organisms in the Fal estuary in southwest England. *Estuaries.*, **10**, 208–219.

Buhler, D.R., R.R. Claeys and B.R. Mate (1975). Heavy metal and chlorinated hydrocarbon residues in California sea lions (*Zalophus californianus californianus*). *J. Fish. Res. Board Can.*, **32**, 2391–2397.

Bulkley, R.V., R.L. Kellogg and L.R. Shannon (1976). Size-related factors associated with dieldrin concentrations in muscle tissue of channel catfish *Ictalurus punctatus. Trans. Amer. Fish. Soc.*, **105**, 301–307.

Bull, K.R., A.F. Dearsley and M.H. Inskip (1981). Growth and mercury content of roach (*Rutilus rutilus*, L.), perch (*Perca fluviatilis*, L.) and pike (*Esox lucius*, L.) living in sewage effluent. *Environ. Pollut. Ser. A.*, **25**, 229–240.

Burdick, G.E., E.J. Harris, H.J. Dean, T.M. Walker, J. Skea and D. Colby (1964). The accumulation of DDT in lake trout and the effect on reproduction. *Trans. Amer. Fish. Soc.*, **93**, 127–136.

Burdon-Jones, C. and G.R.W. Denton (1984). *Metals in marine organisms from the Great Barrier Reef province. Final Report. Part 1. Baseline survey.* Report of the Department of Marine Biology, James Cook University of North Queensland, Queensland, Australia.

Burdon-Jones, C., G.R.W. Denton, G.B. Jones and K.A. McPhie (1982). Regional and seasonal variations of trace metals in tropical Phaeophyceae from North Queensland. *Mar. Environ. Res.*, **7**, 13–30.

Burnett, R. (1971). DDT residues: distribution of concentrations in *Emerita analoga* (Stimpson) along coastal California. *Science.*, **174**, 606–608.

Burns, K.A. and J.L. Smith (1980). Hydrocarbons in Victorian coastal waters. *Aust. J. mar. Freshwat. Res.*, **31**, 251–256.

Burns, K.A. and J.L. Smith (1981). Biological monitoring of ambient water quality: the case for using bivalves as sentinel organisms for monitoring petroleum pollution in coastal waters. *Estuar. cstl. Shelf Sci.*, **13**, 433–443.

Burns, K.A. and J.L. Smith (1982). Hydrocarbons in Victorian coastal ecosystems (Australia): chronic petroleum inputs to Western Port and Port Phillip Bays. *Arch. Environ. Contam. Toxicol.*, **11**, 129–140.

Burns, K.A., J.P. Villeneuve, V.C. Anderlini and S.W. Fowler (1982). Survey of tar, hydrocarbon and metal pollution in the coastal waters of Oman. *Mar. Pollut. Bull.*, **13**, 240–247.

Burton, J.D. and P.J. Statham (1990). Trace metals in seawater. In *Heavy metals in the marine environment,* ed. R.W. Furness and P.S. Rainbow, CRC Press, Boca Raton, Florida. pp. 5–25.

Burton, J.D. and M.L. Young (1980). Trace metals in shelf seas of the British Isles. In *The northwest European shelf seas: the seabed and the sea in motion. II. Physical and chemical oceanography and physical resources,* ed. F.T. Banner, M.B. Collins and K.S. Massie, Elsevier, Amsterdam. pp. 495–516.

Burton, M.A.S. and P.J. Peterson (1979). Metal accumulation by aquatic bryophytes from polluted mine streams. *Environ. Pollut.*, **19**, 39–46.

Bush, B., K.W. Simpson, L. Shane and R.R. Koblintz (1985). PCB congener analysis of water and caddisfly larvae (Insecta: Trichoptera) in the upper Hudson River by glass capillary chromatography. *Bull. Environ. Contam. Toxicol.*, **34**, 96–105.

Butler, P.A. (1966). The problem of pesticides in estuaries. In *A symposium on estuarine fisheries. Trans. Amer. Fish. Soc. Spec. Publ.*, **3**, 110–115.

Butler, P.A. (1969a). The significance of DDT residues in estuarine fauna. In *Chemical fallout*, ed. M.W. Miller and G.G. Berg, Charles C. Thomas, Springfield, Illinois. pp. 205–220.

Butler, P.A. (1969b). Monitoring pesticide pollution. *BioScience.*, **19**, 889–891.

Butler, P.A. (1971). Influence of pesticides on marine ecosystems. *Proc. Roy. Soc. Lond. B.*, **177**, 321–329.

Butler, P.A. (1973). Organochlorine residues in estuarine molluscs, 1965–1972 — National Pesticides Monitoring Program. *Pestic. Monit. J.*, **6**, 238–262.

Butler, P.A. and R.L. Schutzmann (1979). Bioaccumulation of DDT and PCB in tissues of marine fishes. In *Aquatic toxicology*, Eds. L.L. Marking and R.A. Kimerle, American Society for Testing and Materials, Philadelphia. pp. 212–220.

Butler, P.A., L. Andrén, G.J. Bonde, A. Jernelöv and D.J. Reisch (1971). Monitoring organisms. In *Food and Agricultural Organisation Technical Conference on marine pollution and its effects on living resources and fishing, Rome, 1970. Supplement 1: Methods of detection, measurement and monitoring of pollutants in the marine environment*, FAO Fisheries Reports No. 99, Suppl. 1, 101–112.

Butler, P.A., R. Childress and A.J. Wilson (1972). The association of DDT residues with losses in marine productivity. In *Marine pollution and sea life*, Ed. M. Ruivo, pp. 262–266. Fishing News Books Ltd., London.

Cabioch, L., J.-C. Dauvin and F. Gentil (1978). Preliminary observations on pollution of the sea bed and disturbance of sub-littoral communities in Northern Brittany by oil from the *Amoco Cadiz. Mar. Pollut. Bull.*, **9**, 303–307.

Cahn, P.H., J. Foehrenbach and W. Guggino (1978). PCB levels in certain organs of some feral fish from New York State. In *Pollution and physiology of marine animals*, ed. F.J. Vernburg, F.P. Thurberg, A. Calabrese and W. Vernberg, Academic Press, New York, pp. 51–61.

Cain, D.J. and S.N. Luoma (1985). Copper and silver accumulation in transplanted and resident clams (*Macoma balthica*) in south San Francisco Bay. *Mar. Environ. Res.*, **15**, 115–135.

Cain, D.J. and S.N. Luoma (1990). Influence of seasonal growth, age and environmental exposure on Cu and Ag in a bivalve indicator, *Macoma balthica*, in San Francisco Bay. *Mar. Ecol. Prog. Ser.*, **60**, 45–55.

Caines, L.A., A.W. Watt and D.E. Wells (1985). The uptake and release of some trace metals by aquatic bryophytes in acidified waters in Scotland. *Environ. Pollut. Ser.B.*, **10**, 1–18.

Cairns, J. and J.R. Pratt (1987). Ecotoxicological effect indices: a rapidly evolving system. *Water Sci. Technol.*, **19**, 1–12.

Cairns, J. and W.H. van der Schalie (1980). Biological monitoring. Part I — Early warning systems. *Water Res.*, **14**, 1179–1196.

Cajal-Medrano, R. and E.A. Gutierrez-Galindo (1981). Concentration et distribu-

tion du DDT dans les huitres *Crassostrea gigas* et *Ostrea edulis* sur la côte de basse Californie. *Rev. Int. Océanogr. Méd.*, **LXII**, 39–45.
Calabrese, A., J.R. MacInnes, D.A. Nelson, R.A. Greig and P.P. Yevich (1984). Effects of long-term exposure to silver or copper on growth, bioaccumulation and histopathology in the blue mussel *Mytilus edulis*. *Mar. Environ. Res.*, **11**, 253–274.
Calabrese, E.J. (1983). Role of epidemiologic studies in drinking water standards of metals. *Environ. Health. Perspect.*, **52**, 99–106.
Campbell, P.J. and M.B. Jones (1990). Water permeability of *Palaemon longirostris* and other euryhaline caridean prawns. *J. exp. Biol.*, **150**, 145–158.
Cantelmo-Cristini, A., F.E. Hospod and R.J. Lazell (1985). An *in situ* study of the adenylate energy charge of *Corbicula fluminea* in a freshwater system. In *Marine pollution and physiology: recent advances*, ed. F.J. Vernberg, F.P. Thurberg, A. Calabrese and W. Vernberg, University of South Carolina Press, Columbia, South Carolina, pp. 45–62.
Canterford, G.S. and D.R. Canterford (1980). Toxicity of heavy metals to the marine diatom *Ditylum brightwelli* (West) Grunow: correlation between toxicity and metal speciation. *J. mar. biol. Ass. U.K.*, **60**, 243–253.
Canton, J.H., G.J. van Esch, P.A. Greve and A.B.A.M. van Hellemond (1977). Accumulation and elimination of α-hexachlorocyclohexane (α-HCH) by the marine algae *Chlamydomonas* and *Dunaliella*. *Water Res.*, **11**, 111–115.
Capuzzo, J.M. and D.R. Kester (1987). Biological effects of waste disposal: experimental results and predictive assessments. In *Oceanic processes in marine pollution, Vol. 1. Biological processes and wastes in the ocean*, ed. J.M. Capuzzo and D.R. Kester, R.E. Krieger Publ. Co., Malabar, Florida, pp. 3–15.
Carlisle, D.B. (1968). Vanadium and other metals in ascidians. *Proc. Roy. Soc. Lond. B.*, **171**, 31–41.
Carlton, J.T. (1979). Introduced invertebrates of San Francisco Bay. In *San Francisco Bay: the urbanised estuary*, ed. T.J. Conomos, American Association for the Advancement of Science, San Francisco, pp. 427–444.
Carmichael, N.G., K.S. Squibb and B.A. Fowler (1979). Metals in the molluscan kidney: a comparison of two closely related bivalve species (*Argopecten*), using X-ray microanalysis and atomic absorption spectroscopy. *J. Fish. Res. Board Can.*, **36**, 1149–1155.
Carpenter, J.H. and R.J. Huggett (1984). Meaningful chemical measurements in the marine environment — transition metals. In *Concepts in marine pollution measurements*, ed. H.H. White, Maryland Sea Grant College, Maryland, pp. 379–403.
Carvalho, F.P. and S.W. Fowler (1985). Americium adsorption on the surfaces of macrophytic algae. *J. Environ. Radioactivity.*, **2**, 311–317.
Casillas, E., D. Misitano, L.L. Johnson, L.D. Rhodes, T.K. Collier, J.E. Stein, B.B. McCain and U. Varanasi (1991). Inducibility of spawning and reproductive success of female English sole (*Parophrys vetulus*) from urban and nonurban areas of Puget Sound, Washington. *Mar. Environ. Res.*, **31**, 99–122.
CEC (1978). *Criteria (dose/effect relationships) for cadmium*. Report of a Working Group of Experts prepared for the Commission of European Communities, Directorate-General for Social Affairs, Health and Safety Directorate. Pergamon Press, Oxford.

Cember, H., E.H. Curtis and B.G. Blaylock (1978). Mercury bioconcentration in fish: temperature and concentration effects. *Environ. Pollut.*, **17**, 311–319.

Chan, H.M. (1988). Accumulation and tolerance to cadmium, copper, lead and zinc by the green mussel *Perna viridis*. *Mar. Ecol. Prog. Ser.*, **48**, 295–303.

Chan, H.M. (1990). *Aspects of the biology of zinc in crabs with particular emphasis on the shore crab Carcinus maenas (L.)*. Unpublished Ph.D. thesis, University of London.

Chapman, P.M. and E.R. Long (1983). The use of bioassays as a part of a comprehensive approach to marine pollution assessment. *Mar. Pollut. Bull.*, **14**, 81–84.

Chapman, P.M., G.P. Romberg and G.A. Vigers (1982). Design of monitoring studies for priority pollutants. *J. Water Poll. Control Fed.*, **54**, 292–297.

Chapman, P.M., R.N. Dexter and L. Goldstein (1987). Development of monitoring programmes to assess the long-term health of aquatic ecosystems. A model from Puget Sound, USA. *Mar. Pollut. Bull.*, **18**, 521–527.

Chen, P.H. and S.-T. Hsu (1986). PCB poisoning from toxic rice-bran oil in Taiwan. In *PCBs and the environment*, Vol. III, Ed. J.S. Waid, CRC Press, Boca Raton, Florida, pp. 18–31.

Chen, P.H., K.T. Chang and Y.D. Lu (1981). Polychlorinated biphenyls and polychlorinated dibenzofurans in the toxic rice-bran oil that caused PCB poisoning in Taichung. *Bull. Environ. Contam. Toxicol.*, **26**, 489–495.

Cherian, M.G. and R.A. Goyer (1978). Metallothioneins and their role in the metabolism and toxicity of metals. *Life Sciences.*, **23**, 1–10.

Chigno, F.E., R.W. Smith and F.L. Shore (1982). Uptake of arsenic, cadmium, lead and mercury from polluted waters by the water hyacinth *Eichhornia crassipes*. *Environ. Pollut. Ser. A.*, **27**, 31–36.

Chiou, C.T. (1985). Partition coefficients of organic compounds in lipid-water systems and correlations with fish bioconcentration factors. *Environ. Sci. Technol.*, **19**, 57–62.

Claisse, D. (1989). Chemical contamination of French coasts: the results of a ten years mussel watch. *Mar. Pollut. Bull.*, **20**, 523–528.

Clark, R.B. (1989). *Marine pollution*, 2nd edn. Clarendon Press, Oxford.

Clark, T., K. Clark, S. Paterson, D. Mackay and R.J. Norstrom (1988). Wildlife monitoring, modeling and fugacity. *Environ. Sci. Technol.* **22**, 120–127.

Clayton, J.R., S.P. Pavlou and N.F. Breitner (1977). Polychlorinated biphenyls in coastal marine zooplankton: bioaccumulation by equilibrium partitioning. *Environ. Sci. Technol.*, **11**, 676–682.

Clegg, D.E. (1974). Chlorinated hydrocarbon pesticide residues in oysters (*Crassostrea commercialis*) in Moreton Bay, Queensland, Australia, 1970–72. *Pestic. Monit. J.*, **8**, 162–166.

Coates, M., D.W. Connell, J. Bodero, G.J. Miller and R. Back (1986). Aliphatic hydrocarbons in Great Barrier Reef organisms and environment. *Estuar. cstl. Shelf Sci.*, **23**, 99–113.

Cocchieri, R.A., A. Arnese and A.M. Minicucci (1990). Polycyclic aromatic hydrocarbons in marine organisms from Italian central Mediterranean coasts. *Mar. Pollut. Bull.*, **21**, 15–18.

Cognetti, G. (1989). SOS from the Adriatic. *Mar. Pollut. Bull.*, **20**, 578–579.

Collins Longman Atlas (1989). William Collins Sons and Co. Ltd., Scotland.

Conan, G. (1982). The long-term effects of the *Amoco Cadiz* oil spill. *Phil. Trans. Roy. Soc. Lond. B.*, **297**, 323–333.

Connell, D.W. (1981). *Petroleum hydrocarbons in the Hudson-Raritan Estuary: a review.* Working Paper 3, Reference 81–6, from the Marine Sciences Research Center, State University of New York, Stony Brook, New York.

Connell, D.W. (1987). Age to PCB concentration relationship with the striped bass (*Morone saxatilis*) in the Hudson River and Long Island Sound. *Chemosphere.*, **16**, 1469–1474.

Connell, D.W. (1988). Bioaccumulation behaviour of persistent organic chemicals with aquatic organisms. *Rev. Environ. Contam. Toxicol.*, **101**, 117–154.

Connell, D.W. and G.J. Miller (1984). *Chemistry and ecotoxicology of pollution.* John Wiley and Sons, New York.

Connell, J.H. (1985). The consequences of variation in initial settlement versus post-settlement mortality in rocky intertidal communities. *J. exp. mar. Biol. Ecol.*, **93**, 11–45.

Conover, R.J. (1966). Assimilation of organic matter by zooplankton. *Limnol. Oceanogr.*, **11**, 338–354.

Contardi, V., R. Capelli, T. Pellacani and G. Zanicchi (1979). PCBs and chlorinated pesticides in organisms from the Ligurian Sea. *Mar. Pollut. Bull.*, **10**, 307–311.

Conway, H.L. and S.C. Williams (1979). Sorption of cadmium and its effect on growth and the utilization of inorganic carbon and phosphorus of two fresh-water diatoms. *J. Fish. Res. Board Can.*, **36**, 579–586.

Cooley, T.N. and D.F. Martin (1979). Cadmium in naturally occurring water hyacinths. *Chemosphere*, **2**, 75–78.

Cormier, S.M. and R.N. Racine (1990). Histopathology of Atlantic tomcod: a possible monitor of xenobiotics in northeast tidal rivers and estuaries. In *Biomarkers of environmental contamination*, ed. J.F. McCarthy and L.R. Shugart, Lewis Publishers, CRC Press, Boca Raton, Florida, pp. 59–71.

Correa, M. (1987). Physiological effects of metal toxicity on the tropical freshwater shrimp *Macrobrachium carcinus* (Linneo, 1758). *Environ. Pollut.*, **45**, 149–155.

Cory, R.L. and P.V. Dresler (1981). Thermal plumes, trace metals and Asiatic clams of the Potomac River and estuary. *Estuaries*, **4**, 296–303.

Cossa, D., E. Bourget and J. Piuze (1979). Sexual maturation as a source of variation in the relationship between cadmium concentration and body weight of *Mytilus edulis* L. *Mar. Pollut. Bull.*, **10**, 174–176.

Couillard, D. (1982). Evaluation des teneurs en composés organochlores dans le fleuve, l'estuaire et le Golfe Saint-Laurent, Canada. *Environ. Pollut. Ser. B.*, **3**, 239–270.

Courtney, W.A.M. and G.R.W. Denton (1976). Persistence of polychlorinated biphenyls in the hard-clam (*Mercenaria mercenaria*) and the effect upon the distribution of these pollutants in the estuarine environment. *Environ. Pollut.*, **10**, 55–64.

Courtney, W.A.M. and W.J. Langston (1978). Uptake of polychlorinated biphenyl (Aroclor 1254) from sediment and from seawater in two intertidal polychaetes. *Environ. Pollut.*, **15**, 303–309.

Cowan, A.A. (1981). Organochlorine compounds in mussels from Scottish coastal waters. *Environ. Pollut. Ser. B.*, **2**, 129–143.

Cox, J.L. (1970). Low ambient level uptake of ^{14}C-DDT by three species of marine phytoplankton. *Bull. Environ. Contam. Toxicol.*, **5**, 218–221.

Cox, J.L. (1971). DDT residues in sea water and particulate matter in the California Current system. *Fishery Bull. U.S.*, **69**, 443–450.

Crathorne, B. and A.J. Dobbs (1990). Chemical pollution of the aquatic environment by priority pollutants and its control. In *Pollution: causes, effects, and control*, ed. R.M. Harrison, Royal Society of Chemistry, London, pp. 1–18.

Crosby, D.G. and R.K. Tucker (1971). Accumulation of DDT by *Daphnia magna*. *Environ. Sci. Technol.*, **5**, 714–716.

Cross, J.N. and J.E. Hose (1988). Evidence for impaired reproduction in white croaker (*Genyonemus lineatus*) from contaminated areas off Southern California. *Mar. Environ. Res.*, **24**, 185–188.

CSWRCB (1986). *Toxic Substances Monitoring Program, 1984*. Water Quality Monitoring Report No. 86-4-WQ, California State Water Resources Control Board, Sacramento, California, USA.

Curran, J.C., P.J. Holmes and J.E. Yersin (1986). Moored shellfish cages for pollution monitoring: design and operational experience. *Mar. Pollut. Bull.*, **17**, 464–465.

Cutshall, N.H., J.R. Naidu and W.G. Percy (1977). Zinc-65 specific activities in the migratory Pacific hake *Merluccius productus*. *Mar. Biol.*, **40**, 75–80.

Dauvin, J.-C. and F. Gentil (1990). Conditions of the peracarid populations of subtidal communities in Northern Brittany ten years after the *Amoco Cadiz* oil spill. *Mar. Pollut. Bull.*, **21**, 123–130.

Davenport, J. (1977). A study of the effects of copper applied continuously and discontinuously to specimens of *Mytilus edulis* (L.) exposed to steady and fluctuating salinity levels. *J. mar. biol. Ass. U.K.*, **57**, 63–74.

Davenport, J. and A. Manley (1978). The detection of heightened sea-water copper concentrations by the mussel *Mytilus edulis*. *J. mar. biol. Ass. U.K.*, **58**, 843–850.

Davenport, J. and K.J. Redpath (1984). Copper and the mussel, *Mytilus edulis* L. In *Toxins, drugs, and pollutants in marine animals*, ed. P. Bolis *et al.*, Springer-Verlag, Berlin, pp. 176–189.

de Campos, M. and A.E. Olszyna-Marzys (1979). Contamination of human milk with chlorinated pesticides in Guatemala and in El Salvador. *Arch. Environ. Contam. Toxicol.*, **8**, 43–58.

DeFoe, D.L., G.D. Veith and R.W. Carlson (1978). Effects of Aroclor 1248 and 1260 on the fathead minnow. *J. Fish. Res. Board Can.*, **35**, 997–1002.

Degobbis, D. (1989). Increased eutrophication of the Northern Adriatic Sea. Second act. *Mar. Pollut. Bull.*, **20**, 452–457.

del Castilho, P., R.G. Gerritse, J.M. Marquenie and W. Salomons (1984). Speciation of heavy metals and the *in-situ* accumulation by *Dreissena polymorpha:* a new method. In *Complexation of trace metals in natural waters*, ed. C.J.M. Kramer and J.C. Duinker, Martinus Nijhoff/Dr. W. Junk Publishers, The Hague, pp. 445–448.

DeLong, R.L., W.G. Gilmartin and J.G. Simpson (1973). Premature births in California sea lions: association with high organochlorine pollutant residue levels. *Science.*, **181**, 1168–1170.

Denton, G.R.W. and C. Burdon-Jones (1981). Influence of temperature and salinity on the uptake, distribution and depuration of mercury, cadmium and lead by the black-lip oyster *Saccostrea echinata*. *Mar. Biol.*, **64**, 317–326.

Denton, G.R.W. and C. Burdon-Jones (1986). Trace metals in algae from the Great Barrier Reef. *Mar. Pollut. Bull.*, **17**, 98–107.

Depledge, M.H. (1984a). Disruption of circulatory and respiratory activity in shore crabs (*Carcinus maenas* (L.)) exposed to heavy metal pollution. *Comp. Biochem. Physiol.*, **78C**, 445–459.

Depledge, M.H. (1984b). Changes in cardiac activity, oxygen uptake and perfusion indices in *Carcinus maenas* (L.) exposed to crude oil and dispersant. *Comp. Biochem. Physiol.*, **78C**, 461–466.

Depledge, M.H. (1989a). Re-evaluation of metabolic requirements for copper and zinc in decapod crustaceans. *Mar. Environ. Res.*, **27**, 115–126.

Depledge, M.H. (1989b). The rational basis for detection of the early effects of marine pollutants using physiological indicators. *Ambio.*, **18**, 301–302.

Depledge, M.H. (1990). Interactions between heavy metals and physiological processes in estuarine invertebrates. In *Estuarine ecotoxicology*, ed. P.L. Chambers and C.M. Chambers, JAPAGA, Ashford, Ireland, pp. 89–100.

Depledge, M.H. and B.B. Andersen (1990). A computer-aided physiological monitoring system for continuous, long-term recording of cardiac activity in selected invertebrates. *Comp. Biochem. Physiol.*, **96A**, 473–477.

Depledge, M.H. and P. Bjerregaard (1989). Haemolymph protein composition and copper levels in decapod crustaceans. *Helgoländer Meeresunters.*, **43**, 207–223.

Depledge, M.H. and D.J.H. Phillips (1986). Circulation, respiration and fluid dynamics in the gastropod mollusc, *Hemifusus tuba* (Gmelin). *J. exp. mar. Biol. Ecol.*, **95**, 1–13.

Depledge, M.H. and P.S. Rainbow (1990). Models of regulation and accumulation of trace metals in marine invertebrates. *Comp. Biochem. Physiol.*, **97C**, 1–7.

Derban, L.K.A. (1974). Outbreak of food poisoning due to alkyl-mercury fungicide. *Arch. Environ. Hlth.* **28**, 49–52.

Dermott, R.M. and K.R. Lum (1986). Metal concentrations in the annual shell layers of the bivalve *Elliptio complanata*. *Environ. Pollut. Ser. B.*, **12**, 131–143.

Deubert, K.H., P. Rule and I. Corte-Real (1981). PCB residues in *Mercenaria mercenaria* from New Bedford Harbor, 1978. *Bull. Environ. Contam. Toxicol.*, **27**, 683–689.

Dickson, D. (1982). United States: lessons of Love Canal prompt clean up. *Ambio.*, **11**, 47–50.

Diks, D.M. and H.E. Allen (1983). Correlation of copper distribution in a freshwater-sediment system to bioavailability. *Bull. Environ. Contam. Toxicol.*, **30**, 37–43.

D'Itri, F.M. (1972). *The environmental mercury problem*. CRC Press, Cleveland, Ohio.

Dix, T.G., A. Martin, G.M. Ayling, K.C. Wilson and D.A. Ratkowsky (1976). Sand flathead (*Platycephalus bassensis*), an indicator species for mercury pollution in Tasmanian waters. *Mar. Pollut. Bull.*, **6**, 142–144.

Dixon, D.G. and J.B. Sprague (1981). Copper bioaccumulation and hepatoprotein synthesis during acclimation to copper by juvenile rainbow trout. *Aquatic Toxicol.*, **1**, 69–81.

Dixon, D.R. and K.R. Clarke (1982). Sister chromatid exchange: a sensitive method for detecting damage caused by exposure to environmental mutagens in the chromosomes of adult *Mytilus edulis*. *Mar. Biol. Letts.*, **3**, 163–172.

DoE (1985). *Water and the environment. The implementation of Directive 76/464/EEC on pollution caused by certain dangerous substances discharged into the aquatic environment of the Community*. Department of the Environment, London (Circular 18/85) and the Welsh Office, Cardiff (Circular 37/85).

Donazzolo, R., O.H. Merlin, L.M. Vitturi and B. Pavoni (1984). Heavy metal content and lithological properties of recent sediments in the Northern Adriatic. *Mar. Pollut. Bull.*, **15**, 93–101.

Donkin, P., S.V. Mann and E.I. Hamilton (1981). Polychlorinated biphenyl, DDT and dieldrin residues in grey seal (*Halichoerus grypus*) males, females and mother-foetus pairs sampled at the Farne Islands, England, during the breeding season. *Sci. Total Environ.*, **19**, 121–142.

Duffy, J.R. and D. O'Connell (1968). DDT residues and metabolites in Canadian Atlantic coast fish. *J. Fish. Res. Board Can.*, **25**, 189–195.

Duinker, J.C. and M.T.J. Hillebrand (1979). Behaviour of PCB, pentachlorobenzene, hexachlorobenzene, α-HCH, γ-HCH, β-HCH, dieldrin, endrin and *p,p'*-DDD in the Rhine-Meuse Estuary and the adjacent coastal area. *Neth. J. Sea Res.*, **13**, 256–281.

Duinker, J.C., M.T.J. Hillebrand, R.F. Nolting, S. Wellershaus and N.K. Jacobsen (1980). The River Varde Å: processes affecting the behaviour of metals and organochlorines during estuarine mixing. *Neth. J. Sea Res.*, **14**, 237–267.

Duinker, J.C., D.E. Schultz and G. Petrick (1988a). Selection of chlorinated biphenyl congeners for analysis in environmental samples. *Mar. Pollut. Bull.*, **19**, 19–25.

Duinker, J.C., A.H. Knap, K.C. Binkley, G.H. van Dam, A. Darrel-Rew and M.T.J. Hillebrand (1988b). Method to represent the qualitative and quantitative characteristics of PCB mixtures: marine mammal tissues and commercial mixtures as examples. *Mar. Pollut. Bull.*, **19**, 74–79.

Edgren, M., M. Olsson and L. Renberg (1979). Preliminary results on uptake and elimination at two different temperatures of *p,p'*-DDT and two chlorobiphenyls in perch from brackish water. *Ambio.*, **8**, 270–272.

EEC (1976). *Council Directive of 4 May 1976 on pollution caused by certain substances discharged into the aquatic environment of the Community*. Directive 76/464/EEC, European Economic Community, Brussels.

EEC (1980). *3-drins in aquatic environments — quality objectives and limit values*. Paper C 83 02/04/80, European Economic Community, Brussels.

EEC (1982a). *Mercury*. Council Directive 82/176/EEC.

EEC (1982b). *Communication from the Commission to the Council on List 1 of Directive 76/464*. Paper C 176 of 14 July 1982, European Economic Community, Brussels.

EEC (1983a). *Cadmium*. Council Directive 83/513/EEC. European Economic Community, Brussels.

EEC (1983b). *Titanium dioxide*. Council Directive 83/29/EEC. European Economic Community, Brussels.

EEC (1984). *Hexachlorocyclohexane*. Council Directive 84/491/EEC. European Economic Community, Brussels.

EEC (1986a). *Council Directive of 12 June 1986 on limit values and quality objectives for discharges of certain dangerous substances included in List I of the Annex to Directive 76/464/EEC*. European Economic Community, Brussels.

EEC (1986b). *Chromium in water*. Paper C 207 18/08/86, European Economic Community, Brussels.

EEC (1990). *Proposal for a Council Directive amending Directive 76/464/EEC on pollution caused by certain dangerous substances discharged into the aquatic environment of the Community*. 90/C 55/09, European Economic Community, Brussels.

EEC (1991). *Commission Directive of 1 March 1991 adapting to technical progress for the twelfth and thirteenth times Council Directive 67/548/EEC on the approximation of the laws, regulations and administrative provisions relating to the classification, packaging and labelling of dangerous substances*. 91/325/EEC, European Economic Community, Brussels.

Eisenbud, M. (1963). *Environmental radioactivity*. McGraw-Hill Book Company, Inc., New York.

Eisenreich, S.J., G.J. Hollod and T.C. Johnson (1979). Accumulation of polychlorinated biphenyls (PCBs) in surficial Lake Superior sediments. Atmospheric deposition. *Environ. Sci. Technol.*, **13**, 569–573.

Eisler, R. (1979). Behavioural responses of marine poikilotherms to pollutants. *Phil. Trans. Roy. Soc. Lond. B.*, **286**, 507–521.

Eisler, R. (1981). *Trace metal concentrations in marine organisms*. Pergamon Press, New York.

Elder, J.F. and H.C. Mattraw (1984). Accumulation of trace elements, pesticides, and polychlorinated biphenyls in sediment and the clam *Corbicula manilensis* of the Apalachicola River, Florida. *Arch. Environ. Contam. Toxicol.*, **13**, 453–469.

Emara, H.I. (1990). Oil pollution in the southern Arabian Gulf and Gulf of Oman. *Mar. Pollut. Bull.*, **21**, 399–401.

Empain, A. (1976a). Estimation de la pollution par métaux lourds dans la Somme, par l'analyse des bryophytes aquatiques. *Bull. Fr. Pisciculture Belg.*, **48**, 138–142.

Empain, A. (1976b). Les bryophytes aquatiques utilisés comme traceurs de la contamination en métaux lourds des eaux douces. *Mem. Soc. r. Bot. Belg.*, **7**, 141–156.

Engel, D.W. and M. Brouwer (1989). Metallothionein and metallothionein-like proteins: physiological importance. *Adv. Comp. Environ. Physiol.*, **5**, 53–75.

Engel, D.W. and B.A. Fowler (1979). Factors influencing cadmium accumulation and its toxicity to marine organisms. *Environ. Health Perspect.*, **28**, 81–88.

Engelhardt, F.R. (1983). Petroleum effects on marine mammals. *Aquatic Toxicol.*, **4**, 199–217.

Engelhardt, F.R., J.R. Geraci and T.G. Smith (1977). Uptake and clearance of petroleum hydrocarbons in the ringed seal (*Phoca hispida*). *J. Fish. Res. Board Can.*, **34**, 1143–1147.

Ernst, W. (1980). Effects of pesticides and related organic compounds in the sea. *Helgoländ. Wiss. Meeresunters.*, **33**, 301–312.

Essink, K. (1980). Mercury pollution in the Ems Estuary. *Helgoländ. Wiss. Meeresunters.*, **33**, 111–121.

Everard, M. and P. Denny (1984). The transfer of lead by freshwater snails in Ullswater, Cumbria. *Environ. Pollut. Ser. A.*, **35**, 299–314.

Fantin, A.M.B., L. Benedetti, L. Bolognani and E. Ottaviani (1982). The effect of lead pollution on the freshwater gastropod *Viviparus viviparus* L.: biochemical and histochemical features. *Malacologia.*, **22**, 19–21.

FAO/WHO (1972). *Report of the Joint Committee on food additives.* Food and Agriculture Organisation and World Health Organisation. *Wld. Health Org. Tech. Rep. Ser.* 505.

Farrington, J.W., R.W. Risebrough, P.L. Parker, A.C. Davis, B. de Lappe, J.K. Winters, D. Boatwright and N.M. Frew (1982). *Hydrocarbons, polychlorinated biphenyls, and DDE in mussels and oysters from the U.S. coast, 1976–1978 — the mussel watch.* Technical Report WHOI-82-42, Woods Hole Oceanographic Institution, Woods Hole, Massachusetts.

Farrington, J.W., E.D. Goldberg, R.W. Risebrough, J.H. Martin and V.T. Bowen (1983). U.S. 'Mussel Watch' 1976–1978: an overview of the trace-metal, DDE, PCB, hydrocarbon, and artificial radionuclide data. *Environ. Sci. Technol.* **17**, 490–496.

Ferguson, D.E., J.L. Ludke and G.G. Murphy (1966). Dynamics of endrin uptake and release by resistant and susceptible strains of mosquitofish. *Trans. Amer. Fish. Soc.*, **95**, 335–344.

Fisher, J.B., R.L. Petty and W. Lick (1983). Release of polychlorinated biphenyls from contaminated lake sediments: flux and apparent diffusivities of four individual PCBs. *Environ. Pollut. Ser. B.*, **5**, 121–132.

Fisher, N.S., L.B. Graham, E.J. Carpenter and C.F. Wurster (1973). Geographic differences in phytoplankton sensitivity to PCBs. *Nature, Lond.*, **241**, 548–549.

Flegal, A.R. and J.H. Martin (1977). Contamination of biological samples by ingested sediment. *Mar. Pollut. Bull.*, **8**, 90–92.

Florence, T.M. (1983). Evaluation of some physico-chemical techniques for the determination of the fraction of dissolved copper toxic to the marine diatom *Nitzschia closterium. Analytica Chim. Acta.*, **151**, 281–295.

Flores-Baez, B.P. and M.S. Galindo-Bect (1989). DDT in *Mytilus edulis:* statistical considerations and inherent variability. *Mar. Pollut. Bull.*, **20**, 496–499.

Foe, C. and A. Knight (1987). Assessment of the biological impact of point source discharges employing Asiatic clams. *Arch. Environ. Contam. Toxicol.*, **16**, 39–51.

Foehrenbach, J. (1972). Chlorinated pesticides in estuarine organisms. *J. Water Pollut. Control Fed.*, **44**, 619–624.

Folke, J. and J. Birklund (1986). Danish coastal water levels of 2,3,4,6–tetrachlorophenol, pentachlorophenol, and total organohalogens in blue mussels (*Mytilus edulis*). *Chemosphere*, **15**, 895–900.

Folsom, T.R. and D.R. Young (1965). Silver-110m and cobalt-60 in oceanic and coastal organisms. *Nature, Lond.*, **206**, 803–806.

Folsom, T.R., D.R. Young, J.N. Johnson and K.C. Pillai (1963). Manganese-54 and zinc-65 in coastal organisms of California. *Nature, Lond.*, **200**, 327–329.

Förstner, U. (1980). Inorganic pollutants, particularly heavy metals in estuaries. In *Chemistry and biogeochemistry of estuaries*, ed. E. Olausson and I. Cato, John Wiley and Sons, New York, pp. 307–348.

Förstner, U. and Solomons, W. (1983). Trace element speciation in surface waters: interactions with particulate matter. In *Trace element speciation in surface waters and its ecological implications*, Nato Conference Series I, Ecology, Vol. 6, ed. G.G. Leppard, Plenum Press, pp. 245–270.

Förstner, U. and G.T.W. Wittmann (1983). *Metal pollution in the aquatic environment*. Second edition. Springer-Verlag, Berlin.

Foster, P., D.T.E. Hunt and A.W. Morris (1978). Metals in an acid mine stream and estuary. *Sci. Total Environ.*, **9**, 75–86.

Foster, R.B. and J.M. Bates (1978). Use of freshwater mussels to monitor point source industrial discharges. *Environ. Sci. Technol.*, **12**, 958–962.

Fowler, S.W. (1990). Critical review of selected heavy metal and chlorinated hydrocarbon concentrations in the marine environment. *Mar. Environ. Res.*, **29**, 1–64.

Fowler, S.W. and B. Oregioni (1976). Trace metals in mussels from the north-west Mediterranean. *Mar. Pollut. Bull.*, **7**, 26–29.

Frank, R., K. Ronald and H.E. Braun (1973). Organochlorine residues in harp seals (*Pagophilus groenlandicus*) caught in eastern Canadian waters. *J. Fish. Res. Board Can.*, **30**, 1053–1063.

Frank, R., A.E. Armstrong, R.G. Boelens, H.E. Braun and C.W. Douglas (1974). Organochlorine insecticide residues in sediment and fish tissues, Ontario, Canada. *Pestic. Monit. J.*, **7**, 165–180.

Frazier, J.M. (1975). The dynamics of metals in the American oyster, *Crassostrea virginica*. I. Seasonal effects. *Chesapeake Sci.*, **16**, 162–171.

Frederick, L.L. (1975). Comparative uptake of a polychlorinated biphenyl and dieldrin by the white sucker (*Catostomus commersoni*). *J. Fish. Res. Board Can.*, **32**, 1705–1709.

Frenet, M. (1981). The distribution of mercury, cadmium and lead between water and suspended matter in the Loire Estuary as a function of the hydrological regime. *Water Res.*, **15**, 1343–1350.

Friberg, L., M. Piscator, G.F. Nordberg and T. Kjellström (Eds.) (1974). *Cadmium in the environment*. 2nd edn. CRC Press, Cleveland, Ohio.

Fujiki, M. (1972). The transitional condition of Minimata Bay and the neighbouring sea polluted by factory waste water containing mercury. In *6th International Conference on water pollution research*, Jerusalem, paper no. 12.

Gaglione, P. and O. Ravera (1964). Manganese-54 concentration in fall-out, water, and *Unio* mussels of Lake Maggiore, 1960–63. *Nature, Lond.*, **204**, 1215–1216.

Galgani, F. and G. Bocquené (1988). A method for routine detection of organophosphates and carbamates in sea water. *Environ. Tech. Letters*, **10**, 311–322.

Gaskin, D.E., M. Holdrinet and R. Frank (1971). Organochlorine pesticide residues in harbour porpoises from the Bay of Fundy region. *Nature, Lond.*, **233**, 499–500.

Gaskin, D.E., R. Frank, M. Holdrinet, K. Ishida, C.J. Walton and M. Smith (1973). Mercury, DDT, and PCB in harbour seals (*Phoca vitulina*) from the Bay of Fundy and Gulf of Maine. *J. Fish. Res. Board Can.*, **30**, 471–475.

Gault, N.F.S., E.L.C. Tolland and J.G. Parker (1983). Spatial and temporal trends in heavy metal concentrations in mussels from Northern Ireland coastal waters. *Mar. Biol.*, **77**, 307–316.

George, S.G. (1990). Biochemical and cytological assessments of metal toxicity in marine animals. In *Heavy metals in the marine environment*, ed. R.W. Furness and P.S. Rainbow, pp. 123–142. CRC Press, Boca Raton, Florida.

George, S.G. and B.J.S. Pirie (1980). Metabolism of zinc in the mussel *Mytilus edulis* (L.): a combined ultrastructural and biochemical study. *J. mar. biol. Ass. U.K.*, **60**, 575–590.

George, S.G., B.J.S. Pirie and T.L. Coombs (1976). The kinetics of accumulation and excretion of ferric hydroxide in *Mytilus edulis* (L.) and its distribution in the tissues. *J. exp. mar. Biol. Ecol.*, **23**, 71–84.

George, S.G., B.J.S. Pirie, A.R. Cheyne, T.L. Coombs and P.T. Grant (1978). Detoxification of metals by marine bivalves: ultrastructural study of the compartmentation of copper and zinc in the oyster, *Ostrea edulis*. *Mar. Biol.*, **45**, 147–156.

George, S.G., E. Carpene, T.L. Coombs, J. Overnell and A. Youngson (1979). Characterisation of cadmium-binding proteins from mussels, *Mytilus edulis* (L.), exposed to cadmium. *Biochem. Biophys. Acta.*, **580**, 225–233.

Gerhardt, R.E., E.A. Crecelius and J.B. Hudson (1980). Moonshine-related arsenic poisoning. *Arch. Intern. Med.*, **140**, 211–213.

Geyer, H., I. Scheunert and F. Korte (1985). Relationship between the lipid content of fish and their bioconcentration potential of 1,2,4–trichlorobenzene. *Chemosphere*, **14**, 545–555.

Gibbs, P.E. and G.W. Bryan (1980). Copper — the major component of glycerid polychaete jaws. *J. mar. biol. Ass. U.K.*, **60**, 205–214.

Gibbs, P.E., G.W. Bryan and K.P. Ryan (1981). Copper accumulation by the polychaete *Melinna palmata:* an antipredation mechanism? *J. mar. biol. Ass. U.K.*, **61**, 707–722.

Gibbs, P.E., W.J. Langston, G.R. Burt and P.L. Pascoe (1983). *Tharyx marioni* (Polychaeta): a remarkable accumulator of arsenic. *J. mar. biol. Ass. U.K.*, **63**, 313–325.

Gibbs, P.E., G.W. Bryan, P.L. Pascoe and G.R. Burt (1987). The use of the dog-whelk, *Nucella lapillus*, as an indicator of tributyltin (TBT) contamination. *J. mar. biol. Ass. U.K.*, **67**, 507–523.

Gibbs, P.E., P.L. Pascoe and G.R. Burt (1988). Sex change in the female dog-whelk, *Nucella lapillus*, induced by tributyltin from antifouling paints. *J. mar. biol. Ass. U.K.*, **68**, 715–731.

Giesy, J.P. and J.G. Wiener (1977). Frequency distributions of trace metal concentrations in five freshwater fishes. *Trans. Amer. Fish. Soc.*, **106**, 393–403.

Gillespie, N. (1984). Ecological and epidemiological aspects of ciguatera fish poisoning. In *Proceedings of the red tide workshop,* Cronulla, 18–20 June 1984. Australian Department of Science, Canberra.

Godsil, P.J. and W.C. Johnson (1968). Pesticide monitoring of the aquatic biota at the Tule Lake National Wildlife Refuge. *Pestic. Monit. J.*, **1**, 21–26.

Goldberg, E.D. (1975). Synthetic organohalides in the sea. *Proc. Roy. Soc. Lond. B.*, **189**, 277–289.

Goldberg, E.D. (Ed.) (1976). *Strategies for marine pollution monitoring*. John Wiley and Sons, New York.

Goldberg, E.D., V.T. Bowen, J.W. Farrington, G. Harvey, J.H. Martin, P.L. Parker, R.W. Risebrough, W. Robertson, E. Schneider and E. Gamble (1978). The mussel watch. *Environ. Conserv.*, **5**, 101–125.

Goldberg, E.D., J.J. Griffin, V. Hodge, M. Koide and H. Windom (1979). Pollution history of the Savannah River Estuary. *Environ. Sci. Technol.*, **13**, 588–594.

Goldberg, E.D., M. Koide, V. Hodge, A.R. Flegal and J. Martin (1983). U.S. Mussel Watch: 1977–1978 results on trace metals and radionuclides. *Estuar. cstl. Shelf Sci.* **16**, 69–93.

Gommes, R. and H. Muntau (1981). Variations de la composition chimique (polyélements et métaux lourds) entre organes de *Trapa natans* L. et de *Polygonum amphibium* L. *Mem. Ist. Ital. Idrobiol.*, **38**, 331–346.

Gordon, M., G.A. Knauer and J.H. Martin (1980). *Mytilus californianus* as a bioindicator of trace metal pollution: variability and statistical considerations. *Mar. Pollut. Bull.*, **11**, 195–198.

Gordon, R.M., J.H. Martin and G.A. Knauer (1982). Iron in north-east Pacific waters. *Nature, Lond.*, **299**, 611–612.

Gossett, R.W., D.A. Brown and D.R. Young (1983). Predicting the bioaccumulation of organic compounds in marine organisms using octanol/water partition coefficients. *Mar. Pollut. Bull.*, **14**, 387–392.

Göthberg, A. (1983). Intensive fishing — a way to reduce the mercury level in fish. *Ambio*, **12**, 259–261.

Grahl-Nielsen, O. and T. Lygre (1990). Identification of samples of oil related to two spills. *Mar. Pollut. Bull.*, **21**, 176–183.

Grandjean, P. (1981). Blood lead concentrations reconsidered. *Nature, Lond.*, **291**, 188.

Graney, R.L., D.S. Cherry and J. Cairns (1983). Heavy metal indicator potential of the Asiatic clam (*Corbicula fluminea*) in artificial stream systems. *Hydrobiologia*, **102**, 81–88.

Graney, R.L., D.S. Cherry and J. Cairns (1984). The influence of substrate, pH, diet and temperature upon cadmium accumulation in the Asiatic clam (*Corbicula fluminea*) in laboratory artificial streams. *Water Res.*, **18**, 833–842.

Grant, A., J.G. Hateley and N.V. Jones (1989). Mapping the ecological impact of heavy metals on the estuarine polychaete *Nereis diversicolor* using inherited metal tolerance. *Mar. Pollut. Bull.*, **20**, 235–238.

Gray, J.S. (1976). The fauna of the polluted river Tees estuary. *Estuar. cstl. Mar. Sci.*, **4**, 653–676.

Gray, J.S. (1979). Pollution-induced changes in populations. *Phil. Trans. Roy. Soc. Lond. B.*, **286**, 545–561.

Gray, J.S. and T.H. Pearson (1982). Objective selection of sensitive species indicative of pollution-induced change in benthic communities. I. Comparative methodology. *Mar. Ecol. Prog. Ser.*, **9**, 111–119.

Gray, J.S., M. Ascham, M.R. Carr, K.R. Clarke, R.H. Green, T.H. Pearson, R. Rosenberg and R.M. Warwick (1988). Analysis of community attributes of the benthic macrofauna of Frierfjord/Langesundfjord and in a mesocosm experiment. *Mar. Ecol. Prog. Ser.*, **46**, 151–165.

Green, D.R., J.K. Stull and T.C. Heesen (1986). Determination of chlorinated hydrocarbons in coastal waters using a moored *in situ* sampler and transplanted live mussels. *Mar. Pollut. Bull.*, **17**, 324–329.

Griffiths, R.C. (1977). The UN system in the fight against marine pollution. *Fishing News International*, **16**, 41–46.

Gschwend, P.M. and S.-C. Wu (1985). On the constancy of sediment-water partition coefficients of hydrophobic organic pollutants. *Environ. Sci. Technol.*, **19**, 90–96.

Guinness World Data Book (1991). Guinness Publishing Ltd., Middlesex, England.

Guy, R.D., C.L. Chakrabarti and D.C. McBain (1978). An evaluation of extraction techniques for the fractionation of copper and lead in model sediment systems. *Water Res.*, **12**, 21–24.

Habib, S. and M.J. Minski (1982). Incidence and variability of some elements in the non tidal region of the River Thames, and River Kennet, U.K. *Sci. Total Environ.* **22**, 253–273.

Hagino, N. and K. Yoshioka (1961). A study on the cause of itai-itai disease. *J. Jpn. Orthop. Assoc.*, **35**, 812–825.

Haigh, N. (1992). *Manual of environmental policy: the EC and Britain.* Longman Group, Harlow, Essex.

Hallegraeff, G.M. and C. Sumner (1986). Toxic plankton blooms affect shellfish farms. *Aust. Fish.*, **45**, 15–18.

Hamelink, J.L. and R.C. Waybrant (1976). DDE and lindane in a large-scale model lentic ecosystem. *Trans. Amer. Fish. Soc.*, **105**, 124–134.

Hamelink, J.L., R.C. Waybrant and R.C. Ball (1971). A proposal: exchange equilibria control the degree chlorinated hydrocarbons are biologically magnified in lentic environments. *Trans. Amer. Fish. Soc.*, **100**, 207–214.

Hamilton, E.I. (1980). Concentration and distribution of uranium in *Mytilus edulis* and associated materials. *Mar. Ecol. Prog. Ser.*, **2**, 61–73.

Hamilton, E.I. and R.J. Clifton (1980). Concentration and distribution of the transuranium radionuclides $^{239+240}Pu$, ^{238}Pu and ^{241}Am in *Mytilus edulis, Fucus vesiculosus* and surface sediments of Esk estuary. *Mar. Ecol. Prog. Ser.*, **3**, 267–277.

Hannon, M.R., Y.A. Greichus, R.L. Applegate and A.C. Fox (1970). Ecological distribution of pesticides in Lake Poinsett, South Dakota. *Trans. Amer. Fish. Soc.*, **99**, 496–500.

Hansen, D.J. and A.J. Wilson (1970). Significance of DDT residues from the estuary near Pensacola, Fla. *Pestic. Monit. J.*, **4**, 51–56.

Hansen, D.J., S.C. Schimmel and J. Forrester (1975). Effects of Aroclor 1016 on embryos, fry, juveniles, and adults of sheepshead minnows (*Cyprinodon variegatus*). *Trans. Amer. Fish. Soc.*, **104**, 584–588.

Hansen, J.C. (1990). Human exposure to metals through consumption of marine foods: a case study of exceptionally high intake among Greenlanders. In *Heavy metals in the marine environment*, ed. R.W. Furness and P.S. Rainbow, CRC Press, Boca Raton, Florida, pp. 227–243.

Hansen, L.G., W.B. Wiekhorst and J. Simon (1976). Effects of dietary Aroclor 1242 on channel catfish (*Ictalurus punctatus*) and the selective accumulation of PCB components. *J. Fish. Res. Board Can.*, **33**, 1343–1352.

Harada, M. and A.M. Smith (1975). Minimata disease: a medical report. In *Minimata*, ed. W.E. Smith and A.M. Smith, Chatto and Windus Ltd., London, pp. 180–192.

Harding, G.C. (1986). Organochlorine dynamics between zooplankton and their environment, a reassessment. *Mar. Ecol. Prog. Ser.*, **33**, 167–191.

Harding, J.P.C. and B.A. Whitton (1978). Zinc, cadmium and lead in water, sediments and submerged plants of the Derwent Reservoir, northern England. *Water Res.*, **12**, 307–316.

Harding, J.P.C. and B.A. Whitton (1981). Accumulation of zinc, cadmium and lead by field populations of *Lemanea*. *Water Res.*, **15**, 301–319.
Harding, J.P.C., I.G. Burrows and B.A. Whitton (1981). Heavy metals in the Derwent Reservoir catchment, northern England. In *Heavy metals in northern England: environmental and biological aspects*, ed. P.J. Say and B.A. Whitton, Department of Botany, University of Durham, UK., pp. 73–86.
Hardy, J.K. and D.H. O'Keeffe (1985). Cadmium uptake by the water hyacinth: effects of root mass, solute volume, complexers and other metal ions. *Chemosphere.*, **14**, 417–426.
Hardy, J.K. and N.B. Raber (1985). Zinc uptake by the water hyacinth: effects of solution factors. *Chemosphere.*, **14**, 1155–1166.
Hare, L., P.G.C. Campbell, A. Tessier and N. Belzile (1989). Gut sediments in a burrowing mayfly (Ephemeroptera, *Hexagenia limbata*): their contribution to animal trace element burdens, their removal, and the efficacy of a correction for their presence. *Can. J. Fish. Aquat. Sci.*, **46**, 451–456.
Harrington, J.T. (1980). Alcohol, arsenic, and (rapidly) old kidneys. *Arch. Intern. Med.*, **140**, 167–168.
Harrison, F.L. (1978). *Effect of the physicochemical form of trace metals on their accumulation by bivalve molluscs.* Preprint UCRL-81134, submitted to the American Chemical Society Annual Meeting, Miami, Florida, September 1978.
Harrison, H.L., O.L. Loucks, J.W. Mitchell, D.F. Parkhurst, C.R. Tracy, D.G. Watts and V.J. Yannacone (1970). Systems studies of DDT transport. *Science.*, **170**, 503–508.
Harvey, G.R., W.G. Steinhauer and J.M. Teal (1973a). Polychlorobiphenyls in North Atlantic ocean water. *Science.*, **180**, 643–644.
Harvey, G.R., W.G. Steinhauer and J.M. Teal (1973b). Chlorinated hydrocarbons in open ocean Atlantic organisms. In *The changing chemistry of the oceans*, ed. D. Green and D. Jagner, John Wiley & Sons, New York, pp. 177–186.
Harvey, G.R., W.G. Steinhauer and H.P. Miklas (1974a). Decline of PCB concentrations in North Atlantic surface water. *Nature, Lond.*, **252**, 387–388.
Harvey, G.R., H.P. Miklas, V.T. Bowen and W.G. Steinhauer (1974b). Observations on the distribution of chlorinated hydrocarbons in Atlantic Ocean oganisms. *J. mar. Res.*, **32**, 103–118.
Harwood, J. and B. Grenfell (1990). Long term risks of recurrent seal plagues. *Mar. Pollut. Bull.*, **21**, 284–287.
Hatch, T. (1962). Changing objectives in occupational health. *Ind. Hyg. J.* Jan-Feb 17 62: 1–7.
Hattula, M.L., J. Särkkä, J. Janatuinen, J. Paasivirta and A. Roos (1977). Total mercury and methyl mercury contents in fish from Lake Päijänne. *Environ. Pollut.*, **17**, 19–29.
Hattula, M.L., J. Janatuinen, J. Särkkä and J. Paasivirta (1978). A five-year monitoring study of the chlorinated hydrocarbons in the fish of a Finnish lake ecosystem. *Environ. Pollut.*, **15**, 121–139.
Haug, A., S. Melsom and S. Omang (1974). Estimation of heavy metal pollution in two Norwegian fjord areas by analysis of the brown alga *Ascophyllum nodosum*. *Environ. Pollut.*, **7**, 179–192.
Hawker, D.W. and D.W. Connell (1985). Relationships between partition coeffi-

cient, uptake rate constant, clearance rate constant and time to equilibrium for bioaccumulation. *Chemosphere*, **14**, 1205–1219.

Hébrard, J.P. and L. Foulquier (1975). Introduction á l'étude de la fixation du manganèse-54 par *Platyhypnidium riparioides* (Hedw.) Dix. *Rev. Byryol. Lichénol.*, **41**, 35–54.

Heip, C., R.M. Warwick, M.R. Carr, P.M.J. Herman, R. Huys, N. Smol and K. Van Holsbeke (1988). Analyses of community attributes of the benthic meiofauna of Frierfjord/Langesundfjord. *Mar. Ecol. Prog. Ser.*, **46**, 171–180.

Hellawell, J.M. (1986). *Biological indicators of freshwater pollution and environmental management*. Elsevier Applied Science Publishers, London.

Hellawell, J.M. (1988). Toxic substances in rivers and streams. *Environ. Pollut.*, **50**, 61–85.

Helle, E., M. Olsson and S. Jensen (1976a). DDT and PCB levels and reproduction in ringed seal from the Bothnian Bay. *Ambio*, **5**, 188–189.

Helle, E., M. Olsson and S. Jensen (1976b). PCB levels correlated with pathological changes in seal uteri. *Ambio*, **5**, 261–263.

Helle, E., H. Hyvärinen, H. Pyysalo and K. Wickström (1983). Levels of organochlorine compounds in an inland seal population in eastern Finland. *Mar. Pollut. Bull.*, **14**, 256–260.

Hellou, J., G. Stenson, I.-H. Ni and J.F. Payne (1990). Polycyclic aromatic hydrocarbons in muscle tissue of marine mammals from the northwest Atlantic. *Mar. Pollut. Bull.*, **21**, 469–473.

Henderson, C., W.L. Johnson and A. Inglis (1969). Organochlorine insecticide residues in fish (National Pesticide Monitoring Program). *Pestic. Monit. J.* **3**, 145–171.

Henderson, C., A. Inglis and W.L. Johnson (1971). Organochlorine insecticide residues in fish — Fall 1969 National Pesticide Monitoring Program. *Pestic. Monit. J.*, **5**, 1–11.

Herman, S.G., R.L. Garrett and R.L. Rudd (1969). Pesticides and the Western grebe. A study of pesticide survival and trophic concentration at Clear Lake, Lake County, California. In *Chemical fallout: current research on persistent pesticides*, ed. M.W. Miller and G.G. Berg, Charles C. Thomas, Springfield, Illinois, pp. 24–53.

Higgins, H.W. and D.J. Mackey (1987). Role of *Ecklonia radiata* (C. Ag.) J. Agardh in determining trace metal availability in coastal waters. I. Total trace metals. *Aust. J. mar. Freshwat. Res.*, **38**, 307–315.

Hillis, D.M. and J.C. Patton (1982). Morphological and electrophoretic evidence for two species of *Corbicula* (Bivalvia: Corbiculidae) in North America. *Am. Midl. Nat.*, **108**, 74–80.

Hinton, D.E. and D.J. Laurén (1990). Liver structural alterations accompanying chronic toxicity in fishes: potential biomarkers of exposure. In *Biomarkers of environmental contamination*, ed. J.F. McCarthy and L.R. Shugart, Lewis Publishers, CRC Press, Boca Raton, Florida, pp. 17–57.

Hodson, P.V., B.R. Blunt, D.J. Spry and K. Austin (1977). Evaluation of erythrocyte δ-amino levulinic acid dehydratase activity as a short term indicator in fish of harmful exposure to lead. *J. Fish. Res. Board Can.*, **34**, 501–508.

Hodson, P.V., B.R. Blunt and D.J. Spry (1978). Chronic toxicity of water-borne and dietary lead levels to rainbow trout (*Salmo gairdneri*) in Lake Ontario water. *Water Res.*, **12**, 869–878.

Holden, A.V. (1962). A study of the absorption of ^{14}C-labelled DDT from water by fish. *Ann. Appl. Biol.*, **50**, 467–477.

Holden, A.V. (1970). International cooperative study of organochlorine pesticide residues in terrestrial and aquatic wildlife, 1967/1968. *Pestic. Monit. J.*, **4**, 117–135.

Holden, A.V. (1972). Monitoring organochlorine contamination of the marine environment by the analysis of residues in seals. In *Marine pollution and sea life*, ed. M. Ruivo, Fishing News (Books) Ltd., London, pp. 266–272.

Holden, A.V. (1981). Organochlorines — an overview. *Mar. Pollut. Bull.*, **12**, 110–115.

Holden, A.V. and K. Marsden (1967). Organochlorine pesticides in seals and porpoises. *Nature, Lond.*, **216**, 1274–1276.

Holden, A.V. and J.E. Portmann (1970). Monitoring organochlorine residues. *Mar. Pollut. Bull.*, **1**, 41–42.

Holdgate, M.W. (1979). *A perspective of environmental pollution.* Cambridge University Press, Cambridge.

Holmes, P.R. and C.W.Y. Lam (1985). Red tides in Hong Kong waters — response to a growing problem. *Asian Mar. Biol.* **2**, 1–10.

Hopkin, S.P. (1989). *Ecophysiology of metals in terrestrial invertebrates.* Elsevier Applied Science, London.

Hopkin, S.P. and J.A. Nott (1979). Some observations on concentrically structured, intracellular granules in the hepatopancreas of the shore crab *Carcinus maenas* (L.). *J. mar. biol. Ass. U.K.*, **59**, 867–877.

Hose, J.E., J.N. Cross, S.G. Smith and D. Diehl (1989). Reproductive impairment in a fish inhabiting a contaminated coastal environment off southern California. *Environ. Pollut.*, **57**, 139–148.

Howells, G., D. Calamari, J. Gray and P.G. Wells (1990). An analytical approach to assessment of long-term effects of low levels of contaminants in the marine environment. *Mar. Pollut. Bull.*, **21**, 371–375.

Huckabee, J.W., S.A. Janzen, B.G. Blaylock, Y. Talmi and J.J. Beauchamp (1978). Methylated mercury in brook trout (*Salvelinus fontinalis*): absence of an *in vivo* methylating process. *Trans. Amer. Fish. Soc.*, **107**, 848–852.

Huggett, R.J. and M.E. Bender (1980). Kepone in the James River. *Environ. Sci. Technol.*, **14**, 918–923.

Hughes, G.M. and S.F. Perry (1976). Morphometric study of trout gills: a light microscope method suitable for the evaluation of pollutant action. *J. exp. Biol.*, **64**, 447–460.

Hungspreugs, M. and C. Yuangthong (1983). A history of heavy metal pollution in the Upper Gulf of Thailand. *Mar. Pollut. Bull.*, **14**, 465–469.

Hunt, E.G. and A.I. Bischoff (1960). Inimical effects on wildlife of periodic DDD application to Clear Lake. *Calif. Fish and Game*, **46**, 91–106.

Hunter, D. and D.S. Russell (1954). Focal cerebral and cerebellar atrophy in a human subject due to organic mercury compounds. *J. Neurol. Neurosurg. Psychiat.*, **17**, 235–241.

Hurlbert, S.N. (1971). The non-concept of species diversity. *Ecology*, **53**, 577–586.

Huschenbeth, E. and U. Harms (1975). On the accumulation of organochlorine

pesticides, PCB and certain heavy metals in fish and shellfish from Thai coastal and inland waters. *Arch. FischWiss.*, **26**, 109–122.
Hutcheson, M.S. (1974). The effect of temperature and salinity on cadmium uptake by the blue crab, *Callinectes sapidus*. *Chesapeake Sci.*, **15**, 237–241.
Hutzinger, O., S. Safe and V. Zitko (1974). *The chemistry of polychlorinated biphenyls*. CRC Press, Cleveland, Ohio.
Icely, J.D. and J.A. Nott (1980). Accumulation of copper in the 'hepatopancreatic' caeca of *Corophium volutator* (Crustacea: Amphipoda). *Mar. Biol.*, **57**, 193–199.
INSAG (1986). *Post accident review meeting on the Chernobyl accident. Summary report*. International Atomic Energy Agency, Vienna.
Ip, H.M.H. (1983). Breast milk contaminants in Hong Kong. *Bull. Hong Kong Med. Assoc.*, **35**, 16pp.
Ip, H.M.H. and D.J.H. Phillips (1989). Organochlorine chemicals in human breast milk in Hong Kong. *Arch. Environ. Contam. Toxicol.*,, **18**, 490–494.
Ireland, M.P. and R.J. Wootton (1977). Distribution of lead, zinc, copper and manganese in the marine gastropods, *Thais lapillus* and *Littorina littorea*, around the coast of Wales. *Environ. Pollut.*, **12**, 27–41.
ISSG (1990). *The Irish Sea: an environmental review*. Report of the Irish Sea Study Group. Liverpool University Press, Liverpool.
IUCN (1983). *Impact of oil pollution on living resources*. Working Group on oil pollution of the IUCN Commission on ecology in co-operation with the World Wildlife Fund. Commission on ecology papers, No.4. International Union for the Conservation of Nature and Natural Resources, Geneva.
Jeffree, R.A. (in press). A radioecological approach to problems of bioaccumulation. In *Proceedings of the bioaccumulation workshop*, 20–22 Feb. 1991, Sydney, Australia. The Water Board, Sydney.
Jenkins, K.D. and A.Z. Mason (1988). Relationships between subcellular distributions of cadmium and perturbations in reproduction in the polychaete *Neanthes arenaceodentata*. *Aquatic Toxicol.*, **12**, 229–244.
Jenkins, K.D. and B.M. Sanders (1986). Relationships between free cadmium ion activity in seawater, cadmium accumulation and subcellular distribution, and growth in polychaetes. *Environ. Health Perspect.*, **65**, 205–211.
Jensen, A.A. (1983). Chemical contaminants in human milk. *Residue Reviews.*, **89**, 1–128.
Jensen, A.L. (1984). PCB uptake and transfer to humans by lake trout. *Environ. Pollut. Ser. A.*, **34**, 73–82.
Jensen, A.L., S.A. Spigarelli and M.M. Thommes (1982). PCB uptake by five species of fish in Lake Michigan, Green Bay of Lake Michigan, and Cayuga Lake, New York. *Can. J. Fish. Aquat. Sci.*, **39**, 700–709.
Jensen, S. (1972). The PCB story. *Ambio*, **1**, 123–131.
Jernelöv, A. (1969). Conversion of mercury compounds. In *Chemical fallout: current research on persistent pesticides*, ed. M.W. Miller and G.G. Berg, Charles C. Thomas, Springfield, Illinois, pp. 68–74.
Jernelöv, A. (1974). Heavy metals, metalloids, and synthetic organics. In *The sea. Vol. 5: Marine chemistry*, ed. E.D. Goldberg, John Wiley and Sons, New York, pp. 799–815.
Jiminez, B.D., A. Oikari, S.M. Adams, D.E. Hinton and J.F. McCarthy (1990). Hepatic enzymes as biomarkers: interpreting the effects of environmental, phy-

siological and toxicological variables. In *Biomarkers of environmental contamination*, ed. J.F. McCarthy and L.R. Shugart, Lewis Publishers, CRC Press, Boca Raton, Florida, pp.123–143.

Johansson, C., D. Cain and S.N. Luoma (1986). Variability in the fractionation of Cu, Ag and Zn among cytosolic proteins in the bivalve *Macoma balthica*. *Mar. Ecol. Prog. Ser.*, **28**, 87–97.

Johnels, A.G., T. Westermark, W. Berg, P.I. Persson and B. Sjöstrand (1967). Pike (*Esox lucius* L.) and some other aquatic organisms in Sweden as indicators of mercury contamination in the environment. *Oikos*, **18**, 323–333.

Johnson, B.T., C.R. Saunders, H.O. Sanders and R.S. Campbell (1971). Biological magnification and degradation of DDT and aldrin by freshwater invertebrates. *J. Fish. Res. Board Can.*, **28**, 705–709.

Jones, K.C. (1986). The distribution and partitioning of silver and other heavy metals in sediments associated with an acid mine drainage stream. *Environ. Pollut. Ser. B.*, **12**, 249–263.

Jones, W.G. and K.F. Walker (1979). Accumulation of iron, manganese, zinc and cadmium by the Australian freshwater mussel *Velesunio ambiguus* (Philippi) and its potential as a biological monitor. *Aust. J. mar. Freshwat. Res.*, **30**, 741–751.

Jorgensen, C.B. (1990). *Bivalve filter feeding: hydrodynamics, bioenergetics, physiology and ecology*. Olsen and Olsen, Denmark.

Justic, D. (1987). Long-term eutrophication of the Northern Adriatic Sea. *Mar. Pollut. Bull.*, **18**, 281–284.

Kaiser, I.I., P.A. Young and J.D. Johnson (1979). Chronic exposure of trout to waters with naturally high selenium levels: effects on transfer RNA methylation. *J. Fish. Res. Board Can.*, **36**, 689–694.

Kannan, N., S. Tanabe, R. Tatsukawa and D.J.H. Phillips (1989). Persistency of highly toxic coplanar PCBs in aquatic ecosystems: uptake and release kinetics of coplanar PCBs in green-lipped mussels (*Perna viridis* Linnaeus). *Environ. Pollut.*, **56**, 65–76.

Karbe, L., T. Borchardt, R. Dannenberg and E. Meyer (1984). Ten years of experience using marine and freshwater hydroid bioassays. In *Ecotoxicological testing for the marine environment*, Vol. 2, ed. G. Persoone, E. Jaspers and C. Claus, State University Ghent and Institute of Marine Scientific Research, Bredene, Belgium, pp. 99–129.

Kawai, S., M. Fukushima, N. Miyazaki and R. Tatsukawa (1988). Relationship between lipid composition and organochlorine levels in the tissues of striped dolphin. *Mar. Pollut. Bull.*, **19**, 129–133.

Keith, L.H. and W.A. Telliard (1979). Priority pollutants. I — A perspective view. *Environ. Sci. Technol.*, **13**, 416–423.

Kellogg, R.L. and R.V. Bulkley (1976). Seasonal concentrations of dieldrin in water, channel catfish, and catfish-food organisms, Des Moines River, Iowa — 1971–73. *Pestic. Monit. J.*, **9**, 186–194.

Kelso, J.R.M. and R. Frank (1974). Organochlorine residues, mercury, copper and cadmium in yellow perch, white bass and smallmouth bass, Long Point Bay, Lake Erie. *Trans. Amer. Fish. Soc.*, **103**, 577–581.

Kelso, J.R.M., H.R. MacCrimmon and D.J. Ecobichon (1970). Seasonal insecticide residue changes in tissues of fish from the Grand River, Ontario. *Trans. Amer. Fish. Soc.*, **99**, 423–426.

Khan, M.A.Q., A. Kamal, R.J. Wolin and J. Runnels (1972). *In vivo* and *in vitro* epoxidation of aldrin by aquatic food chain organisms. *Bull. Environ. Contam. Toxicol.*, **8**, 219–228.

Kingston, P.F. (1987). Field effects of platform discharges on benthic macrofauna. *Phil. Trans. Roy. Soc. Lond. B.*, **316**, 545–565.

Klauda, R.J., T.H. Peck and G.K. Rice (1981). Accumulation of polychlorinated biphenyls in Atlantic tomcod (*Microgadus tomcod*) collected from the Hudson River estuary, New York. *Bull. Environ. Contam. Toxicol.*, **27**, 829–835.

Klerks, P.L. and J.S. Levington (1989). Rapid evolution of metal resistance in a benthic oligochaete inhabiting a metal-polluted site. *Biol. Bull. Woods Hole, Mass.*, **176**, 135–141.

Klerks, P.L. and J.S. Weis (1987). Genetic adaptation to heavy metals in aquatic organisms: a review. *Environ. Pollut.*, **45**, 173–205.

Klumpp, D.W. and C. Burdon-Jones (1982). Investigations of the potential of bivalve molluscs as indicators of heavy metal levels in tropical marine waters. *Aust. J. mar. Freshwat. Res.*, **33**, 285–300.

Knap, A.H., K.A. Burns, R. Dawson, M. Ehrhardt and K.H. Palmork (1986). Dissolved/dispersed hydrocarbons, tarballs and the surface microlayer: experiences from an IOC/UNEP workshop in Bermuda, December, 1984. *Mar. Pollut. Bull.*, **17**, 313–319.

Knauer, G.A. (1976). Immediate industrial effects on sediment mercury concentrations in a clean coastal environment. *Mar. Pollut. Bull.*, **7**, 112–115.

Knauer, G.A. (1977). Immediate industrial effects on sediment metals in a clean coastal environment. *Mar. Pollut. Bull.*, **8**, 249–254.

Knauer, G.A., J.H. Martin and R.M. Gordon (1982). Cobalt in north-east Pacific waters. *Nature, Lond.*, **297**, 49–51.

Knezovich, J.P., F.L. Harrison and R.G. Wilhelm (1987). The bioavailability of sediment-sorbed organic chemicals: a review. *Water, Air Soil Pollut.*, **32**, 233–245.

Kobayashi, J. (1971). Relation between the 'Itai-Itai' disease and the pollution of river water by cadmium from a mine. In *Advances in water pollution research; proceedings of the 5th International Conference*, San Francisco, pp.1–7.

Kobayashi, N. (1984). Marine ecotoxicological testing with echinoderms. In *Ecotoxicological testing for the marine environment*, Vol. 1. ed. G. Persoone, E. Jaspers and C. Claus, State University Ghent and Institute of Marine Scientific Research, Bredene, Belgium, pp. 341–405.

Koide, M., P.W. Williams and E.D. Goldberg (1981). Am-241/Pu-239 + -240 ratios in the marine environment. *Mar. Environ. Res.*, **5**, 241–246.

Koide, M., D.S. Lee and E.D. Goldberg (1982). Metal and transuranic records in mussel shells, byssal threads and tissues. *Estuar. cstl. Shelf Sci.*, **15**, 679–695.

Krom, M.D., H. Hornung and Y. Cohen (1990). Determination of the environmental capacity of Haifa Bay with respect to the input of mercury. *Mar. Pollut. Bull.*, **21**, 349–354.

Kuratsune, M. (1980). The Yusho rice oil incident. In *Halogenated biphenyls, terphenyls, naphthalenes, dibenzodioxins and related products*, ed. R.D. Kimbrough, Elsevier/North Holland, Amsterdam, pp. 287–302.

Kurelec, B., A. Garg, S. Krča and R.C. Gupta (1990). DNA adducts in marine mussel *Mytilus galloprovincialis* living in polluted and unpolluted environments.

In *Biomarkers of environmental contamination*, ed. J.F. McCarthy and L.R. Shugart, Lewis Publishers, CRC Press, Boca Raton, Florida, pp. 217–227.

Lack, T.J. and D. Johnson (1985). Assessment of the biological effects of sewage sludge at a licensed site off Plymouth. *Mar. Pollut. Bull.*, **16**, 147–152.

Lande, E. (1977). Heavy metal pollution in Trondheimsfjorden, Norway, and the recorded effects on the fauna and flora. *Environ. Pollut.*, **12**, 187–198.

Langston, W.J. (1978a). Accumulation of polychlorinated biphenyls in the cockle *Cerastoderma edule* and the tellin *Macoma balthica*. *Mar. Biol.*, **45**, 265–272.

Langston, W.J. (1978b). Persistence of polychlorinated biphenyls in marine bivalves. *Mar. Biol.*, **46**, 35–40.

Langston, W.J. (1980). Arsenic in U.K. estuarine sediments and its availability to deposit-feeding bivalves. *J. mar. biol. Ass. U.K.*, **60**, 869–881.

Langston, W.J. (1982). Distribution of mercury in British estuarine sediments and its availability to deposit-feeding bivalves. *J. mar. biol. Ass. U.K.*, **62**, 667–684.

Langston, W.J. (1986). Metals in sediments and benthic organisms in the Mersey Estuary. *Estuar. cstl. Shelf Sci.*, **23**, 239–261.

Langston, W.J. (1990). Toxic effects of metals and the incidence of metal pollution in marine ecoystems. In *Heavy metals in the marine environment*, ed. R.W. Furness and P.S. Rainbow, CRC Press, Boca Raton, Florida, pp. 101–122.

Langston, W.J. and M. Zhou (1987). Cadmium accumulation, distribution and elimination in the bivalve *Macoma balthica:* neither metallothionein nor metallothionein-like proteins are involved. *Mar. Environ. Res.*, **21**, 225–237.

Langston, W.J., M.J. Bebianno and M. Zhou (1989). A comparison of metal-binding proteins and cadmium metabolism in the marine molluscs *Littorina littorea* (Gastropoda), *Mytilus edulis* and *Macoma balthica* (Bivalvia). *Mar. Environ. Res.*, **28**, 195–200.

Langston, W.J., G.W. Bryan, G.R. Burt and P.E. Gibbs (1990). Assessing the impact of tin and TBT in estuaries and coastal regions. *Functional Ecol.*, **4**, 433–443.

Latouche, Y.D. and M.C. Mix (1982). The effects of depuration, size and sex on trace metal levels in Bay mussels. *Mar. Pollut. Bull.*, **13**, 27–29.

Laubier, L. (1978). The *Amoco Cadiz* oil spill — lines of study and early observations. *Mar. Pollut. Bull.*, **9**, 285–287.

Lauenstein, G.G., A. Robertson and T.P. O'Connor (1990). Comparison of trace metal data in mussels and oysters from a mussel watch programme of the 1970s with those from a 1980s programme. *Mar. Pollut. Bull.*, **21**, 440–447.

Lavie, B. and E. Nevo (1986). Genetic selection of homozygote allozyme genotypes in marine gastropods exposed to cadmium pollution. *Sci. Total Environ.*, **57**, 91–98.

Lavigne, D.M. and O.J. Schmitz (1990). Global warming and increasing population densities: a prescription for seal plagues. *Mar. Pollut. Bull.*, **21**, 280–284.

Law, R.J., M. Marchand, G. Dahlmann and T.W. Fileman (1987). Results of two bilateral comparisons of the determination of hydrocarbon concentrations in coastal seawater by fluorescence spectroscopy. *Mar. Pollut. Bull.*, **18**, 486–489.

Law, R.J., C.R. Allchin and J. Harwood (1989). Concentrations of organochlorine compounds in the blubber of seals from eastern and north-eastern England, 1988. *Mar. Pollut. Bull.*, **20**, 110–115.

Lederman, T.C. and G.-Y. Rhee (1982). Bioconcentration of a hexachlorobiphenyl in Great Lakes planktonic algae. *Can. J. Fish. Aquat. Sci.*, **39**, 380–387.

Lee, M.C., E.S.K. Chian and R.A. Griffin (1979). Solubility of polychlorinated biphenyls and capacitor fluid in water. *Water Res.*, **13**, 1249–1257.

Lemly, A.D. (1982). Response of juvenile centrarchids to sublethal concentrations of waterborne selenium. I. Uptake, tissue distribution, and retention. *Aquatic Toxicol.*, **2**, 235–252.

Livingstone, D.R. (1985). Responses of the detoxication/toxication enzyme systems of molluscs to organic pollutants and xenobiotics. *Mar. Pollut. Bull.*, **16**, 158–164.

Livingstone, D.R. (1988). Responses of the microsomal NADPH-cytochrome c reductase activity and cytochrome P-450 in digestive glands of *Mytilus edulis* and *Littorina littorea* to environmental and experimental exposure to pollutants. *Mar. Ecol. Prog. Ser.*, **46**, 37–43.

Livingstone, D.R., P. Garcia Martinez, X. Michel, J.F. Narbonne, S. O'Hara, D. Ribera and G.W. Winston (1990). Oxyradical production as a pollution-mediated mechanism of toxicity in the common mussel, *Mytilus edulis* L., and other molluscs. *Functional Ecol.*, **4**, 415–424.

Lobel, P.B. (1986). Role of the kidney in determining the whole soft tissue zinc concentration of individual mussels (*Mytilus edulis*). *Mar. Biol.*, **92**, 355–364.

Lobel, P.B. (1987a). Short-term and long-term uptake of zinc by the mussel, *Mytilus edulis:* a study in individual variability. *Arch. Environ. Contam. Toxicol.*, **16**, 723–732.

Lobel, P.B. (1987b). Intersite, intrasite and inherent variability of the whole soft tissue zinc concentrations of individual mussels *Mytilus edulis:* importance of the kidney. *Mar. Environ. Res.*, **27**, 59–71.

Lobel, P.B. (1987c). Inherent variability in the ratio of zinc to other elements in the kidney of the mussel *Mytilus edulis*. *Comp. Biochem. Physiol.*, **87C**, 47–50.

Lobel, P.B. and D.A. Wright (1983). Frequency distribution of zinc in *Mytilus edulis* (L.). *Estuaries.*, **6**, 154–159.

Lobel, P.B., P. Mogie, D.A. Wright and B.L. Wu (1982). Metal accumulation in four molluscs. *Mar. Pollut. Bull.*, **13**, 170–174.

Lobel, P.B., S.P. Belkhode, S.E. Jackson and H.P. Longerich (1989). A universal method for quantifying and comparing the residual variability of element concentrations in biological tissues using 25 elements in the mussel *Mytilus edulis* as a model. *Mar. Biol.*, **102**, 513–518.

Lobel, P.B., S.P. Belkhode, S.E. Jackson and H.P. Longerich (1990). Recent taxonomic discoveries concerning the mussel *Mytilus:* implications for biomonitoring. *Arch. Environ. Contam. Toxicol.*, **19**, 508–512.

Lobel, P.B., S.P. Belkhode, S.E. Jackson and H.P. Longerich (1991). Sediment in the intestinal tract: a potentially serious source of error in aquatic biological monitoring programs. *Mar. Environ. Res.*, **31**, 163–174.

Lockhart, W.L., J.F. Uthe, A.R. Kenney and P.M. Mehrle (1972). Methylmercury in northern pike (*Esox lucius*): distribution, elimination, and some biochemical characteristics of contaminated fish. *J. Fish. Res. Board Can.*, **29**, 1519–1523.

Loganathan, B.G., S. Tanabe, M. Goto and R. Tatsukawa (1989). Temporal trends of organochlorine residues in lizard goby *Rhinogobius flumineus* from the River Nagaragawa, Japan. *Environ. Pollut.*, **62**, 237–251.

Loganathan, B.G., S. Tanabe, H. Tanaka, S. Watanabe, N. Miyazaki, M. Amano and R. Tatsukawa (1990). Comparison of organochlorine residue levels in the striped dolphin from western North Pacific, 1978–79 and 1986. *Mar. Pollut. Bull.*, **21**, 435–439.

Long, E.R. and M.F. Buchman (1990). A comparative evaluation of selected measures of biological effects of exposure of marine organisms to toxic chemicals. In *Biomarkers of environmental contamination*, ed. J.F. McCarthy and L.R. Shugart, Lewis Publishers, CRC Press, Boca Raton, Florida, pp. 355–394.

Long, E.R. and P.M. Chapman (1985). A Sediment Quality Triad: measures of sediment contamination, toxicity and infaunal community composition in Puget Sound. *Mar. Pollut. Bull.*, **16**, 405–415.

Loring, D.H. (1981). Potential availability of metals in eastern Canadian estuarine and coastal sediments. *Rapp. P.-V. Réun. Cons. Int. Explor. Mer.*, **181**, 93–101.

Lorz, H.W. and B.P. McPherson (1976). Effects of copper or zinc in freshwater on the adaptation to sea water and ATPase activity, and the effects of copper on migratory disposition of coho salmon (*Oncorhynchus kisutch*). *J. Fish. Res. Board Can.*, **33**, 2023–2030.

Lovegrove, S.M. and B. Eddy (1982). Uptake and accumulation of zinc in juvenile rainbow trout, *Salmo gairdneri*. *Env. Biol. Fish.*, **7**, 285–289.

Lowe, D.M. and M.N. Moore (1978). Cytology and quantitative cytochemistry of a proliferative atypical hemocytic condition in *Mytilus edulis* (Bivalvia, Mollusca). *J. natn. Cancer Inst.*, **60**, 1455–1459.

Lunsford, C.A. and C.R. Blem (1982). Annual cycle of Kepone residue and lipid content of the estuarine clam, *Rangia cuneata*. *Estuaries*, **5**, 121–130.

Luoma, S.N. (1977a). Dynamics of biologically available mercury in a small estuary. *Estuar. cstl. Mar. Sci.*, **5**, 643–652.

Luoma, S.N. (1977b). Detection of trace contaminant effects in aquatic ecosystems. *J. Fish. Res. Board Can.*, **34**, 436–439.

Luoma, S.N. (1983). Bioavailability of trace metals to aquatic organisms — a review. *Sci. Total Environ.*, **28**, 1–22.

Luoma, S.N. (1989). Can we determine the biological availability of sediment-bound trace elements? *Hydrobiologia*, **176/177**, 379–396.

Luoma, S.N. (1990). Processes affecting metal concentrations in estuarine and coastal marine sediments. In *Heavy metals in the marine environment*, ed. R.W. Furness and P.S. Rainbow, CRC Press, Boca Raton, Florida, pp. 51–66.

Luoma, S.N. and G.W. Bryan (1978). Factors controlling the availability of sediment-bound lead to the estuarine bivalve *Scrobicularia plana*. *J. mar. biol. Ass. U.K.*, **58**, 793–802.

Luoma, S.N. and G.W. Bryan (1982). A statistical study of environmental factors controlling concentrations of heavy metals in the burrowing bivalve *Scrobicularia plana* and the polychaete *Nereis diversicolor*. *Estuar. cstl. Shelf Sci.*, **15**, 95–108.

Luoma, S.N., G.W. Bryan and W.J. Langston (1982). Scavenging of heavy metals from particulates by brown seaweed. *Mar. Pollut. Bull.*, **13**, 394–396.

Luoma, S.N., D. Cain and C. Johansson (1985). Temporal fluctuations of silver, copper and zinc in the bivalve *Macoma balthica* at five stations in South San Francisco Bay. *Hydrobiologia*, **129**, 109–120.

Luoma, S.N., R. Dagovitz and E. Axtmann (1990). Temporally intensive study of trace metals in sediments and bivalves from a large river-esturanne system: Suisun Bay/Delta in San Francisco Bay. *Sci. Total Environ.*, 97/98, 685–712.

Lyman, L.D., W.A. Tompkins and J.A. McCann (1968). Massachusetts pesticide monitoring study. *Pestic. Monit. J.*, **2**, 109–122.

Lynch, T.R., C.J. Popp and G.Z. Jacobi (1988). Aquatic insects as environmental monitors of trace metal contamination: Red River, New Mexico. *Water, Air, Soil Pollut.*, **42**, 19–31.

MacCrimmon, H.R., C.D. Wren and B.L. Gots (1983). Mercury uptake by lake trout, *Salvelinus namaycush*, relative to age, growth, and diet in Tadenac Lake with comparative data from other PreCambrian Shield lakes. *Can. J. Fish. Aquat. Sci.*, **40**, 114–120.

Macek, K.J. and S. Korn (1970). Significance of the food chain in DDT accumulation by fish. *J. Fish. Res. Board Can.*, **27**, 1496–1498.

Macek, K.J., C.R. Rodgers, D.L. Stalling and S. Korn (1970). The uptake, distribution and elimination of dietary ^{14}C-DDT and ^{14}C-dieldrin in rainbow trout. *Trans. Amer. Fish. Soc.*, **99**, 689–695.

Mackay, D. and A.I. Hughes (1984). Three-parameter equation describing the uptake of organic compounds by fish. *Environ. Sci. Technol.*, **18**, 439–444.

Mackay, N.J., R.J. Williams, J.L. Kacprzac, M.N. Kazacos, A.J. Collins and E.H. Auty (1975). Heavy metals in cultivated oysters (*Crassostrea commercialis = Saccostrea cucullata*) from the estuaries of New South Wales. *Aust. J. mar. Freshwat. Res.*, **26**, 31–46.

Maclean, J.L. (1984). Indo-Pacific red tide occurrences, 1972–1984. In *Toxic red tides and shellfish toxicity in South-east Asia*, ed. A.W. White, M. Anraku and K.K. Hooi. Southeast Asian Fisheries Development Centre, Singapore, and International Development Research Centre, Ottowa. pp. 92–98.

Maclean, J.L. (1989). Indo-Pacific red tides, 1985–1988. *Mar. Pollut. Bull.*, **20**, 304–310.

Maclean, J.L. and R.M. Temprosa (1984). A bibliography on toxic red tides and shellfish poisoning related to the Indo-Pacific region. In *Toxic red tides and shellfish toxicity in South-east Asia*, ed. A.W. White, M. Anraku and K.K. Hooi, Southeast Asian Fisheries Development Centre, Singapore, and International Development Research Centre, Ottowa, pp. 99–102.

Maclean, J.L. and A.W. White (1985). Toxic dinoflagellate blooms in Asia: a growing concern. In *Toxic dinoflagellates*, ed. D.M. Anderson, A.W. White and D.G. Baden, Elsevier, New York, pp. 517–520.

Maclean, R.O. and A.K. Jones (1975). Studies of tolerance to heavy metals in the flora of the rivers Ystwyth and Clarach in Wales. *Freshwat. Biol.*, **5**, 431–444.

MacLeod, J.C. and E. Pessah (1973). Temperature effects on mercury accumulation, toxicity, and metabolic rate in rainbow trout (*Salmo gairdneri*). *J. Fish. Res. Board Can.*, **30**, 485–492.

Maher, W.A. and R.H. Norris (1990). Water quality assessment programs in Australia: deciding what to measure, and how and where to use bioindicators. *Environ. Monit. Assessment.*, **14**, 115–130.

Mance, G. (1987). *Pollution threat of heavy metals in aquatic environments*. Elsevier Applied Science, New York.

Mangum, C.P. (1979). A note on blood and water mixing in large marine gastropods. *Comp. Biochem. Physiol.*, **63A**, 389–391.

Manley, A.R. (1983). The effects of copper on the behaviour, respiration, filtration and ventilation activity of *Mytilus edulis*. *J. mar. biol. Ass. U.K.*, **63**, 205–222.

Manly, R. and W.O. George (1977). The occurrence of some heavy metals in populations of the freshwater mussel *Anodonta anatina* (L.) from the River Thames. *Environ. Pollut.*, **14**, 139–154.

Mantel, L.H. and Farmer, L.L. (1983). Osmotic and ionic regulation. In *The biology of Crustacea*, Vol. 5, ed. L.H. Mantel, Academic Press, New York, pp. 53–161.

Mantoura, R.F.C., A. Dickson and J.P. Riley (1978). The complexation of metals with humic materials in natural waters. *Estuar. cstl. Mar. Sci.*, **6**, 387–408.

Marchand, M., D. Vas and E.K. Duursma (1976). Levels of PCBs and DDTs in mussels from the N.W. Mediterranean. *Mar. Pollut. Bull.*, **7**, 65–69.

Margerison, T., M. Wallace and D. Hallenstein (1980). *The superpoison*. MacMillan, London.

Martin, H. (Ed.) (1963). *Insecticide and fungicide handbook for crop protection*. Blackwell Scientific Publications, Oxford.

Martin, J.H., G.A. Knauer and R.M. Gordon (1983). Silver distributions and fluxes in north-east Pacific waters. *Nature, Lond.*, **305**, 306–309.

Martin, M. (1985). State mussel watch: toxics surveillance in California. *Mar. Pollut. Bull.*, **16**, 140–146.

Martin, M. and W. Castle (1984). Petrowatch: petroleum hydrocarbons, synthetic organic compounds, and heavy metals in mussels from the Monterey Bay area of central California. *Mar. Pollut. Bull.*, **15**, 259–266.

Martin, M. and B.J. Richardson (1991). Long term contaminant monitoring: views from southern and northern hemisphere perspectives. *Mar. Pollut. Bull.*, **22**, 533–537.

Martin, M., K.E. Osborn, P. Billig and N. Glickstein (1981). Toxicities of ten metals to *Crassostrea gigas* and *Mytilus edulis* embryos and *Cancer magister* larvae. *Mar. Pollut. Bull.*, **12**, 305–308.

Martin, M., M.D. Stephenson, D.R. Smith, E.A. Gutierrez-Galindo and G.F. Munoz (1988). Use of silver in mussels as a tracer of domestic wastewater discharge. *Mar. Pollut. Bull.*, **19**, 512–520.

Martin, M.H. and P.J. Coughtrey (1982). *Biological monitoring of heavy metal pollution: land and air*. Applied Science Publishers, London.

Martineau, D., A. Lagace, P. Béland, R. Higgins, D. Armstrong and L.R. Shugart (1988). Pathology of stranded Beluga whales (*Delphinapterus leucas*) from St. Lawrence estuary, Quebec, Canada. *J. Comp. Pathol.*, **98**, 287–311.

Mason, A.Z. and J.A. Nott (1981). The role of intracellular biomineralized granules in the regulation and detoxification of metals in gastropods with special reference to the marine prosobranch *Littorina littorea*. *Aquatic Toxicol.*, **1**, 239–256.

Mason, A.Z., K.D. Jenkins and P.A. Sullivan (1988). Mechanisms of trace metal accumulation in the polychaete *Neanthes arenaceodentata*. *J. mar. biol. Ass. U.K.*, **68**, 61–80.

Mayer, L.M. and L.K. Fink (1980). Granulometric dependence of chromium

accumulation in estuarine sediments in Maine. *Estuar. cstl. Mar. Sci.*, **11**, 491–503.
McCarthy, J.F. (1990). Concluding remarks: implementation of a biomarker-based environmental monitoring program. In *Biomarkers of environmental contamination*, ed. J.F. McCarthy and L.R. Shugart, Lewis Publishers, CRC Press, Boca Raton, Florida, pp. 429–439.
McClurg, T.P. (1984). Trace metals and chlorinated hydrocarbons in Ross seals from Antarctica. *Mar. Pollut. Bull.*, **15**, 384–389.
McEvoy, E.G. (1988). Heavy metals in marine nemerteans. *Hydrobiologia.*, **156**, 135–143.
McGreer, E.R. (1982). Factors affecting the distribution of the bivalve, *Macoma balthica* (L.) on a mudflat receiving sewage effluent, Fraser River estuary, British Columbia. *Mar. Environ. Res.*, **7**, 131–149.
McLeod, M.J. (1986). Electrophoretic variation in North American *Corbicula.* In *Proceedings of the 2nd international Corbicula symposium, Special edition No. 2*, ed. J.C. Britton and R. Prezant. *Am. Malac. Bull.*, pp. 125–133.
McLeese, D.W., C.D. Metcalfe and D.S. Pezzack (1980). Uptake of PCBs from sediment by *Nereis virens* and *Crangon septemspinosa. Arch Environ. Contam. Toxicol.*, **9**, 507–513.
McLusky, D.S., V. Bryant and R. Campbell (1986). The effects of temperature and salinity on the toxicity of heavy metals to marine and estuarine invertebrates. *Oceanogr. mar. Biol. Ann. Rev.*, **24**, 481–520.
Merlini, M. and G. Pozzi (1977). Lead and freshwater fishes: Part 1 — Lead accumulation and water pH. *Environ. Pollut.*, **12**, 167–172.
Mileikovsky, S.A. (1971). Types of larval development in marine bottom invertebrates, their distribution and ecological significance: a re-evaluation. *Mar. Biol.*, **10**, 193–213.
Miller, G.J. and D.W. Connell (1982). Global production and fluxes of petroleum and recent hydrocarbons. *Int. J. Environ. Studies.*, **19**, 273–280.
Miller, M.M., S.P. Wasik, G.-L. Huang, W.-Y. Shiu and D. Mackay (1985). Relationships between octanol-water partition coefficient and aqueous solubility. *Environ. Sci. Technol.*, **19**, 522–529.
Millington, P.J. and K.F. Walker (1983). Australian freshwater mussel *Velesunio ambiguus* (Philippi) as a biological monitor for zinc, iron and manganese. *Aust. J. mar. Freshwat. Res.*, **34**, 873–892.
Miramand, P., P. Germain and J.P. Trilles (1989). Histo-autoradiographic localisation of americium (^{241}Am) in tissues of European lobster *Homarus gammarus* and edible crab *Cancer pagurus* after uptake from labelled seawater. *Mar. Ecol. Prog. Ser.*, **52**, 217–225.
Mochiike, A., T. Matsuo, H. Kanamori, N. Hoshita and I. Sakamoto (1986). Determination of PCQs by HPLC and its application to the analysis of Yusho patient blood and toxic rice oil and to the distribution of synthetic PCQs in mice. *Chemosphere*, **15**, 599–606.
Modin, J.C. (1968). Estuarine monitoring of Mollusca for organochlorine pesticides in the California coastal environment. *Proc. Symp. Moll., Mar. Biol. Ass. India.*, **1968**, 40–41.
Modin, J.C. (1969). Chlorinated hydrocarbon pesticides in California bays and estuaries. *Pestic. Monit. J.*, **3**, 1–7.

Moilanen, R., H. Pyysalo, K. Wickström and R. Linko (1982). Time trends of chlordane, DDT, and PCB concentrations in pike (*Esox lucius*) and Baltic herring (*Clupea harengus*) in the Turku Archipelago, northern Baltic Sea for the period 1971–1982. *Bull. Environ. Contam. Toxicol.*, **29**, 334–340.

Moore, J.W. and D.J. Sutherland (1980). Mercury concentrations in fish inhabiting two polluted lakes in northern Canada. *Water Res.*, **14**, 903–907.

Moore, M.N. (1980). Cytochemical determination of cellular responses to environmental stressors in marine organisms. *Rapp. P.-V. Réun. Cons. Int. Explor. Mer.*, **179**, 7–15.

Moore, M.N. (1985). Cellular responses to pollutants. *Mar. Pollut. Bull.*, **16**, 134–139.

Moore, M.N. (1988a). Cytochemical responses of the lysosomal system and NADPH-ferrihemoprotein reductase in molluscan digestive cells to environmental and experimental exposure to xenobiotics. *Mar. Ecol. Prog. Ser.*, **46**, 81–89.

Moore, M.N. (1988b). Cellular and histopathological effects of a pollutant gradient — summary. *Mar. Ecol. Prog. Ser.*, **46**, 109–110.

Moore, M.N., J. Widdows, J.J. Cleary, R.K. Pipe, P.N. Salkeld, P. Donkin, S.V. Farrar, S.V. Evans and P.E. Thomson (1984). Responses of the mussel *Mytilus edulis* to copper and phenanthrene: interactive effects. *Mar. Environ. Res.*, **14**, 167–183.

Moore, M.N., D.R. Livingstone, J. Widdows, D.M. Lowe and R.K. Pipe (1987a). Molecular, cellular and physiological effects of oil-derived hydrocarbons on molluscs and their use in impact assessment. *Phil. Trans. Roy. Soc. Lond. B.*, **316**, 603–623.

Moore, M.N., R.K. Pipe, S.V. Farrar, S. Thomson and P. Donkin (1987b). Lysosomal and microsomal responses to oil-derived hydrocarbons in *Littorina littorea*. In *Oceanic processes in marine pollution, Vol. 1. Biological processes and wastes in the ocean*, ed. J.M. Capuzzo and D.R. Kester, R.E. Krieger Publ. Co., Malabar, Florida, pp. 89–96.

Moore, P.G. and P.S. Rainbow (1984). Ferritin crystals in the gut caeca of *Stegocephaloides christianiensis* Boeck and other Stegocephalidae (Amphipoda: Gammaridea): a functional interpretation. *Phil. Trans. Roy. Soc. Lond. B.*, **306**, 219–245.

Moore, P.G. and P.S. Rainbow (1987). Copper and zinc in an ecological series of talitroidean Amphipoda (Crustacea). *Oecologia.*, **73**, 120–126.

Morales-Alamo, R. and D.S. Haven (1983). Uptake of Kepone from sediment suspensions and subsequent loss by the oyster *Crassostrea virginica*. *Mar. Biol.*, **74**, 187–201.

Moriarty, F. (1988). *Ecotoxicology: the study of pollutants in ecosystems*, 2nd edn., Academic Press, London.

Moriarty, F. and M.C. French (1977). Mercury in waterways that drain into the Wash, eastern England. *Water Res.*, **11**, 367–372.

Moriarty, F., H.M. Hanson and P. Freestone (1984). Limitations of body burden as an index of environmental contamination: heavy metals in fish *Cottus gobio* L. from the River Ecclesbourne, Derbyshire. *Environ. Pollut. Ser. A.*, **34**, 297–320.

Morris, R.L. and L.G. Johnson (1971). Dieldrin levels in fish from Iowa streams. *Pestic. Monit. J.*, **5**, 12–16.

Morris, R.J., R.J. Law, C.R. Allchin, C.A. Kelly and C.F. Fileman (1989). Metals and organochlorines in dolphins and porpoises of Cardigan Bay, Wales. *Mar. Pollut. Bull.*, **20**, 512–523.

Mortimer, D.C. (1985). Freshwater aquatic macrophytes as heavy metal monitors — the Ottowa River experience. *Environ. Monit. Assessment*, **5**, 311–323.

Morton, B.S. (1985). Marine pollution induced environmental changes in Hong Kong — the Tolo Harbour case study. In *Pollution in the urban environment, POLMET 85*, ed. M.W.H. Chan, R.W.M. Hoare, P.R. Holmes, R.J.S. Law and S.B. Reed, Elsevier Applied Science Publishers, London, pp. 548–558.

Morton, B.S. (1989). Pollution of the coastal waters of Hong Kong. *Mar. Pollut. Bull.*, **20**, 310–318.

Mouvet, C. (1984). Accumulation of chromium and copper by the aquatic moss *Fontinalis antipyretica* L. ex Hedw transplanted in a metal-contaminated river. *Environ. Technol. Letts.*, **5**, 541–548.

Mowrer, J., K. Åswald, G. Burgermeister, L. Machado and J. Tarradellas (1982). PCB in a Lake Geneva ecosystem. *Ambio*, **11**, 355–358.

Murphy, P.G. (1970). Effects of salinity on uptake of DDT, DDE and DDD by fish. *Bull. Environ. Contam. Toxicol.*, **5**, 404–407.

Murphy, P.G. (1971). The effect of size on the uptake of DDT from water by fish. *Bull. Environ. Contam. Toxicol.*, **6**, 20–23.

Murray, J., A.B. Thompson, A. Stagg, R. Hardy, K.J. Whittle and P.R. Mackie (1977). On the origin of hydrocarbons in marine organisms. *Rapp. P.-V. Réun. Cons. Int. Explor. Mer.*, **171**, 84–90.

Nadeau, R.J. and R.A. Davis (1976). Polychlorinated biphenyls in the Hudson River (Hudson Falls — Fort Edward, New York State). *Bull. Environ. Contam. Toxicol.*, **16**, 436–444.

NAS (1975). *Petroleum in the marine environment.* National Academy of Sciences, Washington, DC.

NAS (1979). *Polychlorinated biphenyls.* National Academy of Sciences, Washington, DC.

NAS (1980). *The international mussel watch.* National Academy of Sciences, Washington, DC.

Nasu, Y., M. Kugimoto, O. Tanaka and A. Takimoto (1983). Comparative studies on the absorption of cadmium and copper in *Lemna paucicostata*. *Environ. Pollut. Ser. A.*, **32**, 201–209.

Nebeker, A.V., F.A. Puglisi and D.L. DeFoe (1974). Effect of polychlorinated biphenyl compounds on survival and reproduction of the fathead minnow and flagfish. *Trans. Amer. Fish. Soc.*, **103**, 562–568.

Neff, J.M. (1984). Bioaccumulation of organic micropollutants from sediments and suspended particulates by aquatic animals. *Fresenius Z. Anal. Chem.*, **319**, 132–136.

Nevo, E., T. Shimony and M. Libni (1978). Pollution selection of allozyme polymorphisms in barnacles. *Experientia*, **34**, 1562–1564.

Newell, R.C. (1976). Adaptations to intertidal life. In *Adaptation to environment: essays on the physiology of marine animals*, ed. R.C. Newell, Butterworths, London, pp. 1–82.

Newman, M.C. and A.W. McIntosh (1982). The influence of lead in components of a freshwater ecosystem on molluscan tissue lead concentrations. *Aquatic Toxicol.*, **2**, 1–19.

Newman, M.C., J.J. Alberts and V.A. Greenhut (1985). Geochemical factors complicating the use of *aufwuchs* to monitor bioaccumulation of arsenic, cadmium, chromium, copper and zinc. *Water Res.*, **19**, 1157–1165.

Nichols, F., J.E. Cloern, S.N. Luoma and D.H. Peterson (1986). The modification of an estuary. *Science*, **231**, 567–573.

Nieboer, E. and D.H.S. Richardson (1980). The replacement of the nondescript term 'heavy metals' by a biologically and chemically significant classification of metal ions. *Environ. Pollut. Ser. B.*, **1**, 3–26.

Nieboer, E., D.H.S. Richardson and F.D. Tomassini (1978). Mineral uptake and release by lichens: an overview. *Bryologist*, **81**, 226–246.

Nielsen, S.A. and A. Nathan (1975). Heavy metal levels in New Zealand molluscs. *New Zealand J. mar. Freshwat. Res.*, **9**, 467–481.

Niimi, A.J. and C.Y. Cho (1981). Elimination of hexachlorobenzene (HCB) by rainbow trout (*Salmo gairdneri*), and an examination of its kinetics in Lake Ontario salmonids. *Can. J. Fish. Aquat. Sci.*, **38**, 1350–1356.

Niimi, A.J. and B.G. Oliver (1983). Biological half-lives of polychlorinated biphenyl (PCB) congeners in whole fish and muscle of rainbow trout (*Salmo gairdneri*). *Can. J. Fish. Aquat. Sci.*, **40**, 1388–1394.

Niimi, A.J. and V. Palazzo (1985). Temperature effect on the elimination of pentachorophenol, hexachlorobenzene and Mirex by rainbow trout (*Salmo gairdneri*). *Water Res.*, **19**, 205–207.

Nimmo, D.R., A.J. Wilson and R.R. Blackman (1970). Localization of DDT in the body organs of pink and white shrimp. *Bull. Environ. Contam. Toxicol.*, **5**, 333–341.

Nott, J.A. and W.J. Langston (1989). Cadmium and the phosphate granules in *Littorina littorea*. *J. mar. biol. Ass. U.K.*, **69**, 219–227.

Nott, J.A. and A. Nicolaidou (1990). Transfer of metal detoxification along marine food chains. *J. mar. biol. Ass. U.K.*, **70**, 905–912.

Nriagu, J.O. (1988). A silent epidemic of environmental metal poisoning? *Environ. Pollut.*, **50**, 139–161.

Nugegoda, D. and P.S. Rainbow (1987). The effect of temperature on zinc regulation by the decapod crustacean *Palaemon elegans* Rathke. *Ophelia*, **27**, 17–30.

Nugegoda, D. and P.S. Rainbow (1988a). Effect of a chelating agent (EDTA) on zinc uptake and regulation by *Palaemon elegans* (Crustacea: Decapoda). *J. mar. biol. Ass. U.K.*, **68**, 25–40.

Nugegoda, D. and P.S. Rainbow (1988b). Zinc uptake and regulation by the sublittoral prawn *Pandalus montagui*. *Estuar. cstl. Mar. Sci.*, **26**, 619–632.

Nugegoda, D. and P.S. Rainbow (1989a). Effects of salinity changes on zinc uptake and regulation by the decapod crustaceans *Palaemon elegans* and *Palaemonetes varians*. *Mar. Ecol. Prog. Ser.*, **51**, 57–75.

Nugegoda, D. and P.S. Rainbow (1989b). Salinity, osmolality and zinc uptake in *Palaemon elegans* (Crustacea: Decapoda). *Mar. Ecol. Prog. Ser.*, **55**, 149–157.

O'Brien, P., P.S. Rainbow and D. Nugegoda (1990). The effect of the chelating agent EDTA on the rate of uptake of zinc by *Palaemon elegans* (Crustacea: Decapoda). *Mar. Environ. Res.*, **30**, 155–159.

Ochiai, M. and T. Hanya (1976). Alpha- and gamma-BHC in Tamagawa River water, Japan (September 1968 to September 1969). *Environ. Pollut.*, **11**, 161–166.

O'Grady, K.T. and M.I. Abdullah (1985). Mobility and residence of Zn in brown trout *Salmo trutta:* results of environmentally induced change through transfer. *Environ. Pollut. Ser. A.*, **38**, 109–127.

O'Hara, J. (1973). Cadmium uptake by fiddler crabs exposed to temperature and salinity stress. *J. Fish. Res. Board Can.*, **30**, 846–848.

Okazaki, R.K. and M.H. Panietz (1981). Depuration of twelve trace metals in tissues of the oysters *Crassostrea gigas* and *C. virginica. Mar. Biol.*, **63**, 113–120.

O'Keeffe, D.H., J.K. Hardy and R.A. Rao (1984). Cadmium uptake by the water hyacinth: effects of solution factors. *Environ. Pollut. Ser. A.*, **34**, 133–147.

Olsson, M. and S. Jensen (1975). *Pike as the test organism for mercury, DDT and PCB pollution. A study of the contamination in the Stockholm Archipelago.* Report No. 54, Institute of Freshwater Research, Drottningholm, pp. 83–106.

Olsson, M. and L. Reutergårdh (1986). DDT and PCB pollution trends in the Swedish aquatic environment. *Ambio*, **15**, 103–109.

Olsson, M., S. Jensen and L. Reutergårdh (1978). Seasonal variation of PCB levels in fish — an important factor in planning aquatic monitoring programs. *Ambio*, **7**, 66–69.

Orren, M.J., G.A. Eagle, H.F.-K.O. Hennig and A. Green (1980). Variations in trace metal content of the mussel *Choromytilus meridionalis* (Kr.) with season and sex. *Mar. Pollut. Bull.*, **11**, 253–257.

Osterberg, C., J. Pattullo and W. Pearcy (1964). Zinc-65 in euphausiids as related to Columbia River water off the Oregon coast. *Limnol. Oceanogr.*, **9**, 249–257.

Parker, J.G. and F. Wilson (1975). Incidence of polychlorinated biphenyls in Clyde seaweed. *Mar. Pollut. Bull.*, **6**, 46–47.

Pärt, P. and G. Wikmark (1984). The influence of some complexing agents (EDTA and citrate) on the uptake of cadmium in perfused rainbow trout gills. *Aquatic Toxicol.*, **5**, 277–289.

Pärt, P., O. Svanberg and A. Kiessling (1985). The availability of cadmium to perfused rainbow trout gills in different water qualities. *Water Res.*, **19**, 427–434.

Patarnello, T., R. Guinez and B. Battaglia (1991). Effects of pollution on heterozygosity in the barnacle *Balanus amphitrite* (Cirripedia: Thoracica). *Mar. Ecol. Prog. Ser.*, **70**, 237–243.

Patrick, F.M. and M.W. Loutit (1977). The uptake of heavy metals by epiphytic bacteria on *Alisma plantago-aquatica. Water Res.*, **11**, 699–703.

Patterson, C.C. and D. Settle (1976a). The reduction of orders of magnitude errors in lead analyses of biological materials and natural waters by evaluating and controlling the extent and sources of industrial lead contamination introduced during sample collecting and analysis. In *Accuracy in trace analysis: sampling, sample handling, analysis*, ed. P. LaFleur, US National Bureau of Standards Special Publication No. 422, pp. 321–351.

Patterson, C.C. and D. Settle (1976b). Interlaboratory analyses of references. In *Strategies for marine pollution monitoring*, ed. E.D. Goldberg, John Wiley and Sons, New York, pp. 235–239.

Pavlou, S.P. and R.N. Dexter (1979). Distribution of polychlorinated biphenyls (PCB) in estuarine ecosystems. Testing the concept of equilibrium partitioning in the marine environment. *Environ. Sci. Technol.*, **13**, 65–71.

Peakall, D.B. and J.L. Lincer (1970). Polychlorinated biphenyls: another long-life widespread chemical in the environment. *BioScience*, **20**, 958–964.

Pearson, T.H. and R. Rosenberg (1978). Macrobenthic succession in relation to organic enrichment and pollution of the marine environment. *Oceanogr. mar. biol. Ann. Rev.*, **16**, 229–311.

Pellew, R. (1991). The environmental damage: pollution from oil fires and the oil slick in the Gulf. Paper presented at *The International Conference on the Reconstruction of Kuwait,* the Dorchester Hotel, London, 30–31 May 1991.

Pentreath, R.J. (1976a). The accumulation of inorganic mercury from sea water by the plaice, *Pleuronectes platessa* L. *J. exp. mar. Biol. Ecol.*, **24**, 103–109.

Pentreath, R.J. (1976b). The accumulation of organic mercury from sea water by the plaice, *Pleuronectes platessa* L. *J. exp. mar. Biol. Ecol.*, **24**, 121–132.

Pentreath, R.J. (1976c). The accumulation of mercury from food by the plaice, *Pleuronectes platessa*. L. *J. exp. mar. Biol. Ecol.*, **25**, 51–65.

Pequegnat, J.E., S.W. Fowler and L.F. Small (1969). Estimates of the zinc requirements of marine organisms. *J. Fish. Res. Board Can.*, **26**, 145–150.

Pereira, W.E., F.D. Hostettler and J.B. Rapp (1992). Bioaccumulation of hydrocarbons derived from terrestrial and anthropogenic sources in the Asian clam, *Potamocorbula amurensis*, in San Francisco Bay Estuary. *Mar. Pollut. Bull.*, **24**, 103–109.

Perttila, M., O. Stenman, H. Pyysalo and K. Wickstrom (1986). Heavy metals and organochlorine compounds in seals in the Gulf of Finland. *Mar. Environ. Res.*, **18**, 43–59.

Petering, D.H., M. Goodrich, W. Hodgman, S. Krezoski, D. Weber, C.F. Shaw, R. Spieler and L. Zettergren (1990). Metal-binding proteins and peptides for the detection of heavy metals in aquatic organisms. In *Biomarkers of environmental contamination*, ed. J.F. McCarthy and L.R. Shugart, Lewis Publishers, CRC Press, Boca Raton, Florida, pp. 239–254.

Petrocelli, S.R., A.R. Hanks and J. Anderson (1973). Uptake and accumulation of an organochlorine insecticide (dieldrin) by an estuarine mollusc, *Rangia cuneata*. *Bull. Environ. Contam. Toxicol.*, **10**, 315–320.

Phillips, D.J.H. (1976a). The common mussel *Mytilus edulis* as an indicator of pollution by zinc, cadmium, lead and copper. I. Effects of environmental variables on uptake of metals. *Mar. Biol.*, **38**, 59–69.

Phillips, D.J.H. (1976b). The common mussel *Mytilus edulis* as an indicator of pollution by zinc, cadmium, lead and copper. II. Relationship of metals in the mussel to those discharged by industry. *Mar. Biol.*, **38**, 71–80.

Phillips, D.J.H. (1977). The use of biological indicator organisms to monitor trace metal pollution in marine and estuarine environments — a review. *Environ. Pollut.*, **13**, 281–317.

Phillips, D.J.H. (1978). Use of biological indicator organisms to quantitate organochlorine pollutants in aquatic environments — a review. *Environ. Pollut.*, **16**, 167–229.

Phillips, D.J.H. (1979a). Trace metals in the common mussel, *Mytilus edulis* (L.), and in the alga *Fucus vesiculosus* (L.) from the region of the Sound (Öresund). *Environ. Pollut.*, **18**, 31–43.

Phillips, D.J.H. (1979b). The rock oyster *Saccostrea glomerata* as an indicator of trace metals in Hong Kong. *Mar. Biol.*, **53**, 353–360.

Phillips, D.J.H. (1980). *Quantitative aquatic biological indicators: their use to monitor trace metal and organochlorine pollution*. Applied Science Publishers, London.
Phillips, D.J.H. (1984). Regional monitoring of conservative pollutants in Asian coastal waters: the WESTPAC bio-indicator programme. *Asian Mar. Biol.*, **1**, 1–15.
Phillips, D.J.H. (1985a). Monitoring and control of coastal water quality. In *Pollution in the urban environment, POLMET 85*, ed. M.W.H. Chan, R.W.M. Hoare, P.R. Holmes, R.J.S. Law and S.B. Reed, Elsevier Applied Science Publishers, London, pp. 559–565.
Phillips, D.J.H. (1985b). Organochlorines and trace metals in green-lipped mussels *Perna viridis* from Hong Kong waters: a test of indicator ability. *Mar. Ecol. Prog. Ser.*, **21**, 251–258.
Phillips, D.J.H. (1986). Use of organisms to quantify PCBs in marine and estuarine environments. In *PCBs and the environment, Vol.II*, ed. J.S. Waid, CRC Press, Boca Raton, Florida, pp. 127–181.
Phillips, D.J.H. (1989). Trace metals and organochlorines in the coastal waters of Hong Kong. *Mar. Pollut. Bull.*, **20**, 319–327.
Phillips, D.J.H. (1990a). Arsenic in aquatic organisms: a review, emphasizing chemical speciation. *Aquatic Toxicol.*, **16**, 151–186.
Phillips, D.J.H. (1990b). Use of macroalgae and invertebrates as monitors of metal levels in estuaries and coastal waters. In *Heavy metals in the marine environment*, ed. R.W. Furness and P.S. Rainbow, CRC Press, Boca Raton, Florida, pp. 81–99.
Phillips, D.J.H. (1990c). Organochlorines in green-lipped mussels (*Perna viridis*) from Hong Kong coastal waters. In *Proceedings of the second international marine biological workshop: the marine fauna and flora of Hong Kong and Southern China*, ed. B.S. Morton, Hong Kong University Press, Hong Kong, pp. 1008–1025.
Phillips, D.J.H. (1991). Selected trace elements and the use of biomonitors in subtropical and tropical marine ecosystems. *Rev. Environ. Contam. Toxicol.*, **120**, 105–129.
Phillips, D.J.H. (in press). Bioaccumulation. In *Handbook of ecotoxicology,*, Ed. P. Calow. Blackwell Scientific Publications, Oxford.
Phillips, D.J.H. and M.H. Depledge (1986a). Chemical forms of arsenic in marine organisms, with emphasis on *Hemifusus* species. *Water Sci. Technol.*, **18**, 213–222.
Phillips. D.J.H. and M.H. Depledge (1986b). Distribution of inorganic and total arsenic in tissues of the marine gastropod *Hemifusus ternatanus*. *Mar. Ecol. Prog. Ser.*, **34**, 261–266.
Phillips, D.J.H. and K. Muttarasin (1985). Trace metals in bivalve molluscs from Thailand. *Mar. Environ. Res.*, **15**, 215–234.
Phillips, D.J.H. and P.S. Rainbow (1988). Barnacles and mussels as biomonitors of trace elements: a comparative study. *Mar. Ecol. Prog. Ser.*, **49**, 83–93.
Phillips, D.J.H. and P.S. Rainbow (1989). Strategies of trace metal sequestration in aquatic organisms. *Mar. Environ. Res.*, **28**, 207–210.
Phillips, D.J.H. and D.A. Segar (1986). Use of bio-indicators in monitoring

conservative contaminants: programme design imperatives. *Mar. Pollut. Bull.*, **17**, 10–17.

Phillips, D.J.H. and R.B. Spies (1988). Chlorinated hydrocarbons in the San Francisco estuarine ecosystem. *Mar. Pollut. Bull.*, **19**, 445–453.

Phillips, D.J.H. and S. Tanabe (1989). Aquatic pollution in the Far East. *Mar. Pollut. Bull.*, **20**, 297–303.

Phillips, D.J.H. and W.W.-S. Yim (1981). A comparative evaluation of oysters, mussels and sediments as indicators of trace metals in Hong Kong waters. *Mar. Ecol. Prog. Ser.*, **6**, 285–293.

Phillips, D.J.H., C.T. Ho and L.H. Ng (1982). Trace elements in the Pacific oyster in Hong Kong. *Arch. Environ. Contam. Toxicol.*, **11**, 533–537.

Phillips, G.R. and D.R. Buhler (1978). The relative contributions of methylmercury from food or water to rainbow trout (*Salmo gairdneri*) in a controlled laboratory experiment. *Trans. Amer. Fish. Soc.*, **107**, 853–861.

Pirie, B.J.S., S.G. George, D.G. Lytton and J.D. Thomson (1984). Metal-containing blood cells of oysters: ultrastructure, histochemistry and X-ray microanalysis. *J. mar. biol. Ass. U.K.*, **64**, 115–123.

Popham, J.D. and J.M. D'Auria (1982). Effects of season and seawater concentrations on trace metal concentrations in organs of *Mytilus edulis*. *Arch. Environ. Contam. Toxicol.*, **11**, 273–282.

Popham, J.D. and J.M. D'Auria (1983). Statistical approach for deciding if mussels (*Mytilus edulis*) have been collected from a water body polluted with trace metals. *Environ. Sci. Technol.*, **17**, 576–582.

Portmann, J.E. (1971). Monitoring metals in marine animals. *Mar. Pollut. Bull.*, **2**, 157–159.

Portmann, J.E. (1975). The bioaccumulation and effects of organochlorine pesticides in marine animals. *Proc. Roy. Soc. Lond. B.*, **189**, 291–304.

Powell, J.H. and D.R. Fielder (1983). Temperature and accumulation of DDT by sea mullet (*Mugil cephalus* L.). *Mar. Pollut. Bull.*, **14**, 17–21.

Preston, A. (1975). Monitoring requirements. In *Petroleum and the continental shelf of North-west Europe, Vol. 2. Environmental protection*, ed. H.A. Cole, Applied Science Publishers Ltd., London, pp. 115–120.

Pruell, R.J., C.B. Norwood, R.D. Bowen, W.S. Boothman, P.F. Rogerson, M. Hackett and B.C. Butterworth (1990). Geochemical study of sediment contamination in New Bedford Harbor, Massachusetts. *Mar. Environ. Res.*, **29**, 77–101.

Pullen, J.S.H. and P.S. Rainbow (1991). The composition of pyrophosphate heavy metal detoxification granules in barnacles. *J. exp. mar. Biol. Ecol.*, **150**, 249–266.

Rainbow, P.S. (1985). Accumulation of Zn, Cu and Cd by crabs and barnacles. *Estuar. cstl. Shelf Sci.*, **21**, 669–686.

Rainbow, P.S. (1987). Heavy metals in barnacles. In *Biology of barnacles*, ed. A.J. Southward, A.A. Balkema, Rotterdam, pp. 405–417.

Rainbow, P.S. (1988). The significance of trace metal concentrations in decapods. *Symp. Zool. Soc. Lond.*, **59**, 291–313.

Rainbow, P.S. (1989). Copper, cadmium and zinc concentrations in oceanic amphipod and euphausiid crustaceans, as a source of heavy metals to pelagic seabirds. *Mar. Biol.*, **103**, 513–518.

Rainbow, P.S. (1990). Heavy metals in marine invertebrates. In *Heavy metals in marine environments*, ed. R.W. Furness and P.S. Rainbow, CRC Press, Boca Raton, Florida, pp. 67–79.

Rainbow, P.S. and P.G. Moore (1986). Comparative metal analyses in amphipod crustaceans. *Hydrobiologia.*, **141**, 273–289.

Rainbow, P.S. and P.G. Moore (1990). Seasonal variation in copper and zinc concentrations in three talitrid amphipods (Crustacea). *Hydrobiologia.*, **196**, 65–72.

Rainbow, P.S. and S.L. White (1989). Comparative strategies of heavy metal accumulation by crustaceans: zinc, copper and cadmium in a decapod, an amphipod and a barnacle. *Hydrobiologia.*, **174**, 245–262.

Rainbow, P.S. and S.L. White (1990). Comparative accumulation of cobalt by three crustaceans: a decapod, an amphipod and a barnacle. *Aquatic Toxicol.*, **16**, 113–126.

Rainbow, P.S., P.G. Moore and D. Watson (1989). Talitrid amphipods as biomonitors for copper and zinc. *Estuar. cstl. Shelf Sci.*, **28**, 567–582.

Rainbow, P.S., D.J.H. Phillips and M.H. Depledge (1990). The significance of trace metal concentrations in marine invertebrates: a need for laboratory investigation of accumulation strategies. *Mar. Pollut. Bull.*, **21**, 321–324.

Rainbow, P.S., I. Malik and P. O'Brien (in press). Physico-chemical and physiological effects on the uptake of dissolved zinc and cadmium by the amphipod crustacean *Orchestia gammarellus. Aquatic Toxicol.*

Ramesh, A., S. Tanabe, R. Tatsukawa, A.N. Subramanian, S. Palanichamy, D. Mohan and V.K. Venugopalan (1989). Seasonal variations of organochlorine insecticide residues in air from Porto Novo, South India. *Environ. Pollut.*, **62**, 213–222.

Rao, D.M.R. and A.S. Murty (1982). Toxicity and metabolism of endosulfan in three freshwater catfishes. *Environ. Pollut. Ser. A.*, **27**, 223–231.

Rapaport, R.A., N.R. Urban, P.D. Capel, J.E. Baker, B.B. Looney, S.J. Eisenreich and E. Gorham (1985). 'New' DDT inputs to North America: atmospheric deposition. *Chemosphere*, **14**, 1167–1173.

Ravid, R., J. Ben-Yosef and H. Hornung (1985). PCBs, DDTs and other chlorinated hydrocarbons in marine organisms from the Mediterranean coast of Israel. *Mar. Pollut. Bull.*, **16**, 35–38.

Readers Digest Atlas of the World (1987). Readers Digest Association Ltd., London.

Reay, J.S.S. (1979). The philosophy of monitoring. *Phil. Trans. Roy. Soc. Lond. A.*, **290**, 609–623.

Redpath, K.J. (1985). Growth inhibition and recovery in mussels (*Mytilus edulis*) exposed to low copper concentrations. *J. mar. biol. Ass. U.K.*, **65**, 421–431.

Reijnders, P.J.H. (1980). Organochlorine and heavy metal residues in harbour seals from the Wadden Sea and their possible effects on reproduction. *Neth. J. Sea Res.*, **14**, 30–65.

Reijnders, P.J.H. (1986). Reproductive failure in common seals feeding on fish from polluted coastal waters. *Nature, Lond.*, **324**, 456–457.

Reinert, R.E. (1970). Pesticide concentrations in Great Lakes fish. *Pestic. Monit. J.*, **3**, 233–240.

Reinert, R.E. (1972). Accumulation of dieldrin in an alga (*Scenedesmus obliquus*),

Daphnia magna, and the guppy (*Poecilia reticulata*). *J. Fish. Res. Board Can.*, **29**, 1413–1418.

Reinert, R.E. and H.L. Bergman (1974). Residues of DDT in lake trout (*Salvelinus namaycush*) and coho salmon (*Oncorhynchus kisutch*) from the Great Lakes. *J. Fish. Res. Board Can.*, **31**, 191–199.

Reinert, R.E., L.J. Stone and W.A. Willford (1974). Effect of temperature on accumulation of methylmercuric chloride and *p,p'*DDT by rainbow trout (*Salmo gairdneri*). *J. Fish. Res. Board Can.*, **31**, 1649–1652.

Reish, D.J. (1978). The effects of heavy metals on polychaetous annelids. *Rev. Int. Océanogr. Med.*, **XLIX**, 99–104.

Ribeyre, F. and A. Boudou (1984). Etude expérimentale des processus de décontamination chez *Salmo gairdneri*, après contamination par voie directe avec deux dérivés du mercure ($HgCl_2$ et CH_3HgCl) — analyse des transferts aux niveaux 'organisme' et 'organes'. *Environ. Pollut. Ser. A.*, **35**, 203–228.

Richardson, B.J. and J.S. Waid (1982). Polychlorinated biphenyls (PCBs): an Australian viewpoint on a global problem. *Search.*, **13**, 17–32.

Richardson, C.A. (1989). An analysis of microgrowth bands in the shell of the common mussel *Mytilus edulis*. *J. mar. biol. Ass. U.K.*, **69**, 477–491.

Ridout, P.S., P.S. Rainbow, H.S.J. Roe and H.R. Jones (1989). Concentrations of V, Cr, Mn, Fe, Ni, Co, Cu, Zn, As, Cd in mesopelagic crustaceans from the north east Atlantic Ocean. *Mar. Biol.*, **100**, 465–471.

Risebrough, R.W., B.W. de Lappe and T.T. Schmidt (1976). Bioaccumulation factors of chlorinated hydrocarbons between mussels and seawater. *Mar. Pollut. Bull.*, **7**, 225–228.

Risebrough, R.W., B.W. de Lappe, W. Walker, B.R.T. Simoneit, J. Grimalt, J. Albaigés, J.A.G. Regueiro, A.B.I Nolla and M.M. Fernandez (1983). Application of the mussel watch concept in studies of the distribution of hydrocarbons in the coastal zone of the Ebro Delta. *Mar. Pollut. Bull.*, **14**, 181–187.

Ritz, D.A., R. Swain and N.G. Elliot (1982). Use of the mussel *Mytilus edulis planulatus* (Lamarck) in monitoring heavy metal levels in seawater. *Aust. J. mar. Freshwat. Res.*, **33**, 491–506.

Roberts, D. (1975). Differential uptake of endosulfan by the tissues of *Mytilus edulis*. *Bull. Environ. Contam. Toxicol.*, **13**, 170–176.

Roberts, M.H. (1981). Kepone distribution in selected tissues of blue crabs, *Callinectes sapidus*, collected from the James River and lower Chesapeake Bay. *Estuaries.*, **4**, 313–320.

Robinson, J., A. Richardson, A.N. Crabtree, J.C. Coulson and G.R. Potts (1967). Organochlorine residues in marine organisms. *Nature, Lond.*, **214**, 1307–1311.

Rodgers, D.W. and F.W.H. Beamish (1983). Water quality modifies uptake of waterborne methylmercury by rainbow trout, *Salmo gairdneri*. *Can. J. Fish. Aquat. Sci.*, **40**, 824–828.

Roesijadi, G. (1980/81). The significance of low molecular weight, metallothionein-like proteins in marine invertebrates: current status. *Mar. Environ. Res.*, **4**, 167–179.

Roesijadi, G. (1986). Mercury-binding proteins from the marine mussel, *Mytilus edulis*. *Environ. Health Perspect.*, **65**, 45–48.

Roesijadi, G. (1992). Metallothioneins in metal regulation and toxicity in aquatic animals. *Aquatic Toxicol.* **22**, 81–114.

Rohrer, T.K., J.H. Hartig and J.C. Forney (1981). *Organochlorine and heavy metal residues in coho and chinook salmon of the Great Lakes — 1980*. Report from the State of Michigan Department of Natural Resources to the Great Lakes National Program Office, Chicago, Illinois, August 1981. Michigan Department of Natural Resources, publication no. 3730-0031.

Ronald, K., R.J. Frank, J.L. Dougan, R. Frank and H.E. Braun (1984). Pollutants in harp seals (*Phoca groenlandica*). I. Organochlorines. *Sci. Total Environ.*, **38**, 133-152.

Rygg, B. (1985). Effects of sediment copper on benthic fauna. *Mar. Ecol. Prog. Ser.*, **25**, 83-89.

Saad, M.A.H., M.M. Abu Elamayem, A.H. El-Sebae and I.F. Sharaf (1982). Occurrence and distribution of chemical pollutants in Lake Mariut, Egypt. *Water, Air, Soil Pollut.*, **17**, 245-252.

Salanki, J., K.V. Balogh and E. Berta (1982). Heavy metals in animals of Lake Balaton. *Water Res.*, **16**, 1147-1152.

Samiullah, Y. (1985). Biological effects of marine oil pollution. *Oil Petrochem. Pollut.*, **2**, 235-264.

Samiullah, Y. (1990). *Biological monitoring of environmental contaminants: animals*. Report No. 37, GEMS-Monitoring and Assessment Research Centre, King's College, University of London.

Sanborn, J.R., W.F. Childers and R.L. Metcalf (1975). Uptake of three polychlorinated biphenyls, DDT, and DDE by the green sunfish, *Lepomis cyanellus* Raf. *Bull. Environ. Contam. Toxicol.*, **13**, 209-217.

Sanders, B.M. (1990). Stress proteins: potential as multitiered biomarkers. In *Biomarkers of environmental contamination*, ed. J.F. McCarthy and L.R. Shugart, Lewis Publishers, CRC Press, Boca Raton, Florida, pp. 165-191.

Sanders, B.M. and K.D. Jenkins (1984). Relationships between free cupric ion concentrations in sea water and copper metabolism and growth in crab larvae. *Biol. Bull. Woods Hole, Mass.*, **167**, 704-712.

Sanders, B.M., K.D. Jenkins, W.G. Sunda and J.D. Costlow (1983). Free cupric ion activity in seawater: effects on metallothionein and growth in crab larvae. *Science.*, **222**, 53-55.

Sanders, H.O. and J.H. Chandler (1972). Biological magnification of a polychlorinated biphenyl (Aroclor 1254) from water by aquatic invertebrates. *Bull. Environ. Contam. Toxicol.*, **7**, 257-263.

Sanders, J.G. (1986). Direct and indirect effects of arsenic on the survival and fecundity of estuarine zooplankton. *Can. J. Fish. Aquat. Sci.*, **43**, 694-699.

Sanders, J.G. and P.S. Vermersch (1982). Responses of marine phytoplankton to low levels of arsenate. *J. Plankton Res.*, **4**, 881-893.

Särkkä, J., M.-L. Hattula, J. Janatuinen and J. Paasivirta (1978). Mercury and chlorinated hydrocarbons in plankton of Lake Päijänne, Finland. *Environ. Pollut.*, **16**, 41-49.

Satsmadjis, J. and G.P. Gabrielides (1983). Organochlorines in mussel and shrimp from the Saronikis Gulf (Greece). *Mar. Pollut. Bull.*, **14**, 356-358.

Say, P.J. and B.A. Whitton (1983). Accumulation of heavy metals by aquatic mosses. 1: *Fontinalis antipyretica* Hedw. *Hydrobiologia*, **100**, 245-260.

Say, P.J., J.P.C. Harding and B.A. Whitton (1981). Aquatic mosses as monitors of heavy metal contamination in the River Etherow, Great Britain. *Environ. Pollut. Ser. B.*, **2**, 295–307.

Schnare, D.W., M. Ben and M.G. Shields (1984). Body burden reductions of PCBs, PBBs and chlorinated pesticides in human subjects. *Ambio*, **13**, 378–380.

Schüler, W., H. Brunn and D. Manz (1985). Pesticides and polychlorinated biphenyls in fish from the Lahn River. *Bull. Environ. Contam. Toxicol.*, **34**, 608–616.

Schulz, D.E., G. Petrick and J.C. Duinker (1988). Chlorinated biphenyls in North Atlantic surface and deep water. *Mar. Pollut. Bull.*, **19**, 526–531.

Schulz-Baldes, M. (1972). Toxizität und Anreicherung von Blei bei der Miesmuschel *Mytilus edulis* im Laborexperiment. *Mar. Biol.*, **16**, 226–229.

Schulz-Baldes, M. (1973). Die Miesmuschel *Mytilus edulis* als Indikator für die Bleikonzentration im Weserästuar und in der Deutschen Bucht. *Mar. Biol.*, **21**, 98–102.

Schulz-Baldes, M. (1974). Lead uptake from sea water and food and lead loss in the common mussel *Mytilus edulis*. *Mar. Biol.*, **25**, 177–193.

Scott, D.P. and F.A.J. Armstrong (1972). Mercury concentration in relation to size in several species of freshwater fishes from Manitoba and northwestern Ontario. *J. Fish. Res. Board Can.*, **29**, 1685–1690.

Seed, R. (1975). *Reproduction in Mytilus (Mollusca: Bivalvia) in European waters.* Pubblicazioni della Stazione Zoologica de Napoli, Milan.

Seed, R. (1976). Ecology. In *Marine mussels: their ecology and physiology*, ed. B. Bayne, Cambridge University Press, Cambridge, pp. 13–65.

Sericano, J.L., E.L. Atlas, T.L. Wade and J.M. Brooks (1990). NOAA's Status and Trends Mussel Watch Program: chlorinated pesticides and PCBs in oysters (*Crassostrea virginica*) and sediments from the Gulf of Mexico, 1986–1987. *Mar. Environ. Res.*, **29**, 161–203.

Shaw, D.G., P. Else, T. Gritz, G. Malinky, G. Mapes, D. McIntosh, J. Schwartz, E. Smith and J. Wiggs (1978). Hydrocarbons: natural distribution and dynamics on the Alaskan outer continental shelf. Section VIII. Hydrocarbons of Bering Sea biota. In *Environmental assessment of the Alaskan continental shelf, Vol. VIII, Contaminant baselines*, National Oceanic and Atmospheric Administration, US Department of Commerce, Boulder, Colorado, pp. 559–567.

Sherlock, J.C. (1983). Dietary surveys on a population at Shipham, Somerset. *Sci. Total Environ.*, **29**, 12–32.

Shugart, L.R. (1990). Biological monitoring: testing for genotoxicity. In *Biomarkers of environmental contamination*, ed. J.F. McCarthy and L.R. Shugart, Lewis Publishers, CRC Press, Boca Raton, Florida, pp. 205–216.

Simkiss, K. (1980). Detoxification, calcification and the intracellular storage of ions. In *The mechanisms of biomineralisation in animals and plants, Proceedings of the third International biomineralisation symposium*, ed. M. Omori and N. Watabe, Tokai University Press, Tokyo, pp. 13–18.

Simkiss, K. (1981). Calcium, pyrophosphate and cellular pollution. *Trends Biochem. Sci.*, April 1981, 1–3.

Simkiss, K. (1983). Lipid solubility of heavy metals in saline solutions. *J. mar. biol. Ass. U.K.*, **63**, 1–7.

Simkiss, K. and M.G. Taylor (1989a). Metal fluxes across the membranes of aquatic organisms. *CRC Crit. Revs. Aquatic Sci.*, **1**, 173–188.

Simkiss, K. and M.G. Taylor (1989b). Convergence of cellular systems of metal detoxification. *Mar. Environ. Res.*, **28**, 211–214.

Simkiss, K., M. Taylor and A.Z. Mason (1982). Metal detoxification and bio-accumulation in molluscs. *Mar. Biol. Letts.*, **3**, 187–201.

Simpson, R.D. (1979). Uptake and loss of zinc and lead by mussels (*Mytilus edulis*) and relationships with body weight and reproductive cycle. *Mar. Pollut. Bull.*, **10**, 74–78.

Skaare, J.U., N.H. Markussen, G. Norheim, S. Haugen and G. Holt (1990). Levels of polychlorinated biphenyls, organochlorine pesticides, mercury, cadmium, copper, selenium, arsenic, and zinc in the harbour seal, *Phoca vitulina*, in Norwegian waters. *Environ. Pollut.*, **66**, 309–324.

Skea, J.C., H.J. Simonin, S. Jackling and J. Symula (1981). Accumulation and retention of Mirex by brook trout fed a contaminated diet. *Bull. Environ. Contam. Toxicol.*, **27**, 79–83.

Slater, T.E. (1978). Biochemical studies on liver injury. In *Biochemical mechanisms of liver injury*, ed. T.E. Slater, Academic Press, New York, pp. 2–44.

Smith, D.R., M.D. Stephenson and A.R. Flegal (1986). Trace metals in mussels transplanted to San Francisco Bay. *Environ. Toxicol. Chem.*, **5**, 129–138.

Smith, J.E. (1968). *'Torrey Canyon' pollution and marine life*. Cambridge University Press, Cambridge.

Smith, J.N. and D.H. Loring (1981). Geochronology for mercury pollution in the sediments of the Saguenay Fjord, Quebec. *Environ. Sci. Technol.*, **15**, 944–951.

Smith, R.M. and C.F. Cole (1970). Chlorinated hydrocarbon insecticide residues in winter flounder, *Pseudopleuronectes americanus*, from the Weweantic River estuary, Massachusetts. *J. Fish. Res. Board Can.*, **27**, 2374–2380.

Södergren, A. (1968). Uptake and accumulation of C^{14}-DDT by *Chlorella* sp. (Chlorophyceae). *Oikos.*, **19**, 126–138.

Södergren, A., B. Svensson and S. Ulfstrand (1972). DDT and PCB in south Swedish streams. *Environ. Pollut.*, **3**, 25–36.

Spies, R.B. and D.W. Rice (1988). Effects of organic contaminants on reproduction of starry flounder, *Platichthys stellatus*, in San Francisco Bay, California, USA. II. Reproductive success of fish captured in San Francisco Bay and spawned in the laboratory. *Mar. Biol.*, **98**, 191–200.

Spies, R.B., J.J. Stegeman, D.W. Rice Jr., B. Woodin, P. Thomas, J.E. Hose, J.N. Cross and M. Prieto (1990). Sublethal responses of *Platichthys stellatus* to organic contamination in San Francisco Bay with emphasis on reproduction. In *Biomarkers of environmental contamination*, ed. J.F. McCarthy and L.R. Shugart, Lewis Publishers, CRC Press, Boca Raton, Florida, pp. 87–121.

Spooner, M. (1978). *Amoco Cadiz* oil spill: editorial introduction. *Mar. Pollut. Bull.*, **9**, 281–284.

Sprague, J.B. and J.R. Duffy (1971). DDT residues in Canadian Atlantic fishes and shellfishes in 1967. *J. Fish. Res. Board Can.*, **28**, 59–64.

Sprague, J.B., P.F. Elson and J.R. Duffy (1971). Decrease in DDT residues in young salmon after forest spraying in New Brunswick. *Environ. Pollut.*, **1**, 191–203.

Stauber, J.L. and T.M. Florence (1985). Interactions of copper and manganese: a mechanism by which manganese alleviates copper toxicity to the marine diatom, *Nitzschia closterium* (Ehrenberg) W. Smith. *Aquatic Toxicol.*, **7**, 241–254.

Stebbing, A.R.D. (1976). The effects of low metal levels on a clonal hydroid. *J. mar. biol. Ass. U.K.*, **56**, 977–994.
Stebbing, A.R.D. (1979). An experimental approach to the determinants of biological water quality. *Phil. Trans. Roy. Soc. Lond. B.*, **286**, 465–481.
Stenzel, H.B. (1971). Oysters. In *Treatise on invertebrate palaontology, Part IV, vol. 3, Mollusca 6*, ed. K.C. Moore, Geological Society of America and University of Kansas, Boulder, Colorado, pp. N953–N1224.
Stephenson, M., D. Smith, G. Ichikawa, J. Goetzl and M. Martin (1986). *State Mussel Watch Program: preliminary data report, 1985–1986*. Report of 2 July 1986 from the California Department of Fish and Game to the State Water Resources Control Board, Sacramento, California.
Stoecker, D. (1980). Relationships between chemical defense and ecology in benthic ascidians. *Mar. Ecol. Prog. Ser.*, **3**, 257–265.
Stokes, P.M., S.I. Dreier, M.O. Farkas and R.A.N. McLean (1983). Mercury accumulation by filamentous algae: a promising biological monitoring system for methyl mercury in acid-stressed lakes. *Environ. Pollut. Ser. B.*, **5**, 255–271.
Stout, V.F., F.L. Beezhold and C.R. Houle (1972). DDT residue levels in some U.S. fishery products and some treatments in reducing them. In *Marine pollution and sea life*, ed. M. Ruivo, Fishing News (Books) Ltd., London, pp. 550–553.
Strigini, P. (1983). The Italian chemical industry and the case of Seveso. *UNEP Industry and Environment*. October–December 1983: 16–21.
Strömgren, T. (1982). Effect of heavy metals (Zn, Hg, Cu, Cd, Pb, Ni) on the length growth of *Mytilus edulis*. *Mar. Biol.*, **72**, 69–72.
Strong, C.R. and S.N. Luoma (1981). Variations in the correlation of body size with concentrations of Cu and Ag in the bivalve *Macoma balthica*. *Can. J. Fish. Aquat. Sci.*, **38**, 1059–1064.
Sullivan, P.A., W.E. Robinson and M.P. Morse (1988). Isolation and characterization of granules from the kidney of the bivalve *Mercenaria mercenaria*. *Mar. Biol.*, **99**, 359–368.
Sunda, W.G. and R.R.L. Guillard (1976). The relationship between cupric ion activity and the toxicity of copper to phytoplankton. *J. mar. Res.*, **34**, 511–529.
Sunda, W.G., D.W. Engel and R.M. Thuotte (1978). Effects of chemical speciation on the toxicity of cadmium to grass shrimp *Palaemonetes pugio:* importance of free cadmium ion. *Environ. Sci. Technol.*, **12**, 409–413.
Sundelin, B. and R. Elmgren (1991). Meiofauna of an experimental soft bottom ecosystem — effects of macrofauna and cadmium exposure. *Mar. Ecol. Prog. Ser.*, **70**, 245–255.
Sunila, I. and R. Lindstrom (1985). Survival, growth and shell deformities of copper- and cadmium-exposed mussels (*Mytilus edulis*) in brackish water. *Estuar. cstl. Shelf Sci.*, **21**, 555–565.
Suteau, P., M. Daubeze, M.L. Migand and J.F. Narbonne (1988). PAH-metabolizing enzymes in whole mussels as biochemical tests for chemical pollution monitoring. *Mar. Ecol. Prog. Ser.*, **46**, 45–49.
Suzuki, M., Y. Yamato and T. Akiyama (1974). BHC (1,2,3,4,5,6–hexachlorocyclohexane) residue concentrations and their seasonal variation in aquatic environments in the Kitakyushu district, Japan, 1970–1973. *Water Res.*, **8**, 643–649.
Takeuchi, T. (1972). Distribution of mercury in the environment of Minimata Bay

and the inland Ariake Sea. In *Environmental mercury contamination*, ed. R. Hartung and B.D. Dinaman, Ann Arbor Science, Ann Arbor, pp. 79–81.

Takizawa, Y. (1979). Epidemiology of mercury poisoning. In *The biogeochemistry of mercury in the environment*, ed. J.O. Nriagu, Elsevier/North Holland Biomedical Press, Amsterdam, pp. 325–365.

Talbot, V. (1985). Heavy metal concentrations in the oysters *Saccostrea cuccullata* and *Saccostrea* sp. (probably *S. commercialis*) from the Dampier Archipelago, Western Australia. *Aust. J. mar. Freshwat. Res.*, **36**, 169–175.

Talbot, V. (1986). Seasonal variation of copper and zinc concentrations in the oyster *Saccostrea cuccullata* from the Dampier Archipelago, Western Australia: implications for pollution monitoring. *Sci. Total Environ.*, **57**, 217–230.

Tanabe, S. (1988). PCB problems in the future: foresight from current knowledge. *Environ. Pollut.*, **50**, 5–28.

Tanabe, S. and R. Tatsukawa (1980). Chlorinated hydrocarbons in the North Pacific and Indian Oceans. *J. Oceanogr. Soc. Japan.*, **36**, 217–226.

Tanabe, S., R. Tatsukawa, H. Tanaka, K. Maruyama, N. Miyazaki and T. Fujiyama (1981). Distribution and total burdens of chlorinated hydrocarbons in bodies of striped dolphins (*Stenella coeruleoalba*). *Agric. Biol. Chem.*, **45**, 2569–2578.

Tanabe, S., M. Kawano and R. Tatsukawa (1982a). Chlorinated hydrocarbons in the Antarctic, western Pacific and eastern Indian Oceans. *Trans. Tokyo Univ. Fish.*, **5**, 97–109.

Tanabe, S., K. Maruyama and R. Tatsukawa (1982b). Absorption efficiency and biological half-life of individual chlorobiphenyls in carp (*Cyrinus carpio*) orally exposed to Kanechlor products. *Agric. Biol. Chem.*, **46**, 891–905.

Tanabe, S., R. Tatsukawa, K. Maruyama and N. Miyazaki (1982c). Transplacental transfer of PCBs and chlorinated hydrocarbon pesticides from the pregnant striped dolphin (*Stenella coeruleoalba*) to her fetus. *Agric. Biol. Chem.*, **46**, 1249–1254.

Tanabe, S., H. Hidaka and R. Tatsukawa (1983a). PCBs and chlorinated hydrocarbon pesticides in Antarctic atmosphere and hydrosphere. *Chemosphere*, **12**, 277–288.

Tanabe, S., T. Mori, R. Tatsukawa and M. Miyazaki (1983b). Global pollution of marine mammals by PCBs, DDTs and HCHs (BHCs). *Chemosphere*, **12**, 1269–1275.

Tanabe, S., N. Kannan, An. Subramanian, S. Watanabe and R. Tatsukawa (1987a). Highly toxic coplanar PCBs: occurrence, source, persistency and toxic implications to wildlife and humans. *Environ. Pollut.*, **47**, 147–163.

Tanabe, S., R. Tatsukawa and D.J.H. Phillips (1987b). Mussels as bioindicators of PCB pollution: a case study on uptake and release of PCB isomers and congeners in green-lipped mussels (*Perna viridis*) in Hong Kong waters. *Environ. Pollut.*, **47**, 41–62.

Tanabe, S., N. Kannan, M. Fukushima, T. Okamoto, T. Wakimoto and R. Tatsukawa (1989a). Persistent organochlorines in Japanese coastal waters: an introspective summary from a Far East developed nation. *Mar. Pollut. Bull.*, **20**, 344–352.

Tanabe, S., N. Kannan, T. Wakimoto, R. Tatsukawa, T. Okamoto and Y. Masuda (1989b). Isomer-specific determination and toxic evaluation of potentially hazardous coplanar PCBs, dibenzofurans and dioxins in the tissues of 'Yusho' PCB poisoning victim and in the causal oil. *Toxicol. Environ. Chem.*, **24**, 215–231.

Tarifeño-Silva, E., L.Y. Kawasaki, D.P. Yu, M.S. Gordon and D.J. Chapman (1982). Aquacultural approaches to recycling of dissolved nutrients in secondarily treated domestic wastewaters — III. Uptake of dissolved heavy metals by artificial food chains. *Water Res.*, **16**, 59–65.

Taruski, A.G., C.E. Olney and H.E. Winn (1975). Chlorinated hydrocarbons in cetaceans. *J. Fish. Res. Board Can.*, **32**, 2205–2209.

Tatton, J.O'G. and J.H.A. Ruzicka (1967). Organochlorine pesticides in Antarctica. *Nature, Lond.*, **215**, 346–348.

Tatum, H.E. (1986). Bioaccumulation of polychlorinated biphenyls and metals from contaminated sediment by freshwater prawns, *Macrobrachium rosembergii* and clams, *Corbicula fluminea*. *Arch. Environ. Contam. Toxicol.*, **15**, 171–183.

Tavares, T.M., V.C. Rocha, C. Porte, D. Barceló and J. Albaigés (1988). Application of the mussel watch concept in studies of hydrocarbons, PCBs and DDT in the Brazilian bay of Todos os Santos (Bahia). *Mar. Pollut. Bull.*, **19**, 575–578.

Taylor, D. (1982). Minimata disease. *Environ. Sci. Technol.*, **16:** 81A.,

Taylor, M.G. and K. Simkiss (1984). Inorganic deposits in invertebrate tissues. *Environ. Chem.*, **3**, 102–138.

Tessier, A. and P.G.C. Campbell (1987). Partitioning of trace metals in sediments: relationships with bioavailability. *Hydrobiologia*, **149**, 43–52.

Tessier, A., P.G.C. Campbell, J.C. Auclair and M. Bisson (1984). Relationships between the partitioning of trace metals in sediments and their accumulation in the tissues of the freshwater mollusc *Elliptio complanata* in a mining area. *Can. J. Fish. Aquat. Sci.*, **41**, 1463–1471.

Thain, J.E. and M.J. Waldock (1986). The impact of tributyl tin (TBT) antifouling paints on molluscan fisheries. *Water Sci. Technol.*, **18**, 193–202.

Theobald, N. (1989). Investigation of 'petroleum hydrocarbons' in seawater, using high performance liquid chromatography with fluorescence detection. *Mar. Pollut. Bull.*, **20**, 134–140.

Thomann, R.V. (1981). Equilibrium model of fate of microcontaminants in diverse aquatic food chains. *Can. J. Fish. Aquat. Sci.*, **38**, 280–296.

Thomann, R.V. and J.P. Connolly (1984). Model of PCB in the Lake Michigan lake trout food chain. *Environ. Sci. Technol.*, **18**, 65–71.

Thomas, W.H. and D.L.R. Seibert (1977). Effects of copper on the dominance and the diversity of algae: controlled ecoystem experiment. *Bull. mar. Sci.*, **27**, 23–33.

Thompson, D.R. (1990). Metal levels in marine vertebrates. In *Heavy metals in the marine environment*, ed. R.W. Furness and P.S. Rainbow, CRC Press, Boca Raton, Florida, pp. 143–182.

Thompson, H.C., D.C. Kendall, W.A. Korfmacher, K.L. Rowland, L.G. Rushing, J.J. Chen, J.R. Kominsky, L.M. Smith and D.L. Stalling (1986). Assessment of the contamination of a multibuilding facility by polychlorinated biphenyls, polychlorinated dibenzo-*p*-dioxins, and polychlorinated dibenzofurans. *Environ. Sci. Technol.*, **20**, 597–603.

Thomson, E.A., S.N. Luoma, C.E. Johansson and D.J. Cain (1984). Comparison of sediments and organisms in identifying sources of biologically available trace metal contamination. *Water Res.*, **18**, 755–765.

Thomson, J.D. (1982). Metal concentration changes in growing Pacific oysters, *Crassostrea gigas*, cultivated in Tasmania, Australia. *Mar. Biol.*, **67**, 135–142.
Thomson, J.D., B.J.S. Pirie and S.G. George (1985). Cellular metal distribution in the Pacific Oyster, *Crassostrea gigas* (Thun.) determined by quantitative X-ray microprobe analysis. *J. exp. mar. Biol. Ecol.*, **85**, 37–45.
Thrower, S.J. and I.J. Eustace (1973). Heavy metal accumulation in oysters grown in Tasmanian waters. *Food Technol. Aust.*, November 1973, 546–553.
Tibbetts, P.J.C., S.J. Rowland, L.L. Tovey and R. Large (1982). Investigation of the sources of aliphatic hydrocarbons in the mussel *Mytilus edulis* from North Sea oil production platforms by capillary glc and CGCMS. *Toxicol. Environ. Chem.*, **5**, 177–193.
Times Atlas and Encyclopaedia of the Sea (1989). Times Books Limited, London.
Tokunaga, T., N. Furuta and M. Morimoto (1976). Accumulation of cadmium in *Eichhornia crassipes* Solms. *J. Hyg. Chem.*, **22**, 234–239.
Tong, S.S.C., W.D. Youngs, W.H. Gutenmann and D.J. Lisk (1974). Trace metals in Lake Cayuga trout (*Salvelinus namaycush*) in relation to age. *J. Fish. Res. Board Can.*, **31**, 238–239.
Tooby, T.E. and F.J. Durbin (1975). Lindane residue accumulation and elimination by rainbow trout (*Salmo gairdneri* Richardson) and roach (*Rutilus rutilus* Linnaeus). *Environ. Pollut.*, **8**, 79–89.
Tseng, W.P. (1977). Effects and dose-response relationships of skin cancer and blackfoot disease with arsenic. *Environ. Health Perspect.*, **19**, 109–119.
Tsuchiya, K. (1977). Various effects of arsenic in Japan depending on type of exposure. *Environ. Health Perspect.*, **19**, 35–42.
Tsui, P.T.P. and P.J. McCart (1981). Chlorinated hydrocarbon residues and heavy metals in several fish species from the Cold Lake area in Alberta, Canada. *Intern. J. Environ. Anal. Chem.*, **10**, 277–285.
Tuthill, C., W. Schutte, C.W. Frank, J. Santolucito and G. Potter (1982). Retrospective monitoring: a review. *Environ. Monit. Assessment.*, **1**, 189–211.
Underwood, A.J. and C.H. Peterson (1988). Towards an ecological framework for investigating pollution. *Mar. Ecol. Prog. Ser.*, **46**, 227–234.
Ünlü, M.Y. and S.W. Fowler (1979). Factors affecting the flux of arsenic through the mussel, *Mytilus galloprovincialis. Mar. Biol.*, **51**, 209–219.
USCDC (1985). *Morbidity and mortality weekly report.* US Centers for Disease Control, Vol. 34: No. 36.
US Federal Register (1981). Environmental Protection Agency: Decisions on petitions to remove ethylbenzene, phenol, 2,4-dichlorophenol, trichlorophenol, pentachlorophenol, monochlorophenyl phenyl ether and chlorodifluoromethane from the list of toxic pollutants under Section 307(a)(1) of the Clean Water Act. *US Federal Register.*, **46**, 2267–2278.
van den Broek, W.L.F. (1979). Seasonal levels of chlorinated hydrocarbons and heavy metals in fish and brown shrimps from the Medway Estuary, Kent. *Environ. Pollut.*, **19**, 21–38.
van der Putte, I. and P. Pärt (1982). Oxygen and chromium transfer in perfused gills of rainbow trout (*Salmo gairdneri*) exposed to hexavalent chromium at two different pH levels. *Aquatic Toxicol.*, **2**, 31–45.
van der Putte, I., J. Lubbers and Z. Kolar (1981). Effect of pH on uptake, tissue

distribution and retention of hexavalent chromium in rainbow trout (*Salmo gairdneri*). *Aquatic Toxicol.*, **1**, 3–18.

Viarengo, A. (1985). Biochemical effects of trace metals. *Mar. Pollut. Bull.*, **16**, 153–158.

Viarengo, A. (1989). Heavy metals in marine invertebrates: mechanisms of regulation and toxicity at the cellular level. *CRC Crit. Revs. Aquatic Sci.*, **1**, 295–317.

Viarengo, A., M. Pertica, G. Mancinelli, S. Palmero, G. Zanicchi and M. Orunesu (1982). Evaluation of general and specific stress indices in mussels collected from populations subjected to different levels of heavy metal pollution. *Mar. Environ. Res.*, **6**, 235–243.

Viarengo, A., M. Pertica, G. Mancinelli, G. Zanicchi, J.M. Bouquegneau and M. Orunesu (1984). Biochemical characterization of Cu-thioneins isolated from the tissues of mussels exposed to the metal. *Mol. Physiol.*, **5**, 41–52.

Viarengo, A., G. Mancinelli, G. Martino, M. Pertica. L. Canesi and A. Mazzucotelli (1988). Integrated cellular stress indices in trace metal contamination: critical evaluation in a field study. *Mar. Ecol. Prog. Ser.*, **46**, 65–70.

Vighi, M. (1981). Lead uptake and release in an experimental trophic chain. *Ecotoxicol. Environ. Safety.*, **5**, 177–193.

Villeneuve, J.P., S.W. Fowler and V.C. Anderlini (1987). Organochlorine levels in edible marine organisms from Kuwaiti coastal waters. *Bull. Environ. Contam. Toxicol.*, **38**, 266–270.

Vreeland, V. (1974). Uptake of chlorobiphenyls by oysters. *Environ. Pollut.*, **6**, 135–140.

Waddington, J.I. (1981). The Mediterranean Action Plan: design and implementation. *J. Water Pollut. Control Fed.*, **1981**, 149–162.

Waid, J.S. (Ed.) (1986). *PCBs in the environment*. CRC Press, Boca Raton, Florida.

Waldichuk, M. (1985). Biological availability of metals to marine organisms. *Mar. Pollut. Bull.*, **16**, 7–11.

Waldichuk, M. (1990). Dioxin pollution near pulp mills. *Mar. Pollut. Bull.*, **20**, 365–366.

Walker, C.H. (1990). Kinetic models to predict bioaccumulation of pollutants. *Functional Ecol.*, **4**, 295–301.

Walker, G. (1977). 'Copper' granules in the barnacle *Balanus balanoides*. *Mar. Biol.*, **39**, 343–349.

Walker, G., P.S. Rainbow, P. Foster and D.J. Crisp (1975a). Barnacles: possible indicators of zinc pollution? *Mar. Biol.*, **30**, 57–75.

Walker, G., P.S. Rainbow, P. Foster and D.L. Holland (1975b). Zinc phosphate granules in tissues surrounding the midgut of the barnacle *Balanus balanoides*. *Mar. Biol.*, **33**, 162–166.

Wang, H.-K. and J.M. Wood (1984). Bioaccumulation of nickel by algae. *Environ. Sci. Technol.*, **18**, 106–109.

Wang, K., B. Rott and F. Korte (1982). Uptake and bioaccumulation of three PCBs by *Chlorella fusca*. *Chemosphere*, **11**, 525–530.

Ward, T.J. (1982). Laboratory study of the accumulation and distribution of cadmium in the Sydney rock oyster *Saccostrea commercialis* (I and R). *Aust. J. mar. Freshwat. Res.*, **33**, 33–44.

Ward, T.J. and P.C. Young (1982). Effects of sediment trace metals and particle size on the community structure of epibenthic seagrass fauna near a lead smelter, South Australia. *Mar. Ecol. Prog. Ser.*, **9**, 137–146.

Warren, G.E. and G.E. Davis (1967). Laboratory studies on the feeding, bioenergetics and growth of fish. In *The biological basis of freshwater fish production*, ed. S.D. Gerking, Blackwell, Oxford, pp. 175–214.

Warwick, R.M. (1988). Effects on community structure of a pollutant gradient — summary. *Mar. Ecol. Prog. Ser.*, **46**, 207–211.

Warwick, R.M. and K.R. Clarke (1991). A comparison of some methods for analysing changes in benthic community structure. *J. mar. biol. Ass. U.K.*, **71**, 225–244.

Watanabe, S., T. Shimada, S. Nakamuru, N. Nishiyama, N. Yamashita, S. Tanabe and R. Tatsukawa (1989). Specific profile of liver microsomal cytochrome P-450 in dolphin and whales. *Mar. Environ. Res.*, **27**, 51–65.

Watkins, J. and J.D. Yarbrough (1975). Aldrin and dieldrin uptake in insecticide-resistant and susceptible mosquitofish (*Gambusia affinis*). *Bull. Environ. Contam. Toxicol.*, **14**, 731–737.

Watling, H.E. and R.J. Watling (1976). Trace metals in oysters from the Knysna Estuary. *Mar. Poll. Bull.*, **7**, 45–48.

Watson, D.G., J.J. Davis and W.C. Hanson (1961). Zinc-65 in marine organisms along the Oregon and Washington coasts. *Science.*, **133**, 1826–1828.

Webb, K. (1983). Indonesia has Minimata disease. *World Environ. Rep.*, November 1983, 5.

Wedermeyer, G. and W.T. Yasutake (1977). *Clinical methods for the assessment of the effects of environmental stress on fish health.* Paper 89, Technical Papers of the U.S. Fish and Wildlife Service, Washington, DC, 18.

Weeks, J.M. and P.G. Moore (1991). The effect of synchronous moulting on body copper and zinc concentrations in four species of talitrid amphipods (Crustacea). *J. mar. biol. Ass. U.K.*, **71**, 481–488.

Weeks, J.M. and P.S. Rainbow (1991). The uptake and accumulation of zinc and copper from solution by two species of talitrid amphipods. *J. mar. biol. Ass. U.K.*, **71**, 811–826.

Wehr, J.D. and B.A. Whitton (1983a). Accumulation of heavy metals by aquatic mosses. 2: *Rhynchostegium riparioides. Hydrobiologia*, **100**, 261–284.

Wehr, J.D. and B.A. Whitton (1983b). Accumulation of heavy metals by aquatic mosses. 3: Seasonal changes. *Hydrobiologia*, **100**, 285–291.

Wehr, J.D., P.J. Say and B.A. Whitton (1981). Heavy metals in an industrially polluted river, the Team. In *Heavy metals in northern England: environmental and biological aspects,* ed. P.J. Say and B.A. Whitton, Department of Botany, University of Durham, UK, pp. 99–107.

Wehr, J.D., A. Empain, C. Mouvet, P.J. Say and B.A. Whitton (1983). Methods for processing aquatic mosses used as monitors of heavy metals. *Water Res.*, **17**, 985–992.

Welsh, R.P.H. and P. Denny (1980). The uptake of lead and copper by submerged aquatic macrophytes in two English lakes. *J. Ecol.*, **68**, 443–455.

Wharfe, J.R. and W.L.F. van den Broek (1978). Chlorinated hydrocarbons in macroinvertebrates and fish from the lower Medway Estuary, Kent. *Mar. Pollut. Bull.*, **9**, 76–79.

White, H.H. (1984). Mussel madness: use and misuse of biological monitors of marine pollution. In *Concepts in marine pollution measurements*, ed. H.H. White, Maryland Sea Grant College, Maryland, pp. 325–337.

White, S.L. and P.S. Rainbow (1982). Regulation and accumulation of copper, zinc and cadmium by the shrimp *Palaemon elegans*. *Mar. Ecol. Prog. Ser.*, **8**, 95–101.

White, S.L. and P.S. Rainbow (1984a). Regulation of zinc concentration by *Palaemon elegans* (Crustacea: Decapoda): zinc flux and effects of temperature, zinc concentration and moulting. *Mar. Ecol. Prog. Ser.*, **16**, 135–147.

White, S.L. and P.S. Rainbow (1984b). Zinc flux in *Palaemon elegans* (Crustacea: Decapoda): moulting, individual variation and tissue distribution. *Mar. Ecol. Prog. Ser.*, **19**, 153–166.

White, S.L. and P.S. Rainbow (1985). On the metabolic requirements for copper and zinc in molluscs and crustaceans. *Mar. Environ. Res.*, **16**, 215–229.

White, S.L. and P.S. Rainbow (1986). Accumulation of cadmium by *Palaemon elegans* (Crustacea: Decapoda). *Mar. Ecol. Prog. Ser.*, **32**, 17–25.

White, S.L. and P.S. Rainbow (1987). Heavy metal concentrations in the mesopelagic decapod crustacean *Systellaspis debilis*. *Mar. Ecol. Prog. Ser.*, **37**, 147–151.

Whitehead, N.E. and R.R. Brooks (1969). Aquatic bryophytes as indicators of uranium mineralisation. *Bryologist*, **72**, 501–507.

Whittle, K.J., R. Hardy, P.R. Mackie and A.S. McGill (1982). A quantitative assessment of the sources and fate of petroleum compounds in the marine environment. *Phil. Trans. Roy. Soc. Lond. B.*, **297**, 193–218.

Whitton, B.A., P.J. Say and J.D. Wehr (1981). Use of plants to monitor heavy metals in rivers. In *Heavy metals in northern England: environmental and biological aspects*, ed. P.J. Say and B.A. Whitton, Department of Botany, University of Durham, UK, pp. 135–145.

Whitton, B.A., P.J. Say and B.P. Jupp (1982). Accumulation of zinc, cadmium and lead by the aquatic liverwort *Scapania*. *Environ. Pollut. Ser. B.*, **3**, 299–316.

WHO (1981). *Environmental health criteria 18: arsenic*. World Health Organization, Geneva.

Widdows, J. (1985). Physiological responses to pollution. *Mar. Pollut. Bull.*, **16**, 129–134.

Widdows, J. and B.L. Bayne (1971). Temperature acclimation of *Mytilus edulis* with reference to its energy budget. *J. mar. biol. Ass. U.K.*, **51**, 827–843.

Widdows, J. and D. Johnson (1988). Physiological energetics for *Mytilus edulis:* scope for growth. *Mar. Ecol. Prog. Ser.*, **46**, 113–121.

Widdows, J., D.K. Phelps and W. Galloway (1981). Measurement of physiological condition of mussels transplanted along a pollution gradient in Narragansett Bay. *Mar. Environ. Res.*, **4**, 181–194.

Widdows, J., T. Bakke, B.L. Bayne, P. Donkin, D.R. Livingstone, D.M. Lowe, M.N. Moore, S.V. Evans and S.L. Moore (1982). Responses of *Mytilus edulis* on exposure to the water-accomodated fraction of North Sea oil. *Mar. Biol.*, **67**, 15–31.

Widdows, J., P. Donkin, P.N. Salkeld, J.J. Cleary, D.M. Lowe, S.V. Evans and P.E. Thomson (1984). Relative importance of environmental factors in determining physiological differences between two populations of mussels (*Mytilus edulis*). *Mar. Ecol. Prog. Ser.*, **17**, 33–47.

Wiener, J.G. and J.P. Giesy (1979). Concentrations of Cd, Cu, Mn, Pb, and Zn in fishes in a highly organic softwater pond. *J. Fish. Res. Board Can.*, **36**, 270–279.

Wiese, C.S. and D.A. Griffin (1978). The solubility of Aroclor 1254 in seawater. *Bull. Environ. Contam. Toxicol.*, **19**, 403–411.

Williams, R.J.P. (1981a). Physico-chemical aspects of inorganic element transfer through membranes. *Phil. Trans. Roy. Soc. Ser. B.*, **294**, 57–74.

Williams, R.J.P. (1981b). Natural selection of the chemical elements. *Proc. Roy. Soc. Lond. B.*, **213**, 361–397.

Wilson, A.J. (1976). Effects of suspended material on measurement of DDT in estuarine water. *Bull. Environ. Contam. Toxicol.*, **15**, 515–521.

Wilson, D., B. Finlayson and N. Morgan (1980). Copper, zinc, and cadmium concentrations of resident trout related to acid-mine wastes. *Calif. Fish and Game.*, **67**, 176–186.

Winter, M., M. Thomas, S. Wernick, S. Levin and M.T. Farvar (1976). Analysis of pesticide residues in 290 samples of Guatemalan mother's milk. *Bull. Environ. Contam. Toxicol.*, **16**, 652–657.

Wolff, E. (1990). Evidence for historical changes in global metal pollution: snow and ice samples. In *Heavy metals in the marine environment*, ed. R.W. Furness and P.S. Rainbow, CRC Press, Boca Raton, Florida, pp. 205–217.

Wong, C.S., E. Boyle, K.W. Bruland, J.D. Burton and E.D. Goldberg (Eds.) (1983). *Trace metals in sea water*. Plenum Press, New York.

Wong, P.T.S., Y.K. Chau, O. Kramar and G.A. Bengert (1981). Accumulation and depuration of tetramethyllead by rainbow trout. *Water Res.*, **15**, 621–625.

Woodhead, D.S. (1984). Contamination due to radioactive materials. In *Marine ecology*, Vol. 5, Part 3, ed. O. Kinne, John Wiley & Sons, New York, pp. 1111–1287.

Woodwell, G.M., C.F. Wurster and P.A. Isaacson (1967). DDT residues in an east coast estuary: a case of biological concentration of a persistent insecticide. *Science.*, **156**, 821–824.

Woodwell, G.M., P.P. Craig and H.A. Johnson (1971). DDT in the biosphere: where does it go? *Science.*, **174**, 1101–1107.

Wren, C.D. and H.R. MacCrimmon (1983). Mercury levels in the sunfish, *Lepomis gibbosus*, relative to pH and other environmental variables of Precambrian Shield lakes. *Can. J. Fish. Aquat. Sci.*, **40**, 1737–1744.

Wright, D.A. (1977a). The effect of salinity on cadmium uptake by the tissues of the shore crab *Carcinus maenas* (L.). *J. exp. Biol.*, **67**, 137–142.

Wright, D.A. (1977b). The effect of calcium on cadmium uptake by the shore crab *Carcinus maenas*. *J. exp. Biol.*, **67**, 163–173.

Wright, D.A. and D.J.H. Phillips (1988). Chesapeake and San Francisco Bays: a study in contrasts and parallels. *Mar. Pollut. Bull.*, **19**, 405–413.

Wright, D.A., J.A. Mihursky and H.L. Phelps (1985). Trace metals in Chesapeake Bay oysters: intra-sample variability and its implications for biomonitoring. *Mar. Environ. Res.*, **16**, 181–197.

Wszolek, P.C., D.J. Lisk, T. Wachs and W.D. Youngs (1979). Persistence of polychlorinated biphenyls and 1,1-dichloro-2,2-bis(*p*-chlorophenyl) ethylene (*p,p′*-DDE) with age in lake trout after 8 years. *Environ. Sci. Technol.*, **13**, 1269–1271.

Yamagata, N. and I. Shigematsu (1970). Cadmium pollution in perspective. *Bull. Inst. Public Health, Tokyo*, **19**, 1–24.

Yamashita, M., N. Kinae, I. Kimura, H. Ishida, H. Kumai and G. Nakamura (1990). The croaker (*Nibea mitsukurii*) and the sea catfish (*Plotosus anguillaris*): useful biomarkers of coastal pollution. In *Biomarkers of environmental contamination*, ed. J.F. McCarthy and L.R. Shugart, Lewis Publishers, CRC Press, Boca Raton, Florida, pp. 73–84.

Yamato, Y., M. Suzuki, K. Shimohara and T. Akiyama (1980). Behaviour of HCH (1,2,3,4,5,6–hexachlorocyclohexane) residue in the aquatic environment. *Water Res.*, **14**, 247–251.

Young, D.R. and T.R. Folsom (1967). Loss of Zn^{65} from the California sea-mussel *Mytilus californianus*. *Biol. Bull. mar. biol. Lab. Woods Hole*. **133**, 438–447.

Young, D.R., T.C. Heesen and D.J. McDermott (1976). An offshore biomonitoring system for chlorinated hydrocarbons. *Mar. Pollut. Bull.*, **7**, 156–159.

Young, M.L. (1975). The transfer of ^{65}Zn and ^{59}Fe along a *Fucus serratus* (L.) → *Littorina obtusata* (L.) food chain. *J. mar. biol. Ass. U.K.*, **55**, 583–610.

Youngs, W.D., W.H. Gutenmann and D.J. Lisk (1972). Residues of DDT in lake trout as a function of age. *Environ. Sci. Technol.*, **6**, 451–452.

Zabik, M.E., C. Merrill and M.J. Zabik (1982). Predictability of PCBs in carp harvested in Saginaw Bay, Lake Huron. *Bull. Environ. Contam. Toxicol.*, **28**, 592–598.

Zamuda, C.D. and W.G. Sunda (1982). Bioavailability of dissolved copper to the American oyster *Crassostrea virginica*. I. Importance of chemical speciation. *Mar. Biol.*, **66**, 77–82.

Zaroogian, G.E., J.H. Gentile, J.F. Heltshe, M. Johnson and A.M. Ivanovici (1982). Application of adenine nucleotide measurement for the evaluation of stress in *Mytilus edulis* and *Crassostrea virginica*. *Comp. Biochem. Physiol.*, **71B**, 643–649.

Zauke, G.-P. (1982). Cadmium in Gammaridae (Amphipoda: Crustacea) of the Rivers Werra and Weser — II. Seasonal variation and correlation to temperature and other environmental variables. *Water Res.*, **16**, 785–792.

Zirino, A. and S. Yamamoto (1972). A pH-dependent model for the chemical speciation of copper, zinc, cadmium, and lead in seawater. *Limnol. Oceanogr.*, **17**, 661–671.

Zitko, V. (1971). Polychlorinated biphenyls and organochlorine pesticides in some freshwater and marine fishes. *Bull. Environ. Contam. Toxicol.*, **6**, 464–470.

Zitko, V. (1989). Characterization of PCBs by Principal Component Analysis (PCA of PCB). *Mar. Pollut. Bull.*, **20**, 26–27.

Author Index

Roman type denotes references to authors within the text; *italic* type denotes references to authors within tables; and an asterisk denotes references to authors within figure captions.

Species Index

Roman type denotes references to species within the text; *italic* type denotes references to species within tables; and an asterisk denotes references to species within figure captions.

Subject Index